Group-Pattern Matrices

Group-Pattern Matrices

Roger Chalkley

Professor Emeritus of Mathematics
University of Cincinnati
Cincinnati, Ohio 45221-0025

Available from Amazon.com and other retail outlets

ISBN: 978-0-578-28788-1

Chalkley, Roger 1931-
    Group-Pattern Matrices

Published by Roger Chalkley

Available from Amazon.com and other retail outlets

2020 *Mathematics Subject Classification.* Primary 15A30, 20D45;
Secondary 15A15, 20-04

Key words and phrases. group, group-pattern,
group-pattern matrix

ABSTRACT. The concept of a *group-pattern matrix* is introduced here to serve
as a unifying starting point for numerous studies both old and new. While it
is quite useful, it does not fit conveniently in the 2020 classification.

A *group-pattern* for a group G having order n is specified by the $n \times n$
interior of a multiplication table for G in which each of the n principal diagonal
components is equal to the identity element e of G. Thus, G and a list $\mathscr{L}$ of
the n elements in G provide a pattern for the components of corresponding
group-pattern matrices. The *standard group-pattern matrix* for G and $\mathscr{L}$ is the
group-pattern matrix for G and $\mathscr{L}$ that has $\left(X_1, X_2, \ldots, X_n\right)$ as its first row.

The block-diagonalization of a standard group-pattern matrix A for G
and $\mathscr{L}$ is examined from a constructive viewpoint. Theorems 13.3 and 14.2
show that *there is a nonsingular $n \times n$ matrix M with components in the
field $\mathbb{C}$ of complex numbers such that the matrix $M^{-1}AM$ is a block-diagonal
matrix for which the determinant of each block is an irreducible polynomial in
$X_1, X_2, \ldots, X_n$ over $\mathbb{C}$*. Chapters 10–14 explain how a suitable set of matrix
representations over $\mathbb{C}$ for G yields the number and the sizes of the blocks for
their convenient positioning as well as yields the components for the explicit
construction of M via computer algebra.

The set of group-pattern matrices for G and $\mathscr{L}$ that have components in
a field forms a ring with respect to addition and multiplication of matrices.

All of the m automorphisms for a group G of order n are specified by
an efficient *Mathematica* program on page 34. With reference to a selected
list $\mathscr{L}_0 = \left(g_1, g_2, \ldots, g_n\right)$ for G having $g_1 = e$, the input can be any group-
pattern matrix for G and $\mathscr{L}_0$ that has n distinct components in its first row.
The program uses that to obtain the standard group-pattern matrix Z_0 for G
and $\mathscr{L}_0$. Then, it finds the lists for G with first element e that specify standard
group-pattern matrices equal to Z_0. Those m lists yield $(1, 2, \ldots, n)$ for $\mathscr{L}_0$
as the first item of output followed by $m - 1$ rearrangements of $(1, 2, \ldots, n)$
that leave 1 fixed. When this output is represented by
$\left(\pi_k(1), \pi_k(2), \ldots, \pi_k(n)\right)$, with $\pi_k(1) = 1$, for $1 \le k \le m$,
the m automorphisms $\phi_1, \phi_2, \ldots, \phi_m$ of G are given by
$\phi_k(g_1) = g_{\pi_k(1)}, \phi_k(g_2) = g_{\pi_k(2)}, \ldots, \phi_k(g_n) = g_{\pi_k(n)}$, for $1 \le k \le m$.

Contents

Preface   xiii
   0.1.   Group-patterns and group-pattern matrices   xiii
   0.2.   Standard group-pattern matrices and group-matrices   xiv
   0.3.   Need for the more inclusive *group-pattern matrices*   xv
   0.4.   Content of various chapters   xvi

Chapter 1.   Nontechnical Views of Group-Pattern Matrices   1
   1.1.   Patterns of multiplication tables for 1, ω, ω^2 1
   1.2.   Patterns of multiplication tables for 1, i, -1, $-i$ 2
   1.3.   Patterns for the symmetries of a nonsquare rectangle   2
   1.4.   Patterns of four tables for 1, $-\omega$, ω^2, -1, ω, $-\omega^2$ 3
   1.5.   A pattern for the symmetries of an equilateral triangle   4
   1.6.   A pattern for the symmetries of a square   5
   1.7.   A pattern for the symmetries of a brick-like object   6
   1.8.   A pattern for some symmetries of a tile-like object   8

Chapter 2.   Formulations for Group-Pattern Matrices   11
   2.1.   Definitions and terminology   11
   2.2.   Restriction to lists $\mathscr{L} = (g_1,, g_2, \ldots, g_n)$ where $g_1 = e$ 13
   2.3.   The set $\mathcal{M}[\mathbb{C}, G, \mathscr{L}]$ is closed under matrix multiplication   14
   2.4.   Injective group-pattern matrices   14
   2.5.   Computer check for an injective group-pattern matrix   19
   2.6.   Verification that a given multiplication table specifies a group   21
   2.7.   Representation of G with permutation matrices   22
   2.8.   Objects of historical interest   24

Chapter 3.   Standard Group-Pattern Matrices   27
   3.1.   The standard group-pattern matrix for G and $\mathscr{L}$ 27
   3.2.   Examples of standard group-pattern matrices   27
   3.3.   The number of standard group-pattern matrices for G 30
   3.4.   Program to obtain all of the automorphisms for G 33
   3.5.   Direct product of groups   36
   3.6.   Computer algebra for the standard group-pattern matrices   38

Chapter 4.   Properties for $\mathcal{M}[\mathfrak{F}, G, \mathscr{L}]$ Based on Definition 2.2   41
   4.1.   Closure under addition, multiplication, and scalar multiplication   41
   4.2.   Some algebraic properties of $\mathcal{M}[\mathfrak{F}, G, \mathscr{L}]$ 42
   4.3.   The group ring and group algebra for G over $\mathfrak{F}$ 44
   4.4.   Cofactor matrices for group-pattern matrices   47
   4.5.   Applications of the isomorphism for $\mathbb{F}[G]$ and $\mathcal{M}[\mathbb{F}, G, \mathscr{L}]$ 49

Chapter 5.   Historical Perspective, Part 1   53
    5.1.   Factorization for the determinant
             of a leftward-circulant matrix   53
    5.2.   Factorization for the determinant of a circulant matrix   54
    5.3.   Procedure of Anton Puchta
             for discovering factorizations   55
    5.4.   Verifications by M. Noether
             for formulations of A. Puchta   57
    5.5.   Factorization procedure of Anton Puchta
             for an example in [43]   58
    5.6.   Verification for the preceding example of A Puchta   59
    5.7.   Observations   61

Chapter 6.   Viewpoints about Circulant Matrices   63
    6.1.   Properties of the field $\mathcal{F}$ 63
    6.2.   The matrix M that corresponds to (5.4) for [16] of 1856   64
    6.3.   Circulant and leftward-circulant matrices   67
    6.4.   Additional results   71
    6.5.   Modifications of Theorem 6.4 and Theorem 6.6   72

Chapter 7.   Diagonalizations
                 of Cyclic-Group-Pattern Matrices   73
    7.1.   Context and principal results   73
    7.2.   Diagonalization for the matrix A_{2p} of Example 7.5   76

Chapter 8.   Diagonalizations
                 of Abelian-Group-Pattern Matrices   79
    8.1.   Context and principal results   79
    8.2.   The structure of $\mathcal{M}[\mathcal{F},\,G,\,\mathscr{L}]$ when G is abelian   84
    8.3.   The matrix $(1/n)M^2$ 85
    8.4.   An abelian group of order n
             has precisely n group characters   86
    8.5.   Factors for the determinant
             of an abelian-group-pattern matrix   87
    8.6.   Brief observations   87

Chapter 9.   Supplement to Chapter 8   89

Chapter 10.   Historical Perspective, Part 2   91
    10.1.   Gruppendeterminanten of Dedekind and Frobenius   91
    10.2.   Dedekind's factorization of the Gruppendeterminante for $\mathfrak{S}_3$ 93
    10.3.   A reinterpretation for the preceding results of Dedekind   94
    10.4.   Matrix identity for (10.7) developed by Frobenius   96
    10.5.   Matrix representations of finite groups and some properties   96

Chapter 11.   Viewpoints about Notation   103
    11.1.   The notation for group-matrices   103
    11.2.   A numbering for triples of integers   104
    11.3.   Selection of components for construction of a matrix like L 106
    11.4.   Contextual improvement   107

11.5.   Machine computations for $\lambda(s)$, $\mu(s)$, and $\nu(s)$ 109

Chapter 12.   An Instructive Frobenius Block-Diagonalization   111
  12.1.   The group $\mathfrak{T}$ of tetrahedral rotational symmetries   111
  12.2.   A matrix representation for $\mathfrak{T}$ of degree 3   112
  12.3.   The matrix representations of degree 1 for $\mathfrak{T}$ 113
  12.4.   Block-diagonalization of a group-pattern matrix for $\mathfrak{T}$ 115
  12.5.   Computer-algebra verifications for the preceding section   118
  12.6.   Further details dependent on Chapter 14   121

Chapter 13.   Frobenius Block-Diagonalizations
              for Group-Pattern Matrices   123
  13.1.   The set $\mathfrak{B}$ of Frobenius block-diagonal matrices for G 123
  13.2.   Employment of a matrix M to relate $\mathfrak{A}$ and $\mathfrak{B}$ 124
  13.3.   Immediate consequences of Theorem 13.3   128
  13.4.   Particular situations for Theorem 13.3   131
  13.5.   Use of a block-diagonalization to deduce others   132

Chapter 14. A in $\mathfrak{A}$ and B in $\mathfrak{B}$ when $B = M^{-1}AM$ 135
  14.1.   Introduction   135
  14.2.   Use of n components d_1, d_2, ..., d_n to specify any B in $\mathfrak{B}$ 135
  14.3.   Supplement for Theorem 13.3   136
  14.4.   Complete factorizations for the determinants
          of standard group-pattern matrices   137

Chapter 15.   The Dihedral Group $\mathcal{D}_n$ 139
  15.1.   Introduction   139
  15.2.   Primary group-pattern matrices for $\mathcal{D}_n$ 142
  15.3.   Other group-pattern matrices for $\mathcal{D}_3$, $\mathcal{D}_4$, and $\mathcal{D}_5$ 145
  15.4.   Secondary group-pattern matrices for $\mathcal{D}_n$ 146
  15.5.   The automorphisms for $\mathcal{D}_n$ 149
  15.6.   Matrix representations of degree 1 for $\mathcal{D}_n$ 150

Chapter 16.   Frobenius Block-Diagonalizations
              of Primary Group-Pattern Matrices for $\mathcal{D}_n$ 155
  16.1.   Frobenius block-diagonalization of A_6 for $\mathcal{D}_3$ 155
  16.2.   Frobenius block-diagonalization of A_8 for $\mathcal{D}_4$ 157
  16.3.   Frobenius block-diagonalization of A_{10} for $\mathcal{D}_5$ 159
  16.4.   Details about the matrix representations used for $\mathcal{D}_5$ 162

Chapter 17.   Auxiliary-Pattern Matrices   165
  17.1.   Introduction   165
  17.2.   Auxiliary-pattern matrices and immediate deductions   166
  17.3.   The situation where G is abelian   167
  17.4.   Injective Auxiliary-Pattern Matrices   168
  17.5.   Computer program that implements Theorems 17.8 and 17.9   173
  17.6.   Some lists are not replaceable by ones having $g_1 = e$ 174

Chapter 18.   Standard Auxiliary-Pattern Matrices   177
  18.1.   The standard auxiliary-pattern matrix for G and $\mathscr{L}$ 177
  18.2.   Examples of standard auxiliary-pattern matrices   177

18.3.   Number of standard auxiliary-pattern matrices for G 179
18.4.   Perspective   181

Chapter 19.   Deduction of Automorphisms
                via Standard Auxiliary-Pattern Matrices   183
19.1.   Computer program for automorphisms   185
19.2.   Auxiliary-pattern viewpoint for the quaternion group Q_8 187

Chapter 20.   Unequal Injective Matrices for Dissimilar Groups   189
20.1.   Injective group-pattern matrix for each of two groups   189
20.2.   Injective group-pattern matrices that are auxiliary ones   189
20.3.   Injective auxiliary-pattern matrix for each of two groups   191

Chapter 21.   Particular Advancements of Richard Dedekind   193

Chapter 22.   Particular Advancements of Georg Frobenius   197

Chapter 23.   Group-Pattern Matrices for Q_8 199
23.1.   Application of Theorem 13.3
             for second example of Dedekind   199
23.2.   A Frobenius block-diagonalization based on Section 13.5   203
23.3.   The Frobenius *Gruppendeterminante* for Q_8 205

Appendix A.   Use of Determinants to Factor Polynomials   207

Appendix B.   Change of Viewpoint   209

Appendix C.   Circulant Matrices without Mention of a Group   211
C.1.   A suitable context   211
C.2.   The principal result   212
C.3.   Two consequences from [7]   213
C.4.   An unusual permissible selection for the field $\mathcal{F}$ when $n = 3$ 214
C.5.   Notation for M and L in (C.2) and (C.3)   215
C.6.   Example to illustrate Formula (D.3) on page 217   215

Appendix D.   Definition of a Circulant Matrix in [19]   217
D.1.   Introduction   217
D.2.   Characterization in [19] of a circulant matrix   217

Appendix E.   Result of Richard Baltzer in [2] of 1864   221
E.1.   Introduction   221
E.2.   The resultant of two polynomials having degrees $n - 1$ and n 221
E.3.   Computation of R. Baltzer in a modified context   222

Appendix F.   Automorphisms
                for Developers of Computer Algebra   225
F.1.   The automorphisms for a group G of order n 225

Appendix G.   Computer Algebra for Transformations
                of Homogeneous Linear Differential Equations   229
G.1.   The two basic types of transformations   229
G.2.   Corresponding transformation formulas for *Mathematica*   230

G.3. Identities for several of the principal invariants 231
G.4. Brief observations 233
G.5. Infinitesimal transformations were a great hindrance 233

Appendix H. Reference to *Mathematica* Notebooks 235

Bibliography 237

Index 239

Preface

Let $\mathscr{L} = (g_1, g_2, \ldots, g_n)$ be a list of the n elements in a group G of order n.

0.1. Group-patterns and group-pattern matrices

(1) *The **group-pattern** $\mathcal{P}$ that is specified by G and $\mathscr{L}$ is the $n \times n$ interior matrix of the multiplication table T for G that has $\mathscr{L}$ as its list of column indices and has its n principal diagonal components equal to the identity element e of G.* (Each row index for T is then the inverse in G of the corresponding column index.)

EXAMPLE 0.1. For the list $\mathscr{L}_1 = (e, \alpha, \alpha^2, \alpha^3)$ of the elements in the cyclic group $C_4 = \{e, \alpha, \alpha^2, \alpha^3\}$ of order 4 and the list $\mathscr{L}_2 = (e, u, v, w)$ of elements in the group $C_2 \times C_2 = \{e, u, v, w\}$ of order 4, their corresponding multiplication tables T_1 and T_2 are uniquely given by

$\cdot$	e	α	α^2	α^3
e	e	α	α^2	α^3
α^3	α^3	e	α	α^2
α^2	α^2	α^3	e	α
α	α	α^2	α^3	e

and

$\cdot$	e	u	v	w
e	e	u	v	w
u	u	e	w	v
v	v	w	e	u
w	w	v	u	e

where $\mathscr{L}_1$ and $\mathscr{L}_2$ are their respective lists of column indices. Consequently, the group-patterns $\mathcal{P}_1$ for C_4 with $\mathscr{L}_1$ and $\mathcal{P}_2$ for $C_2 \times C_2$ with $\mathscr{L}_2$ are respectively

$$(0.1) \qquad \begin{bmatrix} e & \alpha & \alpha^2 & \alpha^3 \\ \alpha^3 & e & \alpha & \alpha^2 \\ \alpha^2 & \alpha^3 & e & \alpha \\ \alpha & \alpha^2 & \alpha^3 & e \end{bmatrix} \quad \text{and} \quad \begin{bmatrix} e & u & v & w \\ u & e & w & v \\ v & w & e & u \\ w & v & u & e \end{bmatrix}.$$

Note that the lists $\mathscr{L}_1$ and $\mathscr{L}_2$ have e as their first element and that those lists are the first rows of the group-patterns in (0.1).

(2) *An $n \times n$ matrix A with components in a set S is a **group-pattern matrix** for G and $\mathscr{L}$ when there is a function f from G to S such that A is obtained from the pattern $\mathcal{P}$ for G and $\mathscr{L}$ by replacing each group element x in $\mathcal{P}$ with $f(x)$.*

EXAMPLE 0.2. For the context of Example 0.1, let $a_1, a_2, a_3, a_4, b_1, b_2, b_3, b_4$ be any elements in the field $\mathbb{C}$ of complex numbers and let S be $\mathbb{C}$. Clearly, there are functions from C_4 to $\mathbb{C}$ and from $C_2 \times C_2$ to $\mathbb{C}$ such that (0.1) specifies

$$(0.2) \qquad A_1 = \begin{bmatrix} a_1 & a_2 & a_3 & a_4 \\ a_4 & a_1 & a_2 & a_3 \\ a_3 & a_4 & a_1 & a_2 \\ a_2 & a_3 & a_4 & a_1 \end{bmatrix} \quad \text{and} \quad A_2 = \begin{bmatrix} b_1 & b_2 & b_3 & b_4 \\ b_2 & b_1 & b_4 & b_3 \\ b_3 & b_4 & b_1 & b_2 \\ b_4 & b_3 & b_2 & b_1 \end{bmatrix}.$$

as group-pattern matrices for C_4 with $\mathscr{L}_1$ and for $C_2 \times C_2$ with $\mathscr{L}_2$.

When S is a field $\mathfrak{F}$ and $\mathfrak{R}$ is the set of group-pattern matrices for G and $\mathscr{L}$ having components in $\mathfrak{F}$, the set $\mathfrak{R}$ forms a ring with addition and multiplication of matrices. That is established in Theorem 4.2.

In particular, if $\mathfrak{R}_1$ is the set of matrices having the form of A_1 in (0.2) and if $\mathfrak{R}_2$ is the set of matrices having the form of A_2 in (0.2), then $\mathfrak{R}_1$ and $\mathfrak{R}_2$ are rings with respect to addition and multiplication of matrices.

0.2. Standard group-pattern matrices and group-matrices

(3) *The **standard group-pattern matrix** A for G and $\mathscr{L}$ is the group-pattern matrix for G and $\mathscr{L}$ that has $(X_1, X_2, \ldots, X_n)$ as its first row.*

Thus, if the first row of the pattern $\mathcal{P}$ for G and $\mathscr{L}$ is $(h_1, h_2, \ldots, h_n)$, then A is the group-pattern matrix for G and $\mathscr{L}$ specified by the function f from G to the set $S = \{X_1, X_2, \ldots, X_n\}$ having $f(h_k) = X_k$, for $k = 1, 2, \ldots, n$.

Alternatively, Proposition 2.7 shows that the list $\mathscr{L}$ can be replaced by one whose first element is e. If $\mathscr{L}$ has $g_1 = e$, then $\mathscr{L}$ is the first row of $\mathcal{P}$ and f is the function from G to S having $f(g_k) = X_k$, for $k = 1, 2, \ldots, n$.

EXAMPLE 0.3. For the context of Example 0.1, we use (0.1) to see that

(0.3)
$$\begin{bmatrix} X_1 & X_2 & X_3 & X_4 \\ X_4 & X_1 & X_2 & X_3 \\ X_3 & X_4 & X_1 & X_2 \\ X_2 & X_3 & X_4 & X_1 \end{bmatrix} \quad \text{and} \quad \begin{bmatrix} X_1 & X_2 & X_3 & X_4 \\ X_2 & X_1 & X_4 & X_3 \\ X_3 & X_4 & X_1 & X_2 \\ X_4 & X_3 & X_2 & X_1 \end{bmatrix}$$

are the standard group-pattern matrices for C_4 with $\mathscr{L}_1$ and for $C_2 \times C_2$ with $\mathscr{L}_2$.

(4) *The **group-matrix** A for G and $\mathscr{L}$ is the group-pattern matrix A for G and $\mathscr{L}$ that is defined by the function f from G to the set $S = \{X_{g_1}, X_{g_2}, \ldots, X_{g_n}\}$ where $f(g_k) = X_{g_k}$, for $k = 1, 2, \ldots, n$.*

The equivalent definition that Georg Frobenius gave for a Gruppenmatrix in [**24**] of 1896 is described in Section 10.1; see also page 197.

Motivation for the introduction of group-matrices involves the property that: for any two lists of the elements in G, the corresponding group-matrices have equal determinants as polynomials over $\mathbb{C}$; e.g., see Theorem 10.1 and page 197. Thus, the group G specifies a unique homogeneous polynomial of degree n in the variables of S over $\mathbb{C}$ as the determinant of any group-matrix for G. That polynomial was called the *Gruppen-Determinante* for G by Richard Dedekind.

In use, simplifications were made; e.g., see the matrix X in Subsection 22.0.2 on page 198 where columns were favored. However, with matrices viewed as lists of their rows, it is more natural to rewrite the group-matrix for G and $\mathscr{L}$ as the unique standard group-pattern matrix for G and $\mathscr{L}$.

EXAMPLE 0.4. For the context of Example 0.1, we use (0.1) to see that the group-matrix for C_4 with $\mathscr{L}_1$ and the group-matrix for $C_2 \times C_2$ with $\mathscr{L}_2$ are

(0.4)
$$\begin{bmatrix} X_e & X_\alpha & X_{\alpha^2} & X_{\alpha^3} \\ X_{\alpha^3} & X_e & X_\alpha & X_{\alpha^2} \\ X_{\alpha^2} & X_{\alpha^3} & X_e & X_\alpha \\ X_\alpha & X_{\alpha^2} & X_{\alpha^3} & X_e \end{bmatrix} \quad \text{and} \quad \begin{bmatrix} X_e & X_u & X_v & X_w \\ X_u & X_e & X_w & X_v \\ X_v & X_w & X_e & X_u \\ X_w & X_v & X_u & X_e \end{bmatrix}.$$

The notation of (0.4) naturally simplifies to that of (0.3).

(5) In a letter to Frobenius dated March 25, 1896, Richard Dedekind indicated that, when G is an abelian group (i.e., $yx = xy$ for each x, y in G), he could express the Gruppendeterminante for G as a product of n linear polynomials over $\mathbb{C}$ by using the n group characters for G; e.g., see page 193. Dedekind never published that result. Naturally he wanted an inclusive explanation valid for any finite group.

(6) In a letter dated April 6, 1896, Dedekind presented Frobenius with details for two nonabelian groups whose Gruppendeterminanten had irreducible nonlinear factors; e.g., see pages 194-195. His first example exhibits a group-matrix A for the nonabelian group D_3 of order 6 and shows that the irreducible factors for $\det(A)$ are two unequal linear polynomials and two equal quadratic polynomials over $\mathbb{C}$.

(7) Georg Frobenius modified that first example in [**25**, pages 1007–1008] of 1897. There, he introduced a nonsingular 6×6 matrix L over $\mathbb{C}$ such that the matrix $U = L^{-1}AL$ is a 6×6 block-diagonal matrix with two unequal blocks of size 1×1 and two equal blocks of size 2×2; e.g., see page 198. The determinants of the blocks are the irreducible polynomials that Dedekind provided. Since $\det(U)$ is a product of those irreducible polynomials and $\det(A) = \det(U)$, Frobenius obtained Dedekind's factorization in a manner that inspired much further research.

(8) Frobenius used the 36 components of three matrix representations for the group D_3 to construct the matrix L. In doing that, Frobenius initiated extensive research about matrix representations for finite groups. Focus then shifted from group-matrices to the discovery of results about matrix representations of groups.

(9) Now, after matrix representations have been thoroughly investigated, it is possible to return to the Frobenius block-diagonalization described in Item (7) and extend that technique to any standard group-pattern matrix. That is done in Chapters 10–14. See the special case that Theorem 14.2 provides for Theorem 13.3.

0.3. Need for the more inclusive *group-pattern matrices*

The concept of a *group-matrix* shows that none of the matrices

$$(0.5) \qquad \begin{bmatrix} c_1 \end{bmatrix}, \quad \begin{bmatrix} c_1 & c_2 \\ c_2 & c_1 \end{bmatrix}, \quad \begin{bmatrix} c_1 & c_2 & c_3 \\ c_3 & c_1 & c_2 \\ c_2 & c_3 & c_1 \end{bmatrix}, \quad \begin{bmatrix} c_1 & c_2 & c_3 & c_4 \\ c_4 & c_1 & c_2 & c_3 \\ c_3 & c_4 & c_1 & c_2 \\ c_2 & c_3 & c_4 & c_1 \end{bmatrix}, \quad \cdots$$

can be correctly described as a group-matrix when their components are complex numbers. However, each is a group-pattern matrix.

When Philip J. Davis wrote [**19**] in 1979, he used

$$(0.6) \qquad C = \begin{bmatrix} c_1 & c_2 & c_3 & \cdots & c_{n-1} & c_n \\ c_n & c_1 & c_2 & \cdots & c_{n-2} & c_{n-1} \\ c_{n-1} & c_n & c_1 & \cdots & c_{n-3} & c_{n-2} \\ \vdots & \vdots & \vdots & \ddots & \vdots & \vdots \\ c_3 & c_4 & c_5 & \cdots & c_1 & c_2 \\ c_2 & c_3 & c_4 & \cdots & c_n & c_1 \end{bmatrix}$$

as a visual definition of an $n \times n$ circulant matrix. He completely avoided the concept of a group except for a reference to [**9**] for generalizations. We shall view the matrix C of (0.6) advantageously as a type of group-pattern matrix. See Chapter 6.

To explicitly specify the components of the preceding matrix C, we introduce $g_k = \alpha^{k-1}$, for $1 \le k \le n$, where $\alpha = \cos\left(\frac{2\pi}{n}\right) + i\sin\left(\frac{2\pi}{n}\right)$ with $i^2 = -1$ and $\alpha^n = 1$. The corresponding multiplication table

(0.7)

$\cdot$	$g_1 = 1$	$g_2 = \alpha$	$g_3 = \alpha^2$	$\cdots$	$g_{n-1} = \alpha^{n-2}$	$g_n = \alpha^{n-1}$
$g_1^{-1} = 1$	g_1	g_2	g_3	$\cdots$	g_{n-1}	g_n
$g_2^{-1} = \alpha^{n-1}$	g_n	g_1	g_2	$\cdots$	g_{n-2}	g_{n-1}
$g_3^{-1} = \alpha^{n-2}$	g_{n-1}	g_n	g_1	$\cdots$	g_{n-3}	g_{n-2}
$\vdots$	$\vdots$	$\vdots$	$\vdots$	$\ddots$	$\vdots$	$\vdots$
$g_{n-1}^{-1} = \alpha^2$	g_3	g_4	g_5	$\cdots$	g_1	g_2
$g_n^{-1} = \alpha$	g_2	g_3	g_4	$\cdots$	g_n	g_1

has its interior pattern similar to that of C in (0.6). In fact, the element in the rth row and sth column of (0.7) is $g_r^{-1}g_s = g_k$ if and only if the element in the rth row and sth column of (0.6) is c_k. Let S be a set that contains the components of C and let f be the function from $G = \{g_1, g_2, \ldots, g_n\}$ to $\mathbb{C}$ such that $f(g_k) = c_k$, for $1 \le k \le n$. The component $[C]_{r,s}$ of C in its rth row and sth column is given by

(0.8) $$[C]_{r,s} = f(g_r^{-1}g_s), \quad \text{for } r, s = 1, 2, \ldots, n.$$

Thus, *an $n \times n$ matrix A with components in a set S is a circulant matrix if and only if there is a function f from G to S such that the component $[A]_{r,s}$ of A in its rth row and sth column is given by*

(0.9) $$[A]_{r,s} = f(g_r^{-1}g_s), \quad \text{for } r, s = 1, 2, \ldots, n.$$

0.4. Content of various chapters

Throughout, except for page 214, the components may be elements of a field $\mathbb{F}$ that is either the field $\mathbb{C}$ of complex numbers or a proper field extension of $\mathbb{C}$.

Depending on circumstances, this monograph may be read in various orders. For instance, to supplement an interest in circulant matrices, a reader may prefer to begin directly with Appendices A, B, C, D followed by Chapters 5, 6, 1, and 2. Or, for those whose primary interest may be automorphisms for finite groups, the recommended order would be Chapters 1, 2, 3, Appendix F, and Chapters 18, 19.

Chapter 1 provides numerous examples of group-pattern matrices based solely on multiplication tables. The concept of a group is introduced later in Definition 2.1.

Chapter 2 uses a list $\mathscr{L} = (g_1, g_2, \ldots, g_n)$ of the n elements in a group G of order n to indicate the form

(0.10)

$\cdot$	g_1	g_2	$\cdots$	g_n
g_1^{-1}	$g_1^{-1}g_1$	$g_1^{-1}g_2$	$\cdots$	$g_1^{-1}g_n$
g_2^{-1}	$g_2^{-1}g_1$	$g_2^{-1}g_2$	$\cdots$	$g_2^{-1}g_n$
$\vdots$	$\vdots$	$\vdots$	$\ddots$	$\vdots$
g_n^{-1}	$g_n^{-1}g_1$	$g_n^{-1}g_2$	$\cdots$	$g_n^{-1}g_n$

of the unique multiplication table for G and $\mathscr{L}$ that has $\mathscr{L}$ as the list of its column indices and also has the identity element e of G in each of its principal diagonal positions. Hence, an $n \times n$ matrix A with components in a set S is a *group-pattern matrix* for G and $\mathscr{L}$ if and only if there is a function f from G to S such that the component $\left[A\right]_{r,s}$ of A in its rth row and sth column is given by

(0.11) $$\left[A\right]_{r,s} = f\!\left(g_r^{-1}g_s\right), \quad \text{for } r, s = 1, 2, \ldots, n.$$

The examples of Chapter 1 illustrate (0.10)–(0.11).

A group-pattern matrix is said to be *injective* when the function f for (0.11) is one-to-one. Thus, a group-pattern matrix A is injective if and only if no two components in its first row are equal.

This leads to the problem of deciding whether a given matrix Z is an injective group-pattern matrix. Necessary and sufficient conditions for that are presented in Corollary 2.12 on page 17. The computer program of page 19 is based on that corollary and decides quickly whether Z is an injective group-pattern matrix for some group G and some list $\mathscr{L}$ of the elements in G. When it is one, the program indicates that fact by providing a multiplication table for a group G and list $\mathscr{L}$ relative to which Z is an injective group-pattern matrix. Otherwise, the program prints a sufficient reason for the negative decision.

Chapter 3 assigns a unique standard group-pattern matrix to each group G of order n and each list $\mathscr{L}$ for G having e as its first element. The number $\mathfrak{N}_{\mathfrak{G}}(G)$ of distinct standard group-pattern matrices for G and the various lists for G is shown in Theorem 3.7 on page 30 to be related to the number m of automorphisms for G by the formula $\mathfrak{N}_{\mathfrak{G}}(G) = (n-1)!/m$.

A remarkably efficient *Mathematica* program to obtain all of the automorphisms for a group G of order n is presented on page 34. It runs rapidly on typical desk-top computers when $n \leq 10$. When $n \geq 11$, more than the usual memory capacity of a typical desk-top computer is needed. However, Appendix F indicates how memory requirements can be greatly reduced.

That program of page 34 specifies the m automorphisms for a given group G of order n in terms of a selected list $\mathscr{L}_0 = (g_1, g_2, \ldots, g_n)$ of the n elements in G for which $g_1 = e$. Suitable input is provided by the representation A of any injective group-pattern matrix A for G and $\mathscr{L}_0$. It is advisable to check A as described above for page 19. The program of page 34 uses the input A to obtain the unique standard group-pattern matrix Z_0 for G and $\mathscr{L}_0$. The program then finds each of the lists for G having e as first element such that the standard group-pattern matrix it specifies is equal to Z_0. Then, instead of printing out lists that look like

$$\left(g_{\pi_k(1)}, g_{\pi_k(2)}, \ldots, g_{\pi_k(n)}\right), \quad \text{with } \pi_k(1) = 1, \text{ for } k = 1, 2, \ldots, m,$$

it represents $(g_1, g_2, \ldots, g_n)$ by $(1, 2, \ldots, n)$ and therefore prints out

$$\left(\pi_k(1), \pi_k(2), \ldots, \pi_k(n)\right), \quad \text{with } \pi_k(1) = 1, \text{ for } k = 1, 2, \ldots, m.$$

Consequently, the m automorphisms $\phi_1, \phi_2, \ldots, \phi_m$ of G are given by

$$\phi_k(g_1) = g_{\pi_k(1)}, \ \phi_k(g_2) = g_{\pi_k(2)}, \ \ldots, \ \phi_k(g_n) = g_{\pi_k(n)}, \quad \text{for } = 1, 2, \ldots, m.$$

Chapter 4 uses the context about (0.10)–(0.11) to derive algebraic properties of the group-pattern matrices for G and $\mathscr{L}$ having components in a ring or field.

Chapter 5 includes early examples where the determinants of various special $n \times n$ matrices were discovered to have remarkable factorizations. Formulations of interest were developed for them before group-pattern matrices for abelian groups were available to unify the subject.

Chapters 6–9 examine the situation where $\mathscr{L} = (g_1, g_2, \ldots, g_n)$ is a list of the elements in an abelian group G of order n and the components of all matrices considered are elements of some field $\mathcal{F}$ that contains a primitive nth root ρ of unity having period n. Thus, the powers of ρ specify n distinct roots of $X^n - 1$ in $\mathcal{F}$. The field $\mathcal{F}$ may be $\mathbb{F}$. As the principal result, Theorem 8.3 of page 81 shows how to construct a nonsingular $n \times n$ matrix M whose components are powers of ρ such that: (i) if A is a group-pattern matrix for G and $\mathscr{L}$, then $M^{-1}AM$ is a diagonal matrix; and (ii) if D is an $n \times n$ diagonal matrix, then MDM^{-1} is a group-pattern matrix for G and $\mathscr{L}$. Thus, if A is a group-pattern matrix for G with $\mathscr{L}$ and $D = M^{-1}AM$, then $\det(A) = \det(D)$ and therefore $det(A)$ equals the product of the n diagonal components of D. This viewpoint was presented in [**9**] of 1976.

The preceding result shows that the standard group-pattern matrix A for G and $\mathscr{L}$ is diagonalizable by a nonsingular $n \times n$ matrix M having components in $\mathbb{C}$. Namely, specialize $\mathbb{F}$ to be a field extension of $\mathbb{C}$ that contains the variables $X_1, X_2, \ldots, X_n$. Then, with $\mathcal{F} = \mathbb{F}$ and ρ in $\mathbb{C}$, the components of M are in $\mathbb{C}$ and the matrix $M^{-1}AM$ is a diagonal matrix. Thus, the diagonal components of $D = M^{-1}AM$ are linear polynomials in $X_1, X_2, \ldots, X_n$ over $\mathbb{C}$.

Theorem 2.18 on page 23 shows that a standard group-pattern matrix for a nonabelian group is not diagonalizable.

Chapters 10–14 examine the general situation where the finite group G is not necessarily abelian. Then, for any group G of order n and any list $\mathscr{L}$ for G, the challenge is to find a nonsingular $n \times n$ matrix M that transforms each group-pattern matrix for G and $\mathscr{L}$ having components in $\mathbb{F}$ to a block-diagonal form that is as simple as possible. The theory of group representations guarantees the existence of various matrix representations for G whose components in $\mathbb{C}$ can be selected to form the components of M. Our main contribution is to provide an algorithm that enables computer algebra to select suitable components for M. In this regard, see the definition of M in (13.6) on page 124.

In particular, when $\mathbb{F}$ is selected to include $X_1, X_2, \ldots, X_n$, the components of the standard group-pattern matrix A for G and $\mathscr{L}$ belong to $\mathbb{F}$. Then, the results show that $M^{-1}AM$ is a block-diagonal matrix for which the determinant of each block is an irreducible polynomial in the variables $X_1, X_2, \ldots, X_n$ over $\mathbb{C}$. This enables $\det(A)$ to be expressed as a product of those irreducible polynomials.

Chapters 15–16 provide details about matrix representations for the dihedral groups $\mathcal{D}_3$ of order 6, $\mathcal{D}_4$ of order 8, and $\mathcal{D}_5$ of order 10. Then, that is used to construct a block-diagonalization of standard group-pattern matrices for each of those three groups. Similarly, Chapter 12 does that for the twelfth-order group of rotational symmetries for a regular tetrahedron. Also, Chapter 23 does that for the quaternion group of order 8. These results for five different nonabelian groups serve to illustrate the technique of machine constructions for the matrix M employed in Theorems 13.3 and 14.2.

Chapters 17–18 use a list $\mathscr{L} = (g_1, g_2, \ldots, g_n)$ of the elements in a group G of order n to specify a unique multiplication table

$$
(0.12) \quad
\begin{array}{c|c|c|c|c}
\cdot & g_1 & g_2 & \cdots & g_n \\
\hline
g_1 & g_1\,g_1 & g_1\,g_2 & \cdots & g_1\,g_n \\
\hline
g_2 & g_2\,g_1 & g_2\,g_2 & \cdots & g_2\,g_n \\
\hline
\vdots & \vdots & \vdots & \ddots & \vdots \\
\hline
g_n & g_n\,g_1 & g_n\,g_2 & \cdots & g_n\,g_n
\end{array}
$$

that serves as a pattern to define $n \times n$ *auxiliary-pattern matrices* for G and $\mathscr{L}$. That table also serves to define a unique standard auxiliary-pattern matrix for G and $\mathscr{L}$ when g_1 is the identity element of G.

Chapter 19 provides another efficient algorithm to obtain the automorphisms for a finite group G. It uses properties of standard auxiliary-pattern matrices rather than standard group-pattern matrices. Its arguments are analogous to the ones made in Section 3.4 but are independent of them.

Chapter 20 shows that nonisomorphic groups have different standard group-pattern matrices and different standard auxiliary-pattern matrices.

Chapter 21 includes two letters dated March 25, 1896 and April 6, 1896 from Richard Dedekind to Georg Frobenius. Dedekind's two examples in that second letter are unusually interesting and relevant.

Chapter 22 includes the detailed explanation that Georg Frobenius gave in [**24**, page 1343] about notation for group-matrices. Moreover, it includes the reexamination by Frobenius in [**25**, pages 1007–1008] of Dedekind's first example. Additional detail about that is provided in Chapter 10.

Chapter 23 shows how Theorem 13.3 on page 125 can be used to transform the matrix in the second example of Dedekind to a suitable block-diagonal form.

Appendices A and **B** provide several motivational details.

Appendix C shows how the principal properties of circulant matrices can be satisfactorily developed without the explicit formalism of a group.

Appendix D examines an interesting detail about circulant matrices in [**19**].

Appendix E completes a clever argument used by Richard Baltzer in 1864 to establish the factorization for the determinant of an $n \times n$ circulant matrix. It serves to extend the historical perspective of Section 5.1.

Appendix F supplements the presentation of Section 3.4 about finding the automorphisms of a group G having order n. The explanation is focused. Thus, there is no need to consider $\mathfrak{N}_{\mathfrak{G}}(G)$ or its computation.

Appendix G describes an area of general interest where a failure to develop basic results curtailed an adequate development for over 100 years. The results are directly suitable for systems of computer algebra.

Throughout, various examples use the computer algebra system *Mathematica*. Where that occurs, there is a corresponding *Mathematica* notebook available for downloading according to the directions in **Appendix H** on page 235.

Roger Chalkley
January 29, 2021
Cincinnati

CHAPTER 1

Nontechnical Views of Group-Pattern Matrices

We begin by considering the patterns for the components of square matrices that are provided by various multiplication tables. Throughout, we let i and ω denote complex numbers that satisfy $i^2 = -1$ and $\omega^2 + \omega + 1 = 0$. In view of $\omega^3 = -\omega^2 - \omega = 1$ and $(\omega^2)^3 = 1$, we see that 1, ω, and ω^2 are the three roots of $X^3 - 1 = 0$ while 1, i, -1, and $-i$ are the four roots of $X^4 - 1 = 0$.

1.1. Patterns of multiplication tables for 1, ω, ω^2

The six multiplication tables

·	1	ω	ω^2
1	1	ω	ω^2
ω^2	ω^2	1	ω
ω	ω	ω^2	1

·	ω	ω^2	1
ω^2	1	ω	ω^2
ω	ω^2	1	ω
1	ω	ω^2	1

·	ω^2	1	ω
ω	1	ω	ω^2
1	ω^2	1	ω
ω^2	ω	ω^2	1

·	1	ω^2	ω
1	1	ω^2	ω
ω	ω	1	ω^2
ω^2	ω^2	ω	1

·	ω^2	ω	1
ω	1	ω^2	ω
ω^2	ω	1	ω^2
1	ω^2	ω	1

·	ω	1	ω^2
ω^2	1	ω^2	ω
1	ω	1	ω^2
ω	ω^2	ω	1

for 1, ω, and ω^2 are the only ones that have 1 in each position of the principal diagonal. With this restriction, these six multiplication tables are uniquely specified by the corresponding lists

$$\mathscr{L}_1 = (1,\, \omega,\, \omega^2), \quad \mathscr{L}_3 = (\omega,\, \omega^2,\, 1), \quad \mathscr{L}_5 = (\omega^2,\, 1,\, \omega),$$
$$\mathscr{L}_2 = (1,\, \omega^2,\, \omega), \quad \mathscr{L}_4 = (\omega^2,\, \omega,\, 1), \quad \mathscr{L}_6 = (\omega,\, 1,\, \omega^2)$$

of elements that serve as column indices in the respective tables.

When the assignment $1 \mapsto a_1$, $\omega \mapsto a_2$, and $\omega^2 \mapsto a_3$ is made, the first three multiplication tables specify a pattern for the components of the matrix

$$(1.1) \qquad\qquad A = \begin{bmatrix} a_1 & a_2 & a_3 \\ a_3 & a_1 & a_2 \\ a_2 & a_3 & a_1 \end{bmatrix}.$$

When the assignment $1 \mapsto a_1$, $\omega^2 \mapsto a_2$, and $\omega \mapsto a_3$ is made, the remaining three multiplication tables also specify a pattern for the matrix A in (1.1). Thus, with the agreement that the first row of the matrix is given by (a_1, a_2, a_3), all six of these multiplication tables specify the single matrix A in (1.1).

1.2. Patterns of multiplication tables for 1, i, -1, $-i$

For the complex numbers 1, i, -1, $-i$, the three lists

$$\mathscr{L}_1 = \big(1,\, i,\, -1,\, -i\big), \quad \mathscr{L}_2 = \big(1,\, i,\, -i,\, -1\big) \quad \mathscr{L}_3 = \big(1,\, -1,\, i,\, -i\big)$$

specify the corresponding multiplication tables

$\cdot$	1	i	-1	$-i$
1	1	i	-1	$-i$
$-i$	$-i$	1	i	-1
-1	-1	$-i$	1	i
i	i	-1	$-i$	1

$\cdot$	1	i	$-i$	-1
1	1	i	$-i$	-1
$-i$	$-i$	1	-1	i
i	i	-1	1	$-i$
-1	-1	$-i$	i	1

$\cdot$	1	-1	i	$-i$
1	1	-1	i	$-i$
-1	-1	1	$-i$	i
$-i$	$-i$	i	1	-1
i	i	$-i$	-1	1

that yield patterns for the corresponding matrices

$$A_1 = \begin{bmatrix} a_1 & a_2 & a_3 & a_4 \\ a_4 & a_1 & a_2 & a_3 \\ a_3 & a_4 & a_1 & a_2 \\ a_2 & a_3 & a_4 & a_1 \end{bmatrix}, \quad A_2 = \begin{bmatrix} a_1 & a_2 & a_3 & a_4 \\ a_3 & a_1 & a_4 & a_2 \\ a_2 & a_4 & a_1 & a_3 \\ a_4 & a_3 & a_2 & a_1 \end{bmatrix}, \quad A_3 = \begin{bmatrix} a_1 & a_2 & a_3 & a_4 \\ a_2 & a_1 & a_4 & a_3 \\ a_4 & a_3 & a_1 & a_2 \\ a_3 & a_4 & a_2 & a_1 \end{bmatrix}.$$

Of the $4! = 24$ lists for $\{1,\, i,\, -1,\, -i\}$, eight yield two patterns for A_1, another eight yield two patterns for A_2, and the remaining eight yield two pattern for A_3.

The matrix A_1 has symmetry with respect to its secondary diagonal and A_2 has symmetry with respect to its center.

1.3. Patterns for the symmetries of a nonsquare rectangle

For the nonsquare rectangle

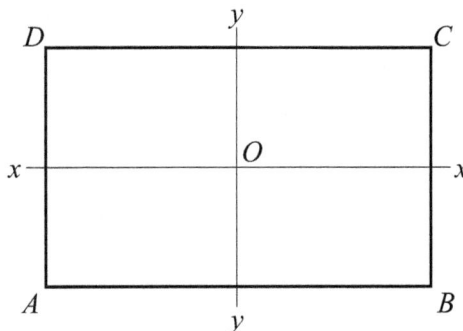

with vertices labeled A, B, C, D as indicated, we use the permutation symbols

$$e = \begin{pmatrix} A\,B\,C\,D \\ A\,B\,C\,D \end{pmatrix}, \quad u = \begin{pmatrix} A\,B\,C\,D \\ D\,C\,B\,A \end{pmatrix}, \quad v = \begin{pmatrix} A\,B\,C\,D \\ B\,A\,D\,C \end{pmatrix}, \quad w = \begin{pmatrix} A\,B\,C\,D \\ C\,D\,A\,B \end{pmatrix}$$

to represent its symmetries. We note that u specifies a reflection in the x-x line, v specifies a reflection in the y-y line, w indicates a rotation about O through 180 degrees, and e is the identity transformation that leaves the rectangle unchanged. Throughout, in computing the product of two juxtaposed permutation symbols, we shall read them from left to right in order to be consistent with references to that usage by Richard Dedekind and Georg Frobenius in Chapters 10, 11, and 23. Thus, we have $uv = w$ because u replaces A, B, C, D with D, C, B, A and v then replaces D, C, B, A with C, D, A, B. In that manner, we obtain

(1.2)

·	e	u	v	w
e	e	u	v	w
u	u	e	w	v
v	v	w	e	u
w	w	v	u	e

as a multiplication table that yields a pattern for the matrix

(1.3)
$$A = \begin{bmatrix} a_1 & a_2 & a_3 & a_4 \\ a_2 & a_1 & a_4 & a_3 \\ a_3 & a_4 & a_1 & a_2 \\ a_4 & a_3 & a_2 & a_1 \end{bmatrix}.$$

We note that A is symmetric with respect to both of its diagonals. Thus, it also has symmetry with respect to its center.

The 24 lists of the four elements e, u, v, w specify 24 multiplication tables having e in each of their diagonal positions. These 24 multiplication tables supply six visibly distinct patterns such that: the matrix A in (1.3) is obtained whenever any one of the six patterns is used to form a 4×4 matrix whose first row is (a_1, a_2, a_3, a_4).

1.4. Patterns of four tables for $1, -\omega, \omega^2, -1, \omega, -\omega^2$

While the six complex numbers $1, -\omega, \omega^2, -1, \omega, -\omega^2$ provide $6! = 720$ lists, we shall merely consider four of them.

The lists

$$\mathscr{L}_1 = \left(1, -1, \ \omega, -\omega, \ \omega^2, -\omega^2\right) \quad \text{and} \quad \mathscr{L}_2 = \left(1, \ \omega, \ \omega^2, -1, -\omega, -\omega^2\right)$$

specify the multiplication tables

·	1	-1	ω	$-\omega$	ω^2	$-\omega^2$
1	1	-1	ω	$-\omega$	ω^2	$-\omega^2$
-1	-1	1	$-\omega$	ω	$-\omega^2$	ω^2
ω^2	ω^2	$-\omega^2$	1	-1	ω	$-\omega$
$-\omega^2$	$-\omega^2$	ω^2	-1	1	$-\omega$	ω
ω	ω	$-\omega$	ω^2	$-\omega^2$	1	-1
$-\omega$	$-\omega$	ω	$-\omega^2$	ω^2	-1	1

·	1	ω	ω^2	-1	$-\omega$	$-\omega^2$
1	1	ω	ω^2	-1	$-\omega$	$-\omega^2$
ω^2	ω^2	1	ω	$-\omega^2$	-1	$-\omega$
ω	ω	ω^2	1	$-\omega$	$-\omega^2$	-1
-1	-1	$-\omega$	$-\omega^2$	1	ω	ω^2
$-\omega^2$	$-\omega^2$	-1	$-\omega$	ω^2	1	ω
$-\omega$	$-\omega$	$-\omega^2$	-1	ω	ω^2	1

whose patterns yield the corresponding matrices

$$A_1 = \begin{bmatrix} a_1 & a_2 & a_3 & a_4 & a_5 & a_6 \\ a_2 & a_1 & a_4 & a_3 & a_6 & a_5 \\ a_5 & a_6 & a_1 & a_2 & a_3 & a_4 \\ a_6 & a_5 & a_2 & a_1 & a_4 & a_3 \\ a_3 & a_4 & a_5 & a_6 & a_1 & a_2 \\ a_4 & a_3 & a_6 & a_5 & a_2 & a_1 \end{bmatrix}, \quad A_2 = \begin{bmatrix} a_1 & a_2 & a_3 & a_4 & a_5 & a_6 \\ a_3 & a_1 & a_2 & a_6 & a_4 & a_5 \\ a_2 & a_3 & a_1 & a_5 & a_6 & a_4 \\ a_4 & a_5 & a_6 & a_1 & a_2 & a_3 \\ a_6 & a_4 & a_5 & a_3 & a_1 & a_2 \\ a_5 & a_6 & a_4 & a_2 & a_3 & a_1 \end{bmatrix}.$$

The lists

$$\mathscr{L}_3 = \left(1, -\omega, \omega^2, -1, \omega, -\omega^2\right) \quad \text{and} \quad \mathscr{L}_4 = \left(1, -\omega, \omega^2, -\omega^2, \omega, -1\right)$$

specify the multiplication tables

·	1	$-\omega$	ω^2	-1	ω	$-\omega^2$
1	1	$-\omega$	ω^2	-1	ω	$-\omega^2$
$-\omega^2$	$-\omega^2$	1	$-\omega$	ω^2	-1	ω
ω	ω	$-\omega^2$	1	$-\omega$	ω^2	-1
-1	-1	ω	$-\omega^2$	1	$-\omega$	ω^2
ω^2	ω^2	-1	ω	$-\omega^2$	1	$-\omega$
$-\omega$	$-\omega$	ω^2	-1	ω	$-\omega^2$	1

·	1	$-\omega$	ω^2	$-\omega^2$	ω	-1
1	1	$-\omega$	ω^2	$-\omega^2$	ω	-1
$-\omega^2$	$-\omega^2$	1	$-\omega$	ω	-1	ω^2
ω	ω	$-\omega^2$	1	-1	ω^2	$-\omega$
$-\omega$	$-\omega$	ω^2	-1	1	$-\omega^2$	ω
ω^2	ω^2	-1	ω	$-\omega$	1	$-\omega^2$
-1	-1	ω	$-\omega^2$	ω^2	$-\omega$	1

whose patterns yield the corresponding matrices

$$A_3 = \begin{bmatrix} a_1 & a_2 & a_3 & a_4 & a_5 & a_6 \\ a_6 & a_1 & a_2 & a_3 & a_4 & a_5 \\ a_5 & a_6 & a_1 & a_2 & a_3 & a_4 \\ a_4 & a_5 & a_6 & a_1 & a_2 & a_3 \\ a_3 & a_4 & a_5 & a_6 & a_1 & a_2 \\ a_2 & a_3 & a_4 & a_5 & a_6 & a_1 \end{bmatrix}, \quad A_4 = \begin{bmatrix} a_1 & a_2 & a_3 & a_4 & a_5 & a_6 \\ a_4 & a_1 & a_2 & a_5 & a_6 & a_3 \\ a_5 & a_4 & a_1 & a_6 & a_3 & a_2 \\ a_2 & a_3 & a_6 & a_1 & a_4 & a_5 \\ a_3 & a_6 & a_5 & a_2 & a_1 & a_4 \\ a_6 & a_5 & a_4 & a_3 & a_2 & a_1 \end{bmatrix}.$$

The submatrices indicated for A_1 and A_2 have subpatterns for their placement. The circulant matrix A_3 is symmetric with respect to its secondary diagonal; and, A_4 has symmetry with respect to its center.

1.5. A pattern for the symmetries of an equilateral triangle

The symmetries of the equilateral triangle

(1.4)

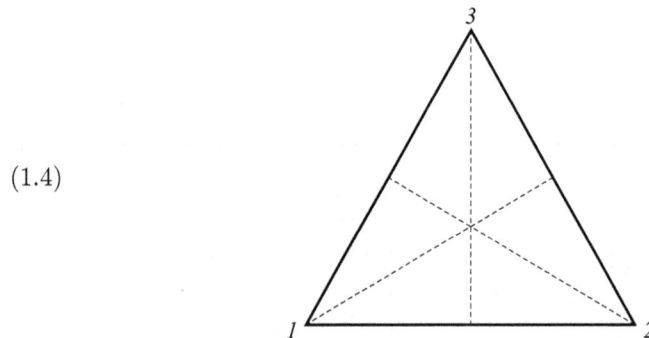

are represented by

$$(1.5) \quad \begin{cases} e = \begin{pmatrix} 1\,2\,3 \\ 1\,2\,3 \end{pmatrix}, & \alpha = \begin{pmatrix} 1\,2\,3 \\ 3\,1\,2 \end{pmatrix}, & \alpha^2 = \begin{pmatrix} 1\,2\,3 \\ 2\,3\,1 \end{pmatrix}, \\ \beta_1 = \begin{pmatrix} 1\,2\,3 \\ 1\,3\,2 \end{pmatrix}, & \beta_2 = \begin{pmatrix} 1\,2\,3 \\ 3\,2\,1 \end{pmatrix}, & \beta_3 = \begin{pmatrix} 1\,2\,3 \\ 2\,1\,3 \end{pmatrix}. \end{cases}$$

We interpret e, α, and α^2 respectively as counterclockwise rotations about the center through 0, 120, and 240 degrees; while, β_1, β_2, and β_3 specify reflections in the altitudes from the vertices 1, 2, and 3. The list $\mathscr{L} = \left(e,\ \alpha,\ \alpha^2,\ \beta_1,\ \beta_2,\ \beta_3\right)$ uniquely specifies the multiplication table

(1.6)

$\cdot$	e	α	α^2	β_1	β_2	β_3
e	e	α	α^2	β_1	β_2	β_3
α^2	α^2	e	α	β_2	β_3	β_1
α	α	α^2	e	β_3	β_1	β_2
β_1	β_1	β_2	β_3	e	α	α^2
β_2	β_2	β_3	β_1	α^2	e	α
β_3	β_3	β_1	β_2	α	α^2	e

as a noncommutative one, and its pattern specifies the corresponding matrix

(1.7)

$$
A = \left[
\begin{array}{ccc|ccc}
a_1 & a_2 & a_3 & a_4 & a_5 & a_6 \\
a_3 & a_1 & a_2 & a_5 & a_6 & a_4 \\
a_2 & a_3 & a_1 & a_6 & a_4 & a_5 \\
\hline
a_4 & a_5 & a_6 & a_1 & a_2 & a_3 \\
a_5 & a_6 & a_4 & a_3 & a_1 & a_2 \\
a_6 & a_4 & a_5 & a_2 & a_3 & a_1
\end{array}
\right].
$$

There are two subpatterns for the indicated 3×3 submatrices of (1.7).

1.6. A pattern for the symmetries of a square

The square

(1.8)

has the eight symmetries represented by

$$
e = \begin{pmatrix} 1\,2\,3\,4 \\ 1\,2\,3\,4 \end{pmatrix}, \quad
\alpha = \begin{pmatrix} 1\,2\,3\,4 \\ 4\,1\,2\,3 \end{pmatrix}, \quad
\alpha^2 = \begin{pmatrix} 1\,2\,3\,4 \\ 3\,4\,1\,2 \end{pmatrix}, \quad
\alpha^3 = \begin{pmatrix} 1\,2\,3\,4 \\ 2\,3\,4\,1 \end{pmatrix},
$$

$$
\beta_1 = \begin{pmatrix} 1\,2\,3\,4 \\ 1\,4\,3\,2 \end{pmatrix}, \quad
\beta_2 = \begin{pmatrix} 1\,2\,3\,4 \\ 4\,3\,2\,1 \end{pmatrix}, \quad
\beta_3 = \begin{pmatrix} 1\,2\,3\,4 \\ 3\,2\,1\,4 \end{pmatrix}, \quad
\beta_4 = \begin{pmatrix} 1\,2\,3\,4 \\ 2\,1\,4\,3 \end{pmatrix}.
$$

We interpret e, α, α^2 and α^3 as counterclockwise rotations about O through 0, 90, 180, and 270 degrees. Also, β_1 is a reflection in the line through vertices 1 and 3, β_2 is a reflection in the x-x line, β_3 is a reflection in the line through vertices 2 and 4, while β_4 is a reflection in the y-y line. With respect to the list

$\mathscr{L} = \left(e,\, \alpha,\, \alpha^2,\, \alpha^3,\, \beta_1,\, \beta_2,\, \beta_3,\, \beta_4\right)$, this yields the multiplication table

(1.9)

$\cdot$	e	α	α^2	α^3	β_1	β_2	β_3	β_4
e	e	α	α^2	α^3	β_1	β_2	β_3	β_4
α^3	α^3	e	α	α^2	β_2	β_3	β_4	β_1
α^2	α^2	α^3	e	α	β_3	β_4	β_1	β_2
α	α	α^2	α^3	e	β_4	β_1	β_2	β_3
β_1	β_1	β_2	β_3	β_4	e	α	α^2	α^3
β_2	β_2	β_3	β_4	β_1	α^3	e	α	α^2
β_3	β_3	β_4	β_1	β_2	α^2	α^3	e	α
β_4	β_4	β_1	β_2	β_3	α	α^2	α^3	e

as a noncommutative one. Its pattern specifies the corresponding matrix

(1.10)

$$A = \left[\begin{array}{cccc|cccc}
a_1 & a_2 & a_3 & a_4 & a_5 & a_6 & a_7 & a_8 \\
a_4 & a_1 & a_2 & a_3 & a_6 & a_7 & a_8 & a_5 \\
a_3 & a_4 & a_1 & a_2 & a_7 & a_8 & a_5 & a_6 \\
a_4 & a_3 & a_2 & a_1 & a_8 & a_5 & a_6 & a_7 \\
\hline
a_5 & a_6 & a_7 & a_8 & a_1 & a_2 & a_3 & a_4 \\
a_6 & a_7 & a_8 & a_5 & a_4 & a_1 & a_2 & a_3 \\
a_7 & a_8 & a_5 & a_6 & a_3 & a_4 & a_1 & a_2 \\
a_8 & a_5 & a_6 & a_7 & a_2 & a_3 & a_4 & a_1
\end{array}\right].$$

Interesting subpatterns yield the indicated 4×4 submatrices of (1.10).

1.7. A pattern for the symmetries of a brick-like object

For the rectangular parallelepiped

(1.11)

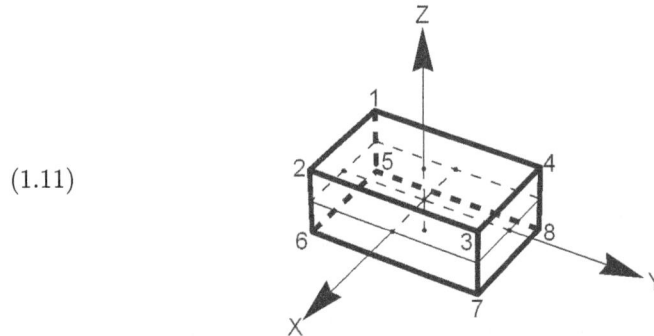

where no two nonparallel edges have the same length, the there symmetries

$$x = \begin{pmatrix} 1\,2\,3\,4\,5\,6\,7\,8 \\ 8\,7\,6\,5\,4\,3\,2\,1 \end{pmatrix}, \quad y = \begin{pmatrix} 1\,2\,3\,4\,5\,6\,7\,8 \\ 6\,5\,8\,7\,2\,1\,4\,3 \end{pmatrix}, \quad z = \begin{pmatrix} 1\,2\,3\,4\,5\,6\,7\,8 \\ 3\,4\,1\,2\,7\,8\,5\,6 \end{pmatrix}$$

correspond respectively to rotations of 180 degrees about the X, Y, and Z axes. We check that a half turn about any one of the three axes followed by a half-turn about a second one of the axes is equivalent to a half-turn about the third axis. Thus, we obtain $x\,y = y\,x = z$, $x\,z = z\,x = y$, and $y\,z = z\,y = x$.

We also have the symmetries

$$m_{xy} = \begin{pmatrix} 1\,2\,3\,4\,5\,6\,7\,8 \\ 5\,6\,7\,8\,1\,2\,3\,4 \end{pmatrix}, \quad m_{xz} = \begin{pmatrix} 1\,2\,3\,4\,5\,6\,7\,8 \\ 4\,3\,2\,1\,8\,7\,6\,5 \end{pmatrix}, \quad m_{yz} = \begin{pmatrix} 1\,2\,3\,4\,5\,6\,7\,8 \\ 2\,1\,4\,3\,6\,5\,8\,7 \end{pmatrix}$$

that correspond respectively to reflections in the xy-plane, the xz-plane, and the yz-plane. We note that (1.11) emphasizes m_{xy} as a reflection in the xy-plane. It is interesting to observe that $m_{xy}m_{xz} = m_{xz}m_{xy} = x$, $m_{xy}m_{yz} = m_{yz}m_{xy} = y$, and $m_{xz}m_{yz} = m_{yz}m_{xz} = z$. The remaining two symmetries are e and m_0 given by

$$e = \begin{pmatrix} 1\,2\,3\,4\,5\,6\,7\,8 \\ 1\,2\,3\,4\,5\,6\,7\,8 \end{pmatrix} \quad \text{and} \quad m_0 = \begin{pmatrix} 1\,2\,3\,4\,5\,6\,7\,8 \\ 7\,8\,5\,6\,3\,4\,1\,2 \end{pmatrix},$$

where e is the identity transformation and, in any order, $m_{xy}m_{yz}m_{xz} = m_0$. Thus, m_0 is a central symmetry produced by the point reflection in the origin.

The list $\mathscr{L}_1 = (e, x, y, z, m_{xy}, m_{xz}, m_{yz}, m_0)$ specifies the multiplication table

(1.12)

$\cdot$	e	x	y	z	m_{xy}	m_{xz}	m_{yz}	m_0
e	e	x	y	z	m_{xy}	m_{xz}	m_{yz}	m_0
x	x	e	z	y	m_{xz}	m_{xy}	m_0	m_{yz}
y	y	z	e	x	m_{yz}	m_0	m_{xy}	m_{xz}
z	z	y	x	e	m_0	m_{yz}	m_{xz}	m_{xy}
m_{xy}	m_{xy}	m_{xz}	m_{yz}	m_0	e	x	y	z
m_{xz}	m_{xz}	m_{xy}	m_0	m_{yz}	x	e	z	y
m_{yz}	m_{yz}	m_0	m_{xy}	m_{xz}	y	z	e	x
m_0	m_0	m_{yz}	m_{xz}	m_{xy}	z	y	x	e

whose computations are simplified by noticing that $m_{xz} = x\,m_{xy}$, $m_{yz} = y\,m_{xy}$, and $m_0 = z\,m_{xy}$. The pattern for (1.12) yields the corresponding matrix

(1.13)
$$A = \left[\begin{array}{cccc|cccc} a_1 & a_2 & a_3 & a_4 & a_5 & a_6 & a_7 & a_8 \\ a_2 & a_1 & a_4 & a_3 & a_6 & a_5 & a_8 & a_7 \\ a_3 & a_4 & a_1 & a_2 & a_7 & a_8 & a_5 & a_6 \\ a_4 & a_3 & a_2 & a_1 & a_8 & a_7 & a_6 & a_5 \\ \hline a_5 & a_6 & a_7 & a_8 & a_1 & a_2 & a_3 & a_4 \\ a_6 & a_5 & a_8 & a_7 & a_2 & a_1 & a_4 & a_3 \\ a_7 & a_8 & a_5 & a_6 & a_3 & a_4 & a_1 & a_2 \\ a_8 & a_7 & a_6 & a_5 & a_4 & a_3 & a_2 & a_1 \end{array}\right].$$

Just as for the matrices

$$\begin{bmatrix} a_1 & a_2 \\ a_2 & a_1 \end{bmatrix} \quad \text{and} \quad \left[\begin{array}{cc|cc} a_1 & a_2 & a_3 & a_4 \\ a_2 & a_1 & a_4 & a_3 \\ \hline a_3 & a_4 & a_1 & a_2 \\ a_4 & a_3 & a_2 & a_1 \end{array}\right],$$

the matrix A in (1.13) is symmetric with respect to both diagonals. Therefore, it also has symmetry with respect to its center.

1.8. A pattern for some symmetries of a tile-like object

The rectangular parallelepiped

(1.14)

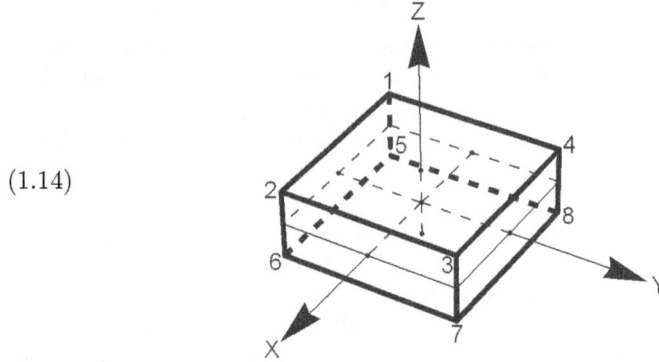

has the lengths of its edges parallel to the X-Y plane equal to each other but unequal to the length of its edges parallel to the Z-axis. The symbols

$$e = \begin{pmatrix} 1\,2\,3\,4\,5\,6\,7\,8 \\ 1\,2\,3\,4\,5\,6\,7\,8 \end{pmatrix}, \quad \alpha = \begin{pmatrix} 1\,2\,3\,4\,5\,6\,7\,8 \\ 4\,1\,2\,3\,8\,5\,6\,7 \end{pmatrix}, \quad m = \begin{pmatrix} 1\,2\,3\,4\,5\,6\,7\,8 \\ 5\,6\,7\,8\,1\,2\,3\,4 \end{pmatrix}$$

correspond respectively to the identity transformation, a rotation about the Z-axis through 90 degrees, and a reflection in the X-Y plane. With $\alpha m = m \alpha$, the list

$$\mathscr{L}_1 = \left(e,\, \alpha, \alpha^2,\, \alpha^3,\, m, m\alpha,\, m\alpha^2,\, m\alpha^3 \right)$$

specifies the commutative multiplication table

(1.15)

$\cdot$	e	α	α^2	α^3	m	$m\alpha$	$m\alpha^2$	$m\alpha^3$
e	e	α	α^2	α^3	m	$m\alpha$	$m\alpha^2$	$m\alpha^3$
α^3	α^3	e	α	α^2	$m\alpha^3$	m	$m\alpha$	$m\alpha^2$
α^2	α^2	α^3	e	α	$m\alpha^2$	$m\alpha^3$	m	$m\alpha$
α	α	α^2	α^3	e	$m\alpha$	$m\alpha^2$	$m\alpha^3$	m
m	m	$m\alpha$	$m\alpha^2$	$m\alpha^3$	e	α	α^2	α^3
$m\alpha^3$	$m\alpha^3$	m	$m\alpha$	$m\alpha^2$	α^3	e	α	α^2
$m\alpha^2$	$m\alpha^2$	$m\alpha^3$	m	$m\alpha$	α^2	α^3	e	α
$m\alpha$	$m\alpha$	$m\alpha^2$	$m\alpha^3$	m	α	α^2	α^3	e

and its pattern yields the corresponding matrix

(1.16)
$$A = \left[\begin{array}{cccc|cccc} a_1 & a_2 & a_3 & a_4 & a_5 & a_6 & a_7 & a_8 \\ a_4 & a_1 & a_2 & a_3 & a_8 & a_5 & a_6 & a_7 \\ a_3 & a_4 & a_1 & a_2 & a_7 & a_8 & a_5 & a_6 \\ a_2 & a_3 & a_4 & a_1 & a_6 & a_7 & a_8 & a_5 \\ \hline a_5 & a_6 & a_7 & a_8 & a_1 & a_2 & a_3 & a_4 \\ a_8 & a_5 & a_6 & a_7 & a_4 & a_1 & a_2 & a_3 \\ a_7 & a_8 & a_5 & a_6 & a_3 & a_4 & a_1 & a_2 \\ a_6 & a_7 & a_8 & a_5 & a_2 & a_3 & a_4 & a_1 \end{array} \right] .$$

The indicated 4×4 submatrices of (1.16) have 4×4 circulant subpatterns.

We observe that the cross section of (1.14) with the X, Y-plane can be identified with the square (1.8) in Section 1.6. Thus, by supplementing the symmetries

$$e = \begin{pmatrix} 1\,2\,3\,4\,5\,6\,7\,8 \\ 1\,2\,3\,4\,5\,6\,7\,8 \end{pmatrix} \quad \text{and} \quad \alpha = \begin{pmatrix} 1\,2\,3\,4\,5\,6\,7\,8 \\ 4\,1\,2\,3\,8\,5\,6\,7 \end{pmatrix}$$

already introduced for (1.14) with the symbolism

$$\beta_2 = \begin{pmatrix} 1\,2\,3\,4\,5\,6\,7\,8 \\ 8\,7\,6\,5\,4\,3\,2\,1 \end{pmatrix} \quad \text{and} \quad \beta_4 = \begin{pmatrix} 1\,2\,3\,4\,5\,6\,7\,8 \\ 6\,5\,8\,7\,2\,1\,4\,3 \end{pmatrix}$$

to represent half-turns about the X and Y axes as well as the symbolism

$$\beta_1 = \begin{pmatrix} 1\,2\,3\,4\,5\,6\,7\,8 \\ 5\,8\,7\,6\,1\,4\,3\,2 \end{pmatrix} \quad \text{and} \quad \beta_3 = \begin{pmatrix} 1\,2\,3\,4\,5\,6\,7\,8 \\ 7\,6\,5\,8\,3\,2\,1\,4 \end{pmatrix}$$

for half-turns about the lines $Y = X$ and $Y = -X$ in the X, Y-plane, we find that the list $\mathscr{L} = \left(e, \alpha, \alpha^2, \alpha^3, \beta_1, \beta_2, \beta_3, \beta_4\right)$ specifies a multiplication table having the same pattern as (1.9) of page 6. Of course, that yields (1.10).

EXERCISES

1. Find a list $\mathscr{L}_1$ of symmetries for (1.14) that specifies a multiplication table whose pattern yields matrices of the type (1.3) on page 3.

2. Find three lists of symmetries for (1.14) that specify three multiplication tables whose patterns yield matrices of the type A_1, A_2, A_3 in Section 1.2 on page 2.

3. The rectangular parallelepiped shown in (1.14) has more symmetries than those listed for (1.15). Specify all of them.

CHAPTER 2

Formulations for Group-Pattern Matrices

2.1. Definitions and terminology

Let G denote a non-vacuous set and let $\cdot$ denote a multiplication for G in the sense that: for any two elements x, y in G, there is unique element in G denoted by $x \cdot y$. This can be described by saying that there is a function from the Cartesian product $G \times G$ to G indicated by $(x, y) \mapsto x \cdot y$, for each x, y in G. Examples of sets having multiplications are given throughout Chapter 1.

DEFINITION 2.1. A set G with a multiplication $\cdot$ is said to be a **group** when:
(i) for each x, y, z in G, the multiplication satisfies $x \cdot (y \cdot z) = (x \cdot y) \cdot z$;
(ii) there is an element e in G such that $e \cdot x = x \cdot e = x$, for each x in G; and,
(iii) for any x in G, there is an element y in G such that $x \cdot y = y \cdot x = e$.
A group G is said to be **abelian** when $v \cdot u = u \cdot v$, for each u, v in G.

The element e is called the **identity element** of G. It is unique because: if e_1 in G has the property $e_1 \cdot x = x \cdot e_1 = x$, for each x in G, then $e_1 = e_1 \cdot e = e$. Also, when $x \cdot y = y \cdot x = e$, the element y is said to be the **inverse** of x and the notation $x^{-1} = y$ is used. It is uniquely specified by x because: if y_1 is an element of G such that $x \cdot y_1 = y_1 \cdot x = e$, then $y_1 = y_1 \cdot e = y_1 \cdot (x \cdot y) = (y_1 \cdot x) \cdot y = e \cdot y = y$. For any x, y in G, we introduce the notation $x\,y = x \cdot y$.

Henceforth, let G be a group for which the number of its elements is a positive integer n. Then, G is said to be a finite group of **order** n and its elements can be denoted in such a manner that they constitute the n elements of the list

$$(2.1) \qquad \mathscr{L} = (g_1, g_2, \ldots, g_n).$$

This list and the multiplication for G uniquely determine the multiplication table

(2.2)

$\cdot$	g_1	g_2	g_3	$\cdots$	g_s	$\cdots$	g_{n-1}	g_n
g_1^{-1}	e	$g_1^{-1}g_2$	$g_1^{-1}g_3$	$\cdots$	$g_1^{-1}g_s$	$\cdots$	$g_1^{-1}g_{n-1}$	$g_1^{-1}g_n$
g_2^{-1}	$g_2^{-1}g_1$	e	$g_2^{-1}g_3$	$\cdots$	$g_2^{-1}g_s$	$\cdots$	$g_2^{-1}g_{n-1}$	$g_2^{-1}g_n$
g_3^{-1}	$g_3^{-1}g_1$	$g_3^{-1}g_2$	e	$\cdots$	$g_3^{-1}g_s$	$\cdots$	$g_3^{-1}g_{n-1}$	$g_3^{-1}g_n$
$\vdots$	$\vdots$	$\vdots$	$\vdots$	$\vdots$	$\vdots$	$\vdots$	$\vdots$	$\vdots$
g_r^{-1}	$g_r^{-1}g_1$	$g_r^{-1}g_2$	$g_r^{-1}g_3$	$\cdots$	$g_r^{-1}g_s$	$\cdots$	$g_r^{-1}g_{n-1}$	$g_r^{-1}g_n$
$\vdots$	$\vdots$	$\vdots$	$\vdots$	$\vdots$	$\vdots$	$\vdots$	$\vdots$	$\vdots$
g_{n-1}^{-1}	$g_{n-1}^{-1}g_1$	$g_{n-1}^{-1}g_2$	$g_{n-1}^{-1}g_3$	$\cdots$	$g_{n-1}^{-1}g_s$	$\cdots$	e	$g_{n-1}^{-1}g_n$
g_n^{-1}	$g_n^{-1}g_1$	$g_n^{-1}g_2$	$g_n^{-1}g_3$	$\cdots$	$g_n^{-1}g_s$	$\cdots$	$g_n^{-1}g_{n-1}$	e

where the column labels are specified by (2.1) and the identity element e of G appears in each of the n principal-diagonal positions of (2.2). Thus, for $1 \le r \le n$, the rth-row label for (2.2) must be g_r^{-1} because its rth-column label is g_r.

When the components of an $n \times n$ matrix A belonging to a set S are positioned according to the pattern provided by (2.2), the element $\left[A\right]_{r,s}$ in the rth row and sth column of A is equal to the element of S that corresponds to the element $g_r^{-1} g_s$ in the rth row and sth column of (2.2). In this situation, there is therefore a function σ from G to S such that $\left[A\right]_{r,s} = \sigma\!\left(g_r^{-1} g_s\right)$, for r, $s = 1, 2, \ldots, n$.

DEFINITION 2.2. For a list $\mathscr{L} = (g_1, g_2, \ldots, g_n)$ of the elements in a group G of order n and an $n \times n$ matrix A with components in a set S, the matrix A is said to be a **group-pattern matrix** for G and $\mathscr{L}$ when there is a function σ from G to S such that the component of A in its rth row and sth column is given by

$$(2.3) \qquad \left[A\right]_{r,s} = \sigma\!\left(g_r^{-1} g_s\right), \quad \text{for } r = 1, 2, \ldots, n \text{ and } s = 1, 2, \ldots, n.$$

A group-pattern matrix given by (2.3) is **injective** if and only if σ is one-to-one. Henceforth, $\mathcal{M}[S, G, \mathscr{L}]$ denotes the set of group-pattern matrices for G and $\mathscr{L}$ that have components in S.

EXAMPLE 2.3. As illustrations of group-pattern matrices, we have the thirteen matrices in Chapter 1 with names selected from among A, A_1, A_2, A_3, A_4. They are injective when no two of a_1, a_2, a_3, $\ldots$ are equal. Associativity of multiplication, i.e. (i) of Definition 2.1, is satisfied throughout Chapter 1 because it either involves multiplication of complex numbers or composition of functions. In each situation, the other group properties are easily checked.

DEFINITION 2.4. A group G is said to be **cyclic** when it contains an element α such that each element of G is expressible as α^k, for some integer k. Any such element α for G is said to be a **generator** for G. Since Exercise 6 at the end of this chapter shows that any two cyclic groups of order n are isomorphic, the notation $\mathcal{C}_n$ is suitable to represent any cyclic group of order n.

EXAMPLE 2.5. Let n denote a positive integer; set $\alpha = \cos\left(\frac{2\pi}{n}\right) + i \sin\left(\frac{2\pi}{n}\right)$, with $i^2 = -1$; and let G be the set consisting of the n distinct complex numbers $g_k = \alpha^{k-1}$, for $1 \leq k \leq n$. Then, with multiplication for complex numbers, G is a cyclic group $\mathcal{C}_n$ of order n and α is a generator. The multiplication table for G and the list $\mathscr{L} = (g_1, g_2, \ldots, g_n)$ of its n elements is given by (0.7) on page xvi. For any function σ from G to a set S having $\sigma(g_k) = a_k$, for $1 \leq k \leq n$, the corresponding group-pattern matrix for G and $\mathscr{L}$ is given by (2.3) as the $n \times n$ circulant matrix

$$(2.4) \qquad A = \begin{bmatrix} a_1 & a_2 & a_3 & \cdots & a_{n-1} & a_n \\ a_n & a_1 & a_2 & \cdots & a_{n-2} & a_{n-1} \\ a_{n-1} & a_n & a_1 & \cdots & a_{n-3} & a_{n-2} \\ \vdots & \vdots & \vdots & \ddots & \vdots & \vdots \\ a_3 & a_4 & a_5 & \cdots & a_1 & a_2 \\ a_2 & a_3 & a_4 & \cdots & a_n & a_1 \end{bmatrix}.$$

Thus, $\mathcal{M}[S, \mathcal{C}_n, \mathscr{L}]$ is the set of $n \times n$ circulant matrices having components in S.

When no two of a_1, a_2, $\ldots$, a_n are equal, the matrix A in (2.4) is injective.

OBSERVATION 2.6. The notation (2.3) was introduced in [**9**] of 1976 to easily derive algebraic properties of $\mathcal{M}[\mathfrak{R}, G, \mathscr{L}]$ and $\mathcal{M}[\mathfrak{F}, G, \mathscr{L}]$ when S is a ring $\mathfrak{R}$ or a field $\mathfrak{F}$. For results of that kind, see Chapter 4 and later chapters.

The problem of recognizing whether a given matrix is an injective group-pattern matrix is one that naturally continues our present context. We begin that study in Section 2.4.

2.2. Restriction to lists $\mathscr{L} = (g_1,, g_2, \ldots, g_n)$ where $g_1 = e$

Chapter 1 provides thirteen examples of group-pattern matrices and for each of them the multiplication table is specified by a list whose first component is the identity element e of the group. The next result shows that: among the lists for G that specify a given group-pattern matrix, there is always at least one such list whose first element is the identity element e for G.

PROPOSITION 2.7. *The interior of each multiplication table (2.2) is specified by at least one list of the form (2.1) in which g_1 is the identity element e of G.*

PROOF. For the multiplication table (2.2) specified by the list in (2.1), we introduce $h_k = g_1^{-1} g_k$, for $1 \le k \le n$, and set

$$(2.5) \qquad \mathscr{L}_1 = \big(h_1, h_2, \ldots, h_n\big) = \big(g_1^{-1} g_1, g_1^{-1} g_2, \ldots, g_1^{-1} g_n\big), \quad \text{where } h_1 = e.$$

For $r, s = 1, 2, \ldots, n$, we have $h_r^{-1} h_s = (g_1^{-1} g_r)^{-1}(g_1^{-1} g_s) = g_r^{-1} g_s$ and obtain

(2.6)

$\cdot$	h_1	h_2	h_3	$\ldots$	h_s	$\ldots$	h_{n-1}	h_n
h_1^{-1}	e	$g_1^{-1}g_2$	$g_1^{-1}g_3$	$\ldots$	$g_1^{-1}g_s$	$\ldots$	$g_1^{-1}g_{n-1}$	$g_1^{-1}g_n$
h_2^{-1}	$g_2^{-1}g_1$	e	$g_2^{-1}g_3$	$\ldots$	$g_2^{-1}g_s$	$\ldots$	$g_2^{-1}g_{n-1}$	$g_2^{-1}g_n$
h_3^{-1}	$g_3^{-1}g_1$	$g_3^{-1}g_2$	e	$\ldots$	$g_3^{-1}g_s$	$\ldots$	$g_3^{-1}g_{n-1}$	$g_3^{-1}g_n$
$\vdots$	$\vdots$	$\vdots$	$\vdots$	$\vdots$	$\vdots$	$\vdots$	$\vdots$	$\vdots$
h_r^{-1}	$g_r^{-1}g_1$	$g_r^{-1}g_2$	$g_r^{-1}g_3$	$\ldots$	$g_r^{-1}g_s$	$\ldots$	$g_r^{-1}g_{n-1}$	$g_r^{-1}g_n$
$\vdots$	$\vdots$	$\vdots$	$\vdots$	$\vdots$	$\vdots$	$\vdots$	$\vdots$	$\vdots$
h_{n-1}^{-1}	$g_{n-1}^{-1}g_1$	$g_{n-1}^{-1}g_2$	$g_{n-1}^{-1}g_3$	$\ldots$	$g_{n-1}^{-1}g_s$	$\ldots$	e	$g_{n-1}^{-1}g_n$
h_n^{-1}	$g_n^{-1}g_1$	$g_n^{-1}g_2$	$g_n^{-1}g_3$	$\ldots$	$g_n^{-1}g_s$	$\ldots$	$g_n^{-1}g_{n-1}$	e

as the multiplication table for G specified by $\mathscr{L}_1$ in (2.5). The interior of the multiplication table (2.6) is the same as that of (2.2) and the list $\mathscr{L}_1$ has $h_1 = e$. This completes the proof. $\square$

EXAMPLE 2.8. For the group $G = \big\{1, \omega, \omega^2\big\}$ of page 1 and the six lists

$$\mathscr{L}_1 = \big(1, \omega, \omega^2\big), \quad \mathscr{L}_3 = \big(\omega, \omega^2, 1\big), \quad \mathscr{L}_5 = \big(\omega^2, 1, \omega\big),$$
$$\mathscr{L}_2 = \big(1, \omega^2, \omega\big), \quad \mathscr{L}_4 = \big(\omega^2, \omega, 1\big), \quad \mathscr{L}_6 = \big(\omega, 1, \omega^2\big)$$

of its elements, the argument above for Proposition 2.7 shows that $\mathscr{L}_3$ and $\mathscr{L}_5$ are redundant to $\mathscr{L}_1$ and it shows that $\mathscr{L}_4$ and $\mathscr{L}_6$ are redundant to $\mathscr{L}_2$. Moreover, the argumet for (1.1) shows that the six sets $\mathcal{M}[S, G, \mathscr{L}_k]$, for $1 \le k \le 6$, are equal. This example provides a simple illustration for Theorem 3.7 and Proposition 3.15.

.

2.3. The set $\mathcal{M}[\mathbb{C}, G, \mathscr{L}]$ is closed under matrix multiplication

THEOREM 2.9. *The set $\mathcal{M}[\mathbb{C}, G, \mathscr{L}]$ of group-pattern matrices for G and $\mathscr{L}$ that have components in the field $\mathbb{C}$ of complex numbers is closed under matrix multiplication and it contains precisely n permutation matrices.*

PROOF. For A_1, A_2 in $\mathcal{M}[\mathbb{C}, G, \mathscr{L}]$, functions σ_1, σ_2 from G to $\mathbb{C}$ exist having

$$(2.7) \quad \left[A_1\right]_{r,s} = \sigma_1\!\left(g_r^{-1} g_s\right) \quad \text{and} \quad \left[A_2\right]_{r,s} = \sigma_2\!\left(g_r^{-1} g_s\right), \quad \text{for } r,\, s = 1,\, 2,\, \ldots,\, n.$$

We use (2.7) to obtain

$$\left[A_1 A_2\right]_{r,s} = \sum_{k=1}^{n} \left[A_1\right]_{r,k} \left[A_2\right]_{k,s} = \sum_{k=1}^{n} \sigma_1\!\left(g_r^{-1} g_k\right) \sigma_2\!\left(g_k^{-1} g_s\right)$$

$$= \sum_{k=1}^{n} \sigma_1\!\left(g_r^{-1} g_k\right) \sigma_2\!\left(\left(g_r^{-1} g_k\right)^{-1}\!\left(g_r^{-1} g_s\right)\right), \quad \text{for } r,\, s = 1,\, 2,\, \ldots,\, n.$$

For each fixed r, as k ranges from 1 to n, both g_k and $g_r^{-1} g_k$ range through the elements of G. Thus, we have

$$\left[A_1 A_2\right]_{r,s} = \sum_{y \in G} \sigma_1(y)\, \sigma_2\!\left(y^{-1}\!\left(g_r^{-1} g_s\right)\right) = \tau\!\left(g_r^{-1} g_s\right), \quad \text{for } r,\, s = 1,\, 2,\, \ldots,\, n,$$

where τ is the function from G to $\mathbb{C}$ defined by

$$\tau(x) = \sum_{y \in G} \sigma_1(y)\, \sigma_2\!\left(y^{-1} x\right), \quad \text{for each } x \text{ in } G.$$

Consequently, the product $A_1 A_2$ is a group-pattern matrix for G and $\mathscr{L}$ having components in $\mathbb{C}$; it is therefore an element of $\mathcal{M}[\mathbb{C}, G, \mathscr{L}]$.

Suppose P is a permutation matrix in $\mathcal{M}[\mathbb{C}, G, \mathscr{L}]$ and let ϕ be the function from G to $\{0, 1\}$ having $\left[P\right]_{r,s} = \phi(g_r^{-1} g_s)$, for $1 \leq r,\, s \leq n$. If $\phi(g_{k_0}) = 1$, then $\phi(g_k) = 0$, for $k \neq k_0$. Thus, there are at most n such matrices in $\mathcal{M}[\mathbb{C}, G, \mathscr{L}]$.

Since the n distinct permutation matrices P_1, P_2, $\ldots$, P_n defined by

$$P_k = \phi_k(g_r^{-1} g_s) \quad \text{with} \quad \phi_k(g_\nu) = \begin{cases} 1, & \text{if } g_\nu = g_k, \\ 0, & \text{if } g_\nu \neq g_k, \end{cases} \quad \text{for } 1 \leq k,\, r,\, s,\, \nu \leq n,$$

are elements of $\mathcal{M}[\mathbb{C}, G, \mathscr{L}]$, this completes the proof. $\qquad\square$

2.4. Injective group-pattern matrices

THEOREM 2.10. *Suppose that Z is an injective group-pattern matrix. Then, the following six assertions about Z are valid.*

    A: *Z is a matrix of size $n \times n$ for some positive integer n;*

    B: *each row of Z and each column of Z has precisely n distinct elements;*

    C: *the number of distinct elements that are components of Z is n;*

    D: *each of the elements in the principal diagonal of Z is equal to $\left[Z\right]_{1,1}$;*

    E: *for $k = 1, 2, \ldots, n$, an $n \times n$ permutation matrix P_k is obtained from Z by replacing each of the n components of Z equal to $\left[Z\right]_{1,k}$ with 1 from $\mathbb{C}$ and replacing each of the $n^2 - n$ components of Z that are unequal to $\left[Z\right]_{1,k}$ with 0 from $\mathbb{C}$; and*

    F: *the set $\mathfrak{P} = \left\{P_1, P_2, \ldots, P_n\right\}$ is closed under matrix multiplication.*

PROOF. We use Definition 2.2 on page 12 to observe that there is a group G of order n as well as a list $\mathscr{L} = (g_1, g_2, \ldots, g_n)$ of its n elements and a one-to-one function σ from G to a set S such that the component $[Z]_{r,s}$ of Z in its rth row and sth column is given by

$$(2.8) \qquad [Z]_{r,s} = \sigma(g_r^{-1}g_s), \quad \text{for } r, s = 1, 2, \ldots, n.$$

(A). In view of (2.8), Z is an $n \times n$ matrix and **A** is satisfied.

(B) For each fixed integer r_0 satisfying $1 \leq r_0 \leq n$, we see that: as s ranges through the n integers from 1 to n, $g_{r_0}^{-1}g_s$ ranges through the n elements of G and $\sigma(g_{r_0}^{-1}g_s)$ ranges through n distinct elements of S. Thus, each row of Z has n distinct elements. Also, for each s_0 satisfying $1 \leq s_0 \leq n$, we note that: as r ranges through the integers from 1 to n, $g_r^{-1}g_{s_0}$ ranges through the n elements of G and $\sigma(g_r^{-1}g_{s_0})$ ranges through n distinct elements of S. Thus, each column of Z has n distinct elements. Consequently, **B** is satisfied.

(C) Due to **B**, Z possesses at least n distinct components that are elements of $\sigma(G)$. Thus, there are precisely n distinct components of Z and **C** is valid.

(D) In terms of the identity element e for G, we use (2.8) to obtain

$$[Z]_{r,r} = \sigma(g_r^{-1}g_r) = \sigma(e) = \sigma(g_1^{-1}g_1) = [Z]_{1,1}, \quad \text{for } r = 1, 2, \ldots, n.$$

Hence, **D** is satisfied.

(E) For $k = 1, 2, \ldots, n$, we introduce $z_k = [Z]_{1,k}$ as the component of Z in its first row and kth column. The results labeled **A**, **B**, and **C** show that each component of Z is an element of $\{z_1, z_2, \ldots, z_n\}$. They also show that: for $1 \leq k \leq n$, the element z_k appears precisely one time as a component in any row of Z and z_k appears precisely one time as a component in any column of Z. Thus, for 0, 1 in $\mathbb{C}$, the $n \times n$ matrix P_k defined, for $1 \leq k, r, s \leq n$, by

$$[P_k]_{r,s} = \begin{cases} 1, & \text{if } [Z]_{r,s} = z_k, \\ 0, & \text{if } [Z]_{r,s} \neq z_k, \end{cases}$$

is an $n \times n$ permutation matrix and **E** is therefore valid.

(F) For $k = 1, 2, \ldots, n$, let ψ_k be the function from $S = \{z_1, z_2, \ldots, z_n\}$ to $\mathbb{C}$ such that

$$\psi_k(x) = \begin{cases} 1, & \text{if } x = z_k, \\ 0, & \text{if } x \neq z_k. \end{cases}$$

Then, for $k = 1, 2, \ldots, n$, we have

$$[P_k]_{r,s} = \psi_k\big([Z]_{r,s}\big) = (\psi_k \circ \sigma)(g_r^{-1}g_s), \quad \text{for } r, s = 1, 2, \ldots, n.$$

Hence, the n permutation matrices $P_1, P_2, \ldots, P_n$ are elements of $\mathcal{M}[\mathbb{C}, G, \mathscr{L}]$. Theorem 2.9 shows that these are the only permutation matrices in $\mathcal{M}[\mathbb{C}, G, \mathscr{L}]$. The product of two permutation matrices is a permutation matrix and, as shown by Theorem 2.9, the product of two elements in $\mathcal{M}[\mathbb{C}, G, \mathscr{L}]$ is an element of $\mathcal{M}[\mathbb{C}, G, \mathscr{L}]$. Hence, the set $\mathfrak{P}$ is closed with respect to matrix multiplication and **F** is therefore satisfied. This completes the proof. $\quad\sqcup$

THEOREM 2.11. *Suppose Z is a matrix about which the assertions* **A**, **B**, **C**, **D**, **E**, **F** *that appear in* Theorem 2.10 *are valid. Then, the set $\mathfrak{P}$ of permutation matrices for* **F** *is a group under matrix multiplication with $P_1 = I_n$; and, $\mathcal{Z}$ is an injective group-pattern matrix with respect to $\mathfrak{P}$ and its list $\mathscr{L}_{\mathfrak{P}} = (P_1, P_2, \ldots, P_n)$.*

PROOF. Condition $\mathbf{F}$ shows that $\mathfrak{P}$ is closed under matrix multiplication; and, matrix multiplication is associative. We use $\mathbf{E}$ with $\mathbf{D}$ to verify that P_1 is the $n \times n$ identity matrix. For $1 \leq k \leq n$, the element P_k of $\mathfrak{P}$ specifies two lists

$$\left(P_1 P_k, P_2 P_k, \ldots, P_n P_k\right) \quad \text{and} \quad \left(P_k P_1, P_k P_2, \ldots, P_k P_n\right)$$

of the n elements for $\mathfrak{P}$ in some order. Thus, there are elements L_k and R_k in $\mathfrak{P}$ that yield $L_k P_k = P_1 = P_k R_k$ and $L_k = L_k P_1 = L_k P_k R_k = P_1 R_k = R_k$. Thus, each element P_k in $\mathfrak{P}$ has an inverse in $\mathfrak{P}$. Consequently, $\mathfrak{P}$ is a group.

The n distinct components of Z may not be elements of $\mathbb{C}$. Thus, we first direct our attention to a matrix having the pattern of Z but possessing components in $\mathbb{C}$.

Let S_Z denote the set whose elements are the n distinct components of Z. We use conditions $\mathbf{A}$, $\mathbf{B}$, $\mathbf{C}$ to see that $S_Z = \{z_1, z_2, \ldots, z_n\}$, where z_k is the kth component $z_k = [Z]_{1,k}$ in the first row of Z, for $k = 1, 2, \ldots, n$. Let $a_1, a_2, \ldots, a_n$ denote n distinct complex numbers. In terms of $\mathbb{C}_A = \{a_1, a_2, \ldots, a_n\}$, we define ψ as the function from S_Z to $\mathbb{C}_A$ having $\psi(z_k) = a_k$, for $k = 1, 2, \ldots, n$. Let A be the $n \times n$ matrix having components in $\mathbb{C}$ that is defined by

$$(2.9) \qquad [A]_{r,s} = \psi\big([Z]_{r,s}\big), \quad \text{for } r, s = 1, 2, \ldots, n,$$

and let $\mathcal{A}, \mathcal{B}, \mathcal{C}, \mathcal{D}, \mathcal{E}, \mathcal{F}$ denote the assertions about A obtained by replacing Z with A throughout each of $\mathbf{A}, \mathbf{B}, \mathbf{C}, \mathbf{D}, \mathbf{E}, \mathbf{F}$. We apply (2.9) to see that each of $\mathcal{A}, \mathcal{B}, \mathcal{C}, \mathcal{D}, \mathcal{E}, \mathcal{F}$ is a valid assertion about A.

Since the components of A are complex numbers, (2.9) and $\mathcal{E}$ yield

$$(2.10) \qquad A = a_1 P_1 + a_2 P_2 + \cdots + a_n P_n.$$

The multiplication table of the form (2.2) on page 11 for the group $\mathfrak{P}$ with respect to the list $\mathscr{L}_{\mathfrak{P}} = (P_1, P_2, \ldots, P_n)$ of its elements can be written as

(2.11)

$\cdot$	P_1	P_2	P_3	$\ldots$	P_s	$\ldots$	P_n
P_1^T	P_1	P_2	P_3	$\ldots$	P_s	$\ldots$	P_n
P_2^T	P_2^T	P_1	$P_2^T P_3$	$\ldots$	$P_2^T P_s$	$\ldots$	$P_2^T P_n$
P_3^T	P_3^T	$P_3^T P_2$	P_1	$\ldots$	$P_3^T P_s$	$\ldots$	$P_3^T P_n$
$\vdots$	$\vdots$	$\vdots$	$\vdots$	$\vdots$	$\vdots$	$\vdots$	$\vdots$
P_r^T	P_r^T	$P_r^T P_2$	$P_r^T P_3$	$\ldots$	$P_r^T P_s$	$\ldots$	$P_r^T P_n$
$\vdots$	$\vdots$	$\vdots$	$\vdots$	$\vdots$	$\vdots$	$\vdots$	$\vdots$
P_n^T	P_n^T	$P_n^T P_2$	$P_n^T P_3$	$\ldots$	$P_n^T P_s$	$\ldots$	P_1

because the inverse of a permutation matrix is equal to its transpose.

For $k = 1, 2, \ldots, n$, to prove that P_k is a group-pattern matrix for the group $\mathfrak{P}$ with respect to the list $\mathscr{L}_{\mathfrak{P}}$ of its elements, we need to specify a function f_k from $\mathfrak{P}$ to $\mathbb{C}$ such that $[P_k]_{r,s} = f_k\big(P_r^T P_s\big)$, for $r, s = 1, 2, \ldots, n$.

For $1 \leq k \leq n$, let f_k be the function from $\mathfrak{P}$ to $\mathbb{C}$ defined, for $1 \leq s \leq n$, by

$$(2.12) \qquad f_k(P_s) = \begin{cases} 1, & \text{if } P_s = P_k, \\ 0, & \text{if } P_s \neq P_k, \end{cases}$$

and, for $k = 1, 2, \ldots, n$, let Q_k be the $n \times n$ matrix with components in $\mathbb{C}$ having

$$(2.13) \qquad [Q_k]_{r,s} = f_k\big(P_r^T P_s\big), \quad \text{for } r, s = 1, 2, \ldots, n.$$

We use (2.13), (2.12), and properties of $\mathfrak{P}$ as a group to verify that Q_k is an $n \times n$ permutation matrix for $1 \le k \le n$.

For $1 \le k,\, r,\, s \le n$, suppose that $\left[Q_k\right]_{r,s} = 1$. Then, we have $P_k = P_r^T P_s$ and

$$\left[P_k\right]_{r,s} = \left[P_r^T P_s\right]_{r,s} = \sum_{\mu=1}^{n} \left[P_r^T\right]_{r,\mu}\left[P_s\right]_{\mu,s} = \sum_{\mu=1}^{n} \left[P_r\right]_{\mu,r}\left[P_s\right]_{\mu,s}.$$

Since we have $\left[P_s\right]_{\mu,s} \ne 0$ only when $\mu = 1$ and $\left[P_s\right]_{1,s} = 1$, we find that

$$\left[P_k\right]_{r,s} = \left[P_r\right]_{1,r} = 1.$$

Consequently, whenever a component of the $n \times n$ permutation matrix Q_k is 1, the corresponding component of the $n \times n$ permutation matrix P_k is 1. Since an $n \times n$ permutation matrix has precisely n components equal to 1, each of the other $n^2 - n$ components of P_k are equal to equal 0. This yields

(2.14) $P_k = Q_k, \quad \text{for } k = 1,\, 2,\, \ldots,\, n.$

Thus, each of the matrices $P_1,\, P_2,\, \ldots,\, P_n$ is a group-pattern matrix with respect to the list $\mathscr{L}_{\mathfrak{P}}$ for $\mathfrak{P}$. In particular, (2.14) and (2.13) yield

(2.15) $\left[P_k\right]_{r,s} = f_k\left(P_r^T P_s\right), \quad \text{for } r,\, s = 1,\, 2,\, \ldots,\, n.$

We employ $A = a_1 P_1 + a_2 P_2 + \cdots + a_n P_n$ from (2.10) with (2.15) to deduce

(2.16) $\left[A\right]_{r,s} = \left[\sum_{k=1}^{n} a_k P_k\right]_{r,s} = \sum_{k=1}^{n} a_k\left[P_k\right]_{r,s} = \sum_{k=1}^{n} a_k f_k\left(P_r^T P_s\right) = \sigma\left(P_r^{-1} P_s\right),$

$$\text{for } r,\, s = 1,\, 2,\, \ldots,\, n,$$

where σ is the function from $\mathfrak{P}$ to $\mathbb{C}$ defined in terms of (2.12) and the components $a_1,\, a_2, \ldots,\, a_n$ of A in its first row by $\sigma = a_1 f_1 + a_2 f_2 + \cdots + a_n f_n$. In view of (2.12), we have $\sigma(P_k) = a_k$, for $1 \le k \le n$. Since $a_1,\, a_2, \ldots,\, a_n$ are distinct elements, σ is one-to-one. Thus, A is an injective group-pattern matrix.

For the matrix Z and the function ϕ from $\mathbb{C}_A$ to S_Z defined by $\phi(a_k) = z_k$, for $k = 1,\, 2,\, \ldots,\, n$, we rewrite (2.9) and use (2.16) to obtain

$$\left[Z\right]_{r,s} = \phi\left(\left[A\right]_{r,s}\right) = \phi\left(\sigma\left(P_r^{-1} P_s\right)\right) = (\phi \circ \sigma)\left(P_r^{-1} P_s\right), \quad \text{for } 1 \le r,\, s \le n,$$

where $\phi \circ \sigma$ is a one-to-one function from $\mathfrak{P}$ to S_Z. Consequently Z is an injective group-pattern matrix. This completes the proof. $\square$

COROLLARY 2.12. *A matrix Z is an injective group-pattern matrix if and only if it satisfies the conditions* **A, B, C, D, E, F** *of Theorem 2.10.*

PROOF. This is a direct consequence of Theorems 2.10 and 2.11. $\square$

EXAMPLE 2.13. To illustrate Theorem 2.11, we consider the matrix

(2.17) $A = \left[\begin{array}{ccc|ccc} a_1 & a_2 & a_3 & a_4 & a_5 & a_6 \\ a_3 & a_1 & a_2 & a_5 & a_6 & a_4 \\ a_2 & a_3 & a_1 & a_6 & a_4 & a_5 \\ \hline a_4 & a_5 & a_6 & a_1 & a_2 & a_3 \\ a_5 & a_6 & a_4 & a_3 & a_1 & a_2 \\ a_6 & a_4 & a_5 & a_2 & a_3 & a_1 \end{array}\right],$

where $a_1, a_2, \ldots, a_6$ are distinct elements in some set S. Clearly, the matrix A satisfies conditions **A**, **B**, **C**, and **D** of Theorem 2.10. For **E**, we have

$$
P_1 = \begin{bmatrix} 1 & 0 & 0 & 0 & 0 & 0 \\ 0 & 1 & 0 & 0 & 0 & 0 \\ 0 & 0 & 1 & 0 & 0 & 0 \\ 0 & 0 & 0 & 1 & 0 & 0 \\ 0 & 0 & 0 & 0 & 1 & 0 \\ 0 & 0 & 0 & 0 & 0 & 1 \end{bmatrix}, \quad
P_2 = \begin{bmatrix} 0 & 1 & 0 & 0 & 0 & 0 \\ 0 & 0 & 1 & 0 & 0 & 0 \\ 1 & 0 & 0 & 0 & 0 & 0 \\ 0 & 0 & 0 & 0 & 1 & 0 \\ 0 & 0 & 0 & 0 & 0 & 1 \\ 0 & 0 & 0 & 1 & 0 & 0 \end{bmatrix}, \quad
P_3 = \begin{bmatrix} 0 & 0 & 1 & 0 & 0 & 0 \\ 1 & 0 & 0 & 0 & 0 & 0 \\ 0 & 1 & 0 & 0 & 0 & 0 \\ 0 & 0 & 0 & 0 & 0 & 1 \\ 0 & 0 & 0 & 1 & 0 & 0 \\ 0 & 0 & 0 & 0 & 1 & 0 \end{bmatrix},
$$

$$
P_4 = \begin{bmatrix} 0 & 0 & 0 & 1 & 0 & 0 \\ 0 & 0 & 1 & 0 & 0 & 1 \\ 0 & 0 & 0 & 0 & 1 & 0 \\ 1 & 0 & 0 & 0 & 0 & 0 \\ 0 & 0 & 1 & 0 & 0 & 0 \\ 0 & 1 & 0 & 0 & 0 & 0 \end{bmatrix}, \quad
P_5 = \begin{bmatrix} 0 & 0 & 0 & 0 & 1 & 0 \\ 0 & 0 & 0 & 1 & 0 & 0 \\ 0 & 0 & 0 & 0 & 0 & 1 \\ 0 & 1 & 0 & 0 & 0 & 0 \\ 1 & 0 & 0 & 0 & 0 & 0 \\ 0 & 0 & 1 & 0 & 0 & 0 \end{bmatrix}, \quad
P_6 = \begin{bmatrix} 0 & 0 & 0 & 0 & 0 & 1 \\ 0 & 0 & 0 & 0 & 1 & 0 \\ 0 & 0 & 0 & 1 & 0 & 0 \\ 0 & 0 & 1 & 0 & 0 & 0 \\ 0 & 1 & 0 & 0 & 0 & 0 \\ 1 & 0 & 0 & 0 & 0 & 0 \end{bmatrix}.
$$

To check that A satisfies **F**, we use the multiplication table

(2.18)

$\cdot$	P_1	P_2	P_3	P_4	P_5	P_6
P_1	P_1	P_2	P_3	P_4	P_5	P_6
P_3	P_3	P_1	P_2	P_5	P_6	P_4
P_2	P_2	P_3	P_1	P_6	P_4	P_5
P_4	P_4	P_5	P_6	P_1	P_2	P_3
P_5	P_5	P_6	P_4	P_3	P_1	P_2
P_6	P_6	P_4	P_5	P_2	P_3	P_1

to see that $\mathfrak{P} = \{P_1, P_2, P_3, P_4, P_5, P_6\}$ is closed. Therefore, Theorem 2.11 shows that $\mathfrak{P}$ forms a group and A in (2.17) is an injective group-pattern matrix with respect to the group $\mathfrak{P}$ and its list $\mathscr{L}_{\mathfrak{P}} = (P_1, P_2, P_3, P_4, P_5, P_6)$.

EXAMPLE 2.14. To illustrate Corollary 2.12, we consider the matrix

$$
A = \begin{bmatrix} 1 & 2 & 3 & 4 & 5 \\ 2 & 1 & 5 & 3 & 4 \\ 4 & 3 & 1 & 5 & 2 \\ 5 & 4 & 2 & 1 & 3 \\ 3 & 5 & 4 & 2 & 1 \end{bmatrix},
$$

which clearly satisfies conditions **A**, **B**, **C**, and **D**. For **E**, we have

$$
P_1 = \begin{bmatrix} 1 & 0 & 0 & 0 & 0 \\ 0 & 1 & 0 & 0 & 0 \\ 0 & 0 & 1 & 0 & 0 \\ 0 & 0 & 0 & 1 & 0 \\ 0 & 0 & 0 & 0 & 1 \end{bmatrix}, \quad
P_2 = \begin{bmatrix} 0 & 1 & 0 & 0 & 0 \\ 1 & 0 & 0 & 0 & 0 \\ 0 & 0 & 0 & 0 & 1 \\ 0 & 0 & 1 & 0 & 0 \\ 0 & 0 & 0 & 1 & 0 \end{bmatrix}, \quad
P_3 = \begin{bmatrix} 0 & 0 & 1 & 0 & 0 \\ 0 & 0 & 0 & 1 & 0 \\ 0 & 1 & 0 & 0 & 0 \\ 0 & 0 & 0 & 0 & 1 \\ 1 & 0 & 0 & 0 & 0 \end{bmatrix},
$$

$$
P_4 = \begin{bmatrix} 0 & 0 & 0 & 1 & 0 \\ 0 & 0 & 0 & 0 & 1 \\ 1 & 0 & 0 & 0 & 0 \\ 0 & 1 & 0 & 0 & 0 \\ 0 & 0 & 1 & 0 & 0 \end{bmatrix}, \quad
P_5 = \begin{bmatrix} 0 & 0 & 0 & 0 & 1 \\ 0 & 0 & 1 & 0 & 0 \\ 0 & 0 & 0 & 1 & 0 \\ 1 & 0 & 0 & 0 & 0 \\ 0 & 1 & 0 & 0 & 0 \end{bmatrix}.
$$

Here, $P_2 P_3$ is not in $\mathfrak{P}$; **F** is not satisfied; and A is not a group-pattern matrix.

2.5. Computer check for an injective group-pattern matrix

When an $n \times n$ matrix Z is given, the program presented here enables one to easily check whether Z is an injective group pattern matrix. It is a direct application of Theorems 2.10 and 2.11. We use a version of *Mathematica* such as [48] to enter in a *Mathematica* notebook the input statement

```
injectiveGroupPatternQ[Z_] :=
  Module[{n, a, r, c, list, rule, Per, S, t, i, j, k, f, pi},
  If[MatrixQ[Z], {}, (Print["It is not a matrix"]; Abort[])];
  If[Dimensions[Z][[1]] == Dimensions[Z][[2]], {},
      (Print["It is not a square matrix"];  Abort[])];
  n = Dimensions[Z][[1]];
  Do[r[k] = Length[Union[Z[[k]]]], {k, 1, n}];
  Do[c[k] = Length[Union[Transpose[Z][[k]]]], {k, 1, n}];
  list = Join[Table[r[k] == n, {k, 1, n}],
              Table[c[k] == n, {k, 1, n}]];
  If[Apply[And, list], {},
      (Print["Some row or column has fewer than ", n,
              " distinct elements."]; Abort[])];
  If[Length[Union[Flatten[Z]]] == n, {},
      (Print["It does not have precisely ", n,
              " distinct elements."]; Abort[])];
  Do[rule[k] = Table[If[i == k, Z[[1, i]] -> 1, Z[[1, i]] -> 0],
      {i, 1, n}], {k, 1, n}];
  Do[Per[k] = Z /. rule[k], {k, 1, n}];
  If[TrueQ[
    Per[1] == IdentityMatrix[n]], {}, (Print[
    "Principal diagonal elements are not equal."]; Abort[])];
  S = Table[Per[k], {k, 1, n}];
  Do[t[i, j] = MemberQ[S, Per[i].Per[j]], {i, 1, n}, {j, 1, n}];
  If[Apply[And, Flatten[Table[t[i, j], {i, 1, n}, {j, 1, n}]]], {},
      (Print[ "Closure for multiplication is violated."]; Abort[])];
  Print["It is an injective group-pattern matrix."];
  Do[If[TrueQ[Transpose[Per[i]].Per[j] == Per[k]],
      f[i, j] = P[k], {}], {i, 1, n}, {j, 1, n}, {k, 1, n}];
  Do[r1[i] = Table[f[i, j], {j, 1, n}], {i, 1, n}]; Clear[P];
  Do[If[TrueQ[Per[j] == Transpose[Per[i]]], pi[i] = j, {}],
      {i, 1, n}, {j, 1, n}];
  Do[r2[i] = Prepend[r1[i], (P[pi[i]])], {i, 1, n}];
  r2[0] = Prepend[Table[P[j], {j, 1, n}], "\[CenterDot]"];
  table = Table[r2[i], {i, 0, n}];
  Print["Its multiplication table for (P[1], P[2], ..., P[",
      n, "]) is:"];
  Print[""]; Print[Grid[table, Frame -> All]]; Print[""];
  Print["where"];
  Do[(Print[""]; Print["P[", k, "] = ", Per[k] // MatrixForm]),
      {k, 1, n}] ]
```

and evaluate it. Then, after the evaluation of a *Mathemataica* representation Z for Z, the evaluation of `injectiveGroupPatternQ[Z]` will either verify that Z is an injective group-pattern matrix for a group and list that it explicitly specifies or it will give a sufficient reason why Z is not an injective group-pattern matrix.

To check that the input of page 19 is free of misprints, we copied it and pasted it into a *Mathematica* notebook that can be downloaded from the web address

$\quad\quad$ `http://homepages.uc.edu/~chalklr/group-pattern.html`

along with the details for the following example. See Appendix H.

EXAMPLE 2.15. The downloadable *Mathematica* notebook mentioned in the preceding paragraph uses a convenient representation to input matrices like

$$(2.19) \quad\quad A = \left[\begin{array}{ccc|ccc|ccc} a_1 & a_2 & a_3 & a_4 & a_5 & a_6 & a_7 & a_8 & a_9 \\ a_3 & a_1 & a_2 & a_6 & a_4 & a_5 & a_9 & a_7 & a_8 \\ a_2 & a_3 & a_1 & a_5 & a_6 & a_4 & a_8 & a_9 & a_7 \\ \hline a_7 & a_8 & a_9 & a_1 & a_2 & a_3 & a_4 & a_5 & a_6 \\ a_9 & a_7 & a_8 & a_3 & a_1 & a_2 & a_6 & a_4 & a_5 \\ a_8 & a_9 & a_7 & a_2 & a_3 & a_1 & a_5 & a_6 & a_4 \\ \hline a_4 & a_5 & a_6 & a_7 & a_8 & a_9 & a_1 & a_2 & a_3 \\ a_6 & a_4 & a_5 & a_9 & a_7 & a_8 & a_3 & a_1 & a_2 \\ a_5 & a_6 & a_4 & a_8 & a_9 & a_7 & a_2 & a_3 & a_1 \end{array}\right] .$$

When a *Mathematica* representation `A` for A in (2.19) is evaluated, we find that the output for the evaluation of `injectiveGroupPatternQ[A]` begins with

$\quad\quad$ "It is an injective group-pattern matrix."

and then provides a list $(P_1, P_2, \ldots, P_9)$ of nine explicitly presented permutation matrices having size 9×9 that form a group $\mathfrak{P}$ relative to which A in (2.19) is an injective group-pattern matrix. Moreover, this output also provides a multiplication table for these permutation matrices and it looks like

$$(2.20)$$

$\cdot$	P_1	P_2	P_3	P_4	P_5	P_6	P_7	P_8	P_9
P_1	P_1	P_2	P_3	P_4	P_5	P_6	P_7	P_8	P_9
P_3	P_3	P_1	P_2	P_6	P_4	P_5	P_9	P_7	P_8
P_2	P_2	P_3	P_1	P_5	P_6	P_4	P_8	P_9	P_7
P_7	P_7	P_8	P_9	P_1	P_2	P_3	P_4	P_5	P_6
P_9	P_9	P_7	P_8	P_3	P_1	P_2	P_6	P_4	P_5
P_8	P_8	P_9	P_7	P_2	P_3	P_1	P_5	P_6	P_4
P_4	P_4	P_5	P_6	P_7	P_8	P_9	P_1	P_2	P_3
P_6	P_6	P_4	P_5	P_9	P_7	P_8	P_3	P_1	P_2
P_5	P_5	P_6	P_4	P_8	P_9	P_7	P_2	P_3	P_1

in which P_1 is the identity element for G. In particular, P_2 and P_4 yield

$$(P_2)^0(P_4)^0 = P_1, \quad\quad (P_2)^1(P_4)^0 = P_2, \quad\quad (P_2)^2(P_4)^0 = P_3,$$
$$(P_2)^0(P_4)^1 = P_4, \quad\quad (P_2)^1(P_4)^1 = P_5, \quad\quad (P_2)^2(P_4)^1 = P_6,$$
$$(P_2)^0(P_4)^2 = P_7, \quad\quad (P_2)^1(P_4)^2 = P_8, \quad\quad (P_2)^2(P_4)^2 = P_9,$$

as well as $P_2P_4 = P_4P_2$, $(P_2)^3 = P_1$, and $(P_4)^3 = P_1$. In particular, the sets $\mathfrak{P}_1 = \{(P_2)^0, (P_2)^1, (P_2)^2\}$ and $\mathfrak{P}_2 = \{(P_4)^0, (P_4)^1, (P_4)^2\}$ form cyclic subgroups each of order 3. This shows that the group $\mathfrak{P}$ for A in (2.19) has the structure of the internal direct product of the two cyclic subgroups $\mathfrak{P}_1$ and $\mathfrak{P}_2$ each of order 3.

2.6. Verification that a given multiplication table specifies a group

For the set $Q_8 = \{1,\ -1,\ i,\ -i,\ j,\ -j,\ k,\ -k\}$ of eight elements about which no details are provided other than that it has the multiplication table

(2.21)

$\cdot$	1	i	-1	$-i$	j	k	$-j$	$-k$
1	1	i	-1	$-i$	j	k	$-j$	$-k$
$-i$	$-i$	1	i	-1	$-k$	j	k	$-j$
-1	-1	$-i$	1	i	$-j$	$-k$	j	k
i	i	-1	$-i$	1	k	$-j$	$-k$	j
$-j$	$-j$	k	j	$-k$	1	$-i$	-1	i
$-k$	$-k$	$-j$	k	j	i	1	$-i$	-1
j	j	$-k$	$-j$	k	-1	i	1	$-i$
k	k	j	$-k$	$-j$	$-i$	-1	i	1

for those elements, we introduce the corresponding matrix

(2.22)

$$A_* = \begin{bmatrix} y_1 & y_2 & y_3 & y_4 & y_5 & y_6 & y_7 & y_8 \\ y_4 & y_1 & y_2 & y_3 & y_8 & y_5 & y_6 & y_7 \\ y_3 & y_4 & y_1 & y_2 & y_7 & y_8 & y_5 & y_6 \\ y_2 & y_3 & y_4 & y_1 & y_6 & y_7 & y_8 & y_5 \\ y_7 & y_6 & y_5 & y_8 & y_1 & y_4 & y_3 & y_2 \\ y_8 & y_7 & y_6 & y_5 & y_2 & y_1 & y_4 & y_3 \\ y_5 & y_8 & y_7 & y_6 & y_3 & y_2 & y_1 & y_4 \\ y_6 & y_5 & y_8 & y_7 & y_4 & y_3 & y_2 & y_1 \end{bmatrix},$$

whose pattern is specified by (2.21). After evaluating a representation A for A_* in a *Mathematica* notebook such as [**48**], we evaluate `injectiveGroupPatternQ[A]`. The output shows that A_* is an injective group-pattern matrix for a group of explicitly given 8×8 permutation matrices whose group properties are given by

(2.23)

$\cdot$	P_1	P_2	P_3	P_4	P_5	P_6	P_7	P_8
P_1	P_1	P_2	P_3	P_4	P_5	P_6	P_7	P_8
P_4	P_4	P_1	P_2	P_3	P_8	P_5	P_6	P_7
P_3	P_3	P_4	P_1	P_2	P_7	P_8	P_5	P_6
P_2	P_2	P_3	P_4	P_1	P_6	P_7	P_8	P_5
P_7	P_7	P_6	P_5	P_8	P_1	P_4	P_3	P_2
P_8	P_8	P_7	P_6	P_5	P_2	P_1	P_4	P_3
P_5	P_5	P_8	P_7	P_6	P_3	P_2	P_1	P_4
P_6	P_6	P_5	P_8	P_7	P_4	P_3	P_2	P_1

.

Since there is a one-to-one correspondence between the entries in the tables (2.21) and (2.23) (including their row and column labels), we conclude that the set Q_8 forms a group when the multiplication for it is defined by (2.21).

We note that: when the elements of G are regarded as quaternions with their usual symbolism, their multiplication as quaternions is provided by (2.21). For that reason, the group Q_8 with (2.21) is standardly termed the *quaternion group*.

2.7. Representation of G with permutation matrices

Let $\mathscr{L} = (g_1, g_2, \ldots, g_n)$ be a list of the elements in a group G of order n. For each g in G, let $\widehat{g}$ be the function from G to a field $\mathfrak{F}$ that is defined by

$$(2.24) \qquad \widehat{g}(x) = \begin{cases} 1, & \text{if } x = g, \\ 0, & \text{if } x \neq g, \end{cases}, \quad \text{for each } x \text{ in } G.$$

For $1 \leq k \leq n$, let P_k be the $n \times n$ group-pattern matrix for G and $\mathscr{L}$ having

$$(2.25) \qquad [P_k]_{r,s} = \widehat{g}_k(g_r^{-1} g_s), \quad \text{for } r, s = 1, 2, \ldots, n.$$

THEOREM 2.16. *Each of $P_1, P_2, \ldots, P_n$ is an $n \times n$ permutation matrix and, with multiplication of matrices, the set $\mathfrak{P} = \{P_1, P_2, \ldots, P_n\}$ forms a group that is isomorphic to G. Moreover, an isomorphism F of G onto $\mathfrak{P}$ is given by*

$$(2.26) \qquad F(g_k) = P_k, \quad \text{for } k = 1, 2, \ldots, n.$$

PROOF. For k, r, s restricted by $1 \leq k, r, s \leq n$, we observe that: for each k and r there is a unique s such that $\widehat{g}_k(g_r^{-1} g_s) = 1$, and for each k and s there is a unique r such that $\widehat{g}_k(g_r^{-1} g_s) = 1$. Thus, each P_k is a permutation matrix.

Clearly, F is onto. Suppose that g_i, g_j in G satisfy $g_i \neq g_j$. Then, there are integers r, s such that $1 \leq r, s \leq n$ and $g_r^{-1} g_s = g_i$. This yields

$$[P_i]_{r,s} = \widehat{g}_i(g_r^{-1} g_s) = \widehat{g}_i(g_i) = 1 \neq 0 = \widehat{g}_j(g_i) = \widehat{g}_j(g_r^{-1} g_s) = [P_j]_{r,s}$$

and $F(g_i) = P_i \neq P_j = F(g_j)$. Thus, F is one-to-one.

For any g_i, g_j in G, we set $g_k = g_i g_j$ and observe, for $1 \leq r, s \leq n$, that

$$(2.27) \quad [F(g_i)\, F(g_j)]_{r,s} = [P_i P_j]_{r,s} = \sum_{k=1}^{n} [P_i]_{r,k}\, [P_j]_{k,s} = \sum_{k=1}^{n} \widehat{g}_i(g_r^{-1} g_k)\, \widehat{g}_j(g_k^{-1} g_s).$$

We apply (2.24) to see that: for $1 \leq k \leq n$, the condition $\widehat{g}_i(g_r^{-1} g_k) \neq 0$ is satisfied if and only if $g_r^{-1} g_k = g_i$ or equivalently $g_k^{-1} = g_i^{-1} g_r^{-1}$. Thus, (2.27) yields

$$[F(g_i)\, F(g_j)]_{r,s} = \widehat{g}_j(g_i^{-1} g_r^{-1} g_s) = \begin{cases} 1, & \text{if } g_i^{-1} g_r^{-1} g_s = g_j, \\ 0, & \text{if } g_i^{-1} g_r^{-1} g_s \neq g_j, \end{cases}$$

$$= \begin{cases} 1, & \text{if } g_r^{-1} g_s = g_i g_j = g_k, \\ 0, & \text{if } g_r^{-1} g_s \neq g_i g_j = g_k, \end{cases} = \widehat{g}_k(g_r^{-1} g_s) = [P_k]_{r,s} = [F(g_k)]_{r,s} = [F(g_i g_j)]_{r,s}.$$

Hence, we have

$$(2.28) \qquad F(g_i g_j) = F(g_i)\, F(g_j), \quad \text{for each } g_i, g_j \text{ in } G.$$

We use (2.26), (2.25), (2.24), and the $n \times n$ identity matrix I_n to see that

$$[F(e)]_{r,s} = \widehat{e}(g_r^{-1} g_s) = \begin{cases} 1, & \text{if } r = s, \\ 0, & \text{if } r \neq s, \end{cases} = [I_n]_{r,s}, \quad \text{for } r, s = 1, 2, \ldots, n.$$

This yields $F(e) = I_n$. Thus, I_n is in $\mathfrak{P}$. For P and Q in $\mathfrak{P}$, we have $P = F(g)$ and $Q = F(h)$, for some g, h in G. That gives $PQ = F(g)F(h) = F(gh)$. Thus, $\mathfrak{P}$ is closed under matrix multiplication. If $h = g^{-1}$, then $PQ = I_n$ and $P^{-1} = Q$. Thus, $\mathfrak{P}$ is a group and (2.28) shows that F is a group isomorphism of G onto $\mathfrak{P}$. This completes the proof. $\qquad \square$

EXAMPLE 2.17. For the elements $g_1 = e$, $g_2 = \rho$, $g_3 = \rho^2$, with $\rho^3 = e$ in a cyclic group G of order 3, the list $\mathscr{L} = (g_1, g_2, g_3)$ yields the multiplication table

(2.29)

$\cdot$	g_1	g_2	g_3
g_1^{-1}	g_1	g_2	g_3
g_2^{-1}	g_3	g_1	g_2
g_3^{-1}	g_2	g_3	g_1

The corresponding 3×3 permutation matrices of (2.24), (2.25), and (2.29) are

$$P_1 = \begin{bmatrix} 1 & 0 & 0 \\ 0 & 1 & 0 \\ 0 & 0 & 1 \end{bmatrix}, \quad P_2 = \begin{bmatrix} 0 & 1 & 0 \\ 0 & 0 & 1 \\ 1 & 0 & 0 \end{bmatrix}, \quad P_3 = \begin{bmatrix} 0 & 0 & 1 \\ 1 & 0 & 0 \\ 0 & 1 & 0 \end{bmatrix}.$$

With respect to matrix multiplication, the matrices P_1, P_2, P_3 form a group $\mathfrak{P}$. An isomorphism F of G onto $\mathfrak{P}$ is given by $F(g_k) = P_k$, for $k = 1, 2, 3$.

THEOREM 2.18. *For a group G of order n with $\mathscr{L} = (g_1, g_2, \ldots, g_n)$ as a list of its elements, let $\mathfrak{R}$ be the polynomial ring in the variables $X_1, X_2, \ldots, X_n$ over a field $\mathfrak{F}$, let σ be the function form G to $\mathfrak{R}$ having $\sigma(g_k) = X_k$, for $1 \le k \le n$, and let A be the $n \times n$ group-pattern matrix for G and $\mathscr{L}$ defined by*

(2.30) $$[A]_{r,s} = \sigma(g_r^{-1} g_s), \quad \text{for } r, s = 1, 2, \ldots, n.$$

Suppose that M is a nonsingular $n \times n$ matrix over $\mathfrak{F}$ such that $D = M^{-1}AM$ is a diagonal matrix. Then, G is an abelian group.

PROOF. For $1 \le k \le n$, we define $\widehat{g_k}$ by (2.24) and employ that formula to deduce $\left(X_1 \widehat{g_1} + X_2 \widehat{g_2} + \cdots + X_n \widehat{g_n}\right)(g_k) = X_k \widehat{g_k}(g_k) = X_k = \sigma(g_k)$ and

(2.31) $$\sigma = X_1 \widehat{g_1} + X_2 \widehat{g_2} + \cdots + X_n \widehat{g_n}.$$

Using (2.30), (2.31), and the permutation matrices P_k defined by (2.25), we find that

$$\begin{aligned} [A]_{r,s} &= \sigma(g_r^{-1} g_s) = X_1 \widehat{g_1}(g_r^{-1} g_s) + X_2 \widehat{g_2}(g_r^{-1} g_s) + \cdots + X_n \widehat{g_n}(g_r^{-1} g_s) \\ &= X_1 [P_1]_{r,s} + X_2 [P_2]_{r,s} + \cdots + X_n [P_n]_{r,s} \\ &= [X_1 P_1 + X_2 P_2 + \cdots + X_n P_n]_{r,s}, \quad \text{for } r, s = 1, 2, \ldots, n, \end{aligned}$$

and $A = X_1 P_1 + X_2 P_2 + \cdots + X_n P_n$. For the diagonal matrix D, this gives

(2.32) $$D = M^{-1}AM = X_1(M^{-1}P_1 M) + X_2(M^{-1}P_2 M) + \cdots + X_n(M^{-1}P_n M).$$

For $k = 1, 2, \ldots, n$, let D_k denote the diagonal matrix that is obtained from D by replacing each X_k in D with 1 and each X_i in D with 0 when $1 \le i \le n$ and $i \ne k$. In view of (2.32), this yields

$$D_k = M^{-1}P_k M \quad \text{and} \quad P_k = MD_k M^{-1}, \quad \text{for } 1 \le k \le n.$$

We use the isomorphism F in Theorem 2.16 and commutativity of multiplication for diagonal matrices to verify, for any integers i, j satisfying $1 \le i, j \le n$, that

$$\begin{aligned} F(g_i g_j) &= F(g_i) F(g_j) = P_i P_j - (MD_i M^{-1})(MD_j M^{-1}) = MD_i D_j M^{-1} \\ &= MD_j D_i M^{-1} = (MD_j M^{-1})(MD_i M^{-1}) = P_j P_i = F(g_j) F(g_i) \\ &= F(g_j g_i) \end{aligned}$$

and $g_i g_j = g_j g_i$. Thus, G is abelian. This completes the proof. $\square$

2.8. Objects of historical interest

In his treatise [**36**, Vol. 2, pp. 400–412, Chapter XIV, Circulants, up to 1860], Thomas Muir introduced in 1911 the term *circulants* to refer to the determinants of the two types of matrices with components in $\mathbb{C}$ given by

$$(2.33) \quad \begin{bmatrix} a_1 \end{bmatrix}, \quad \begin{bmatrix} a_1 & a_2 \\ a_2 & a_1 \end{bmatrix}, \quad \begin{bmatrix} a_1 & a_2 & a_3 \\ a_3 & a_1 & a_2 \\ a_2 & a_3 & a_1 \end{bmatrix}, \quad \begin{bmatrix} a_1 & a_2 & a_3 & a_4 \\ a_4 & a_1 & a_2 & a_3 \\ a_3 & a_4 & a_1 & a_2 \\ a_2 & a_3 & a_4 & a_1 \end{bmatrix}, \quad \cdots$$

and by

$$(2.34) \quad \begin{bmatrix} b_1 \end{bmatrix}, \quad \begin{bmatrix} b_1 & b_2 \\ b_2 & b_1 \end{bmatrix}, \quad \begin{bmatrix} b_1 & b_2 & b_3 \\ b_2 & b_3 & b_1 \\ b_3 & b_1 & b_2 \end{bmatrix}, \quad \begin{bmatrix} b_1 & b_2 & b_3 & b_4 \\ b_2 & b_3 & b_4 & b_1 \\ b_3 & b_4 & b_1 & b_2 \\ b_4 & b_1 & b_2 & b_3 \end{bmatrix}, \quad \cdots .$$

The term **circulant matrices** in [**1**] of 1963, [**7**] of 1975, and [**19**] of 1979 to describe the matrices of (2.33) became widely accepted. By adopting terminology in [**19**, page 69], we use **leftward-circulant matrices** for those in (2.34).

EXERCISES

1. For an integer $n \geq 3$ of your choosing, let A_1 and A_2 denote $n \times n$ matrices obtained from the $n \times n$ matrix in (2.33) by replacing its components $a_1, a_2, \ldots, a_n$ with corresponding integers. Also, let B_1 and B_2 denote $n \times n$ matrices obtained from the $n \times n$ matrix in (2.34) by replacing its components $b_1, b_2, \ldots, b_n$ with corresponding integers. Use a system of computer algebra to verify that the matrix products $A_1 A_2$ and $B_1 B_2$ have the pattern of the $n \times n$ matrix in (2.33) while the products $A_1 B_2$ and $B_1 A_2$ have the pattern of the $n \times n$ matrix in (2.34). Moreover, for the indicated matrix product of $2n \times 2n$ matrices, check that

$$(2.35) \quad \left[\begin{array}{c|c} A_1 & B_1 \\ \hline B_1 & A_1 \end{array}\right] \left[\begin{array}{c|c} A_2 & B_2 \\ \hline B_2 & A_2 \end{array}\right] = \left[\begin{array}{c|c} A_3 & B_3 \\ \hline B_3 & A_3 \end{array}\right],$$

where $A_3 = A_1 A_2 + B_1 B_2$ and $B_3 = A_1 B_2 + B_1 A_2$. For general results about this, see Section 6.3 beginning on page 67 and see Observation 15.3 on page 144.

As an example, when the selections $n = 3$ and

$$A_1 = \begin{bmatrix} 1 & 2 & 3 \\ 3 & 1 & 2 \\ 2 & 3 & 1 \end{bmatrix}, A_2 = \begin{bmatrix} 4 & 5 & 0 \\ 0 & 4 & 5 \\ 5 & 0 & 4 \end{bmatrix}, B_1 = \begin{bmatrix} 4 & 5 & 0 \\ 5 & 0 & 4 \\ 0 & 4 & 5 \end{bmatrix}, B_2 = \begin{bmatrix} 1 & 2 & 3 \\ 2 & 3 & 1 \\ 3 & 1 & 2 \end{bmatrix}$$

are made, we find that $A_2 A_1 = A_1 A_2$,

$$A_1 A_2 = \begin{bmatrix} 19 & 13 & 22 \\ 22 & 19 & 13 \\ 13 & 22 & 19 \end{bmatrix}, B_1 B_2 = \begin{bmatrix} 14 & 23 & 17 \\ 17 & 14 & 23 \\ 23 & 17 & 14 \end{bmatrix}, B_2 B_1 = \begin{bmatrix} 14 & 17 & 23 \\ 23 & 14 & 17 \\ 17 & 23 & 14 \end{bmatrix},$$

$$B_2 A_1 \neq A_1 B_2 = \begin{bmatrix} 14 & 11 & 11 \\ 11 & 11 & 14 \\ 11 & 14 & 11 \end{bmatrix}, \quad \text{and} \quad A_2 B_1 \neq B_1 A_2 = \begin{bmatrix} 16 & 40 & 25 \\ 40 & 25 & 16 \\ 25 & 16 & 40 \end{bmatrix}.$$

For (2.35), we have $A_3 = \begin{bmatrix} 33 & 36 & 39 \\ 39 & 33 & 36 \\ 36 & 39 & 33 \end{bmatrix}$ and $B_3 = \begin{bmatrix} 30 & 51 & 36 \\ 51 & 36 & 30 \\ 36 & 30 & 51 \end{bmatrix}.$

2. Use the *Mathematica* input command of page 19 to conclude that the matrix

$$(2.36) \quad Z_1 = \left[\begin{array}{ccccc|ccccc} a_1 & a_2 & a_3 & a_4 & a_5 & b_1 & b_2 & b_3 & b_4 & b_5 \\ a_5 & a_1 & a_2 & a_3 & a_4 & b_2 & b_3 & b_4 & b_5 & b_1 \\ a_4 & a_5 & a_1 & a_2 & a_3 & b_3 & b_4 & b_5 & b_1 & b_2 \\ a_3 & a_4 & a_5 & a_1 & a_2 & b_4 & b_5 & b_1 & b_2 & b_3 \\ a_2 & a_3 & a_4 & a_5 & a_1 & b_5 & b_1 & b_2 & b_3 & b_4 \\ \hline b_1 & b_2 & b_3 & b_4 & b_5 & a_1 & a_2 & a_3 & a_4 & a_5 \\ b_2 & b_3 & b_4 & b_5 & b_1 & a_5 & a_1 & a_2 & a_3 & a_4 \\ b_3 & b_4 & b_5 & b_1 & b_2 & a_4 & a_5 & a_1 & a_2 & a_3 \\ b_4 & b_5 & b_1 & b_2 & b_3 & a_3 & a_4 & a_5 & a_1 & a_2 \\ b_5 & b_1 & b_2 & b_3 & b_4 & a_2 & a_3 & a_4 & a_5 & a_1 \end{array}\right]$$

is an injective group-pattern matrix. Its group can be identified with the group of symmetries for a regular pentagon. Namely, see Proposition 15.2 on page 142. We note that the indicated 5×5 submatrices of Z_1 are the 5×5 matrices in (2.33) and (2.34). Proposition 15.2 is similarly illustrated by the 4×4 submatrices of A in (1.10) on page 6, by the 3×3 submatrices of A in (1.7) on page 5, and by the the 2×2 submatrices of A in (1.3) on page 3. Of course the 1×1 submatrices of $\left[\begin{array}{c|c} a_1 & b_1 \\ \hline b_1 & a_1 \end{array}\right]$ are the 1×1 matrices of (2.33) and (2.34).

3. Use the *Mathematica* input command of page 19 to conclude that the matrix

$$(2.37) \quad Z_2 = \left[\begin{array}{cccccccc|cccccccc} a & b & c & d & e & f & g & h & i & j & k & l & m & n & o & p \\ b & a & d & c & f & e & h & g & j & i & l & k & n & m & p & o \\ c & d & a & b & g & h & e & f & k & l & i & j & o & p & m & n \\ d & c & b & a & h & g & f & e & l & k & j & i & p & o & n & m \\ e & f & g & h & a & b & c & d & m & n & o & p & i & j & k & l \\ f & e & h & g & b & a & d & c & n & m & p & o & j & i & l & k \\ g & h & e & f & c & d & a & b & o & p & m & n & k & l & i & j \\ h & g & f & e & d & c & b & a & p & o & n & m & l & k & j & i \\ \hline i & j & k & l & m & n & o & p & a & b & c & d & e & f & g & h \\ j & i & l & k & n & m & p & o & b & a & d & c & f & e & h & g \\ k & l & i & j & o & p & m & n & c & d & a & b & g & h & e & f \\ l & k & j & i & p & o & n & m & d & c & b & a & h & g & f & e \\ m & n & o & p & i & j & k & l & e & f & g & h & a & b & c & d \\ n & m & p & o & j & i & l & k & f & e & h & g & b & a & d & c \\ o & p & m & n & k & l & i & j & g & h & e & f & c & d & a & b \\ p & o & n & m & l & k & j & i & h & g & f & e & d & c & b & a \end{array}\right]$$

is an injective group-pattern matrix for a group G in which each element is equal to its inverse. (Since that implies G is abelian, we can use the fundamental theorem for finite abelian groups to see that G has the structure of the direct product of four cyclic groups each of order 2.) We note that Z_2 is symmetric with respect to each of its two diagonals and with respect to its center. Also, each of its four 8×8 submatrices has the pattern of A in (1.13) on page 7.

Mathematica input for the matrix Z_2 can be downloaded from a notebook at
http://homepages.uc.edu/~chalklr/group-pattern.html
where its name corresponds to the number of this page. See Appendix H.

4. Is the matrix

$$
M_1 = \begin{bmatrix}
b_1 & b_2 & b_3 & a_1 & a_2 & a_3 \\
b_2 & b_3 & b_1 & a_3 & a_1 & a_2 \\
b_3 & b_1 & b_2 & a_2 & a_3 & a_1 \\
a_1 & a_2 & a_3 & b_1 & b_2 & b_3 \\
a_3 & a_1 & a_2 & b_2 & b_3 & b_1 \\
a_2 & a_3 & a_1 & b_3 & b_1 & b_2
\end{bmatrix}
$$

an injective group-pattern matrix?

5. Is the matrix

$$
M_2 = \begin{bmatrix}
1 & 2 & 3 & 4 & 5 & 6 \\
2 & 1 & 6 & 5 & 4 & 3 \\
3 & 5 & 1 & 6 & 2 & 4 \\
4 & 6 & 5 & 1 & 3 & 2 \\
6 & 4 & 2 & 3 & 1 & 5 \\
5 & 3 & 4 & 2 & 6 & 1
\end{bmatrix}
$$

an injective group-pattern matrix?

6. Let G be a cyclic group of order n with generator α and let H be a cyclic group of order n with generator β. For the function F from G to H defined by $F(\alpha^k) = \beta^k$, for $0 \le k \le n-1$, prove that F is a one-to-one function from G onto H that satisfies $F(xy) = F(x)F(y)$, for each x, y in G.

7. Let G be a non-vacuous set with a multiplication $\cdot$ subject to:
(i) for each x, y, z in G, the multiplication satisfies $x \cdot (y \cdot z) = (x \cdot y) \cdot z$;
(ii) there is an element e in G such that $e \cdot x = x \cdot e = x$, for each x in G; and,
(iii)$'$ for any x in G, there is an element y in G such that $x \cdot y = e$.
Here, only (iii)$'$ differs from (iii) in Definition 2.1. Prove that G is a group.

CHAPTER 3

Standard Group-Pattern Matrices

3.1. The standard group-pattern matrix for G and $\mathscr{L}$

Throughout, let X_1, X_2, X_3, ... denote distinct variables.

DEFINITION 3.1. For a finite group G of order n and a list $\mathscr{L} = \big(g_1, \, g_2, \, \ldots, \, g_n\big)$ of the n elements in G with the special property that g_1 is the identity element $g_1 = e$ of G, there is a unique $n \times n$ matrix $Stan_{\mathfrak{G}}\big(G, \, \mathscr{L}\big)$ defined by

$$(3.1) \qquad \big[Stan_{\mathfrak{G}}\big(G, \, \mathscr{L}\big)\big]_{r,s} = f_{\mathscr{L}}\big(g_r^{-1}g_s\big), \quad \text{for } r, \, s = 1, \, 2, \, \ldots, \, n,$$

where $f_{\mathscr{L}}$ is the function from G to $\big\{X_1, \, X_2, \, \ldots, \, X_n\big\}$ having

$$(3.2) \qquad f_{\mathscr{L}}(g_k) = X_k, \quad \text{for } k = 1, \, 2, \, \ldots, \, n.$$

The matrix $Stan_{\mathfrak{G}}\big(G, \, \mathscr{L}\big)$ is the ***standard group-pattern matrix*** for G and $\mathscr{L}$.

Due to $g_1 = e$, we use (3.1) and (3.2) to see that the first row of $Stan_{\mathfrak{G}}\big(G, \, \mathscr{L}\big)$ consists of the components $\big[Stan_{\mathfrak{G}}\big(G, \, \mathscr{L}\big)\big]_{1,k} = f_{\mathscr{L}}(g_k) = X_k$, for $1 \le k \le n$. Also, each element in the principal diagonal of $Stan_{\mathfrak{G}}\big(G, \, \mathscr{L}\big)$ equals X_1.

The same G and $\mathscr{L}$ specify the set $\mathcal{M}[S, \, G, \, \mathscr{L}]$ of group-pattern matrices for G and $\mathscr{L}$ having components in a set S as described in Definition 2.2 on page 12. The matrix $Stan_{\mathfrak{G}}\big(G, \, \mathscr{L}\big)$ serves as a standard pattern for $\mathcal{M}[S, \, G, \, \mathscr{L}]$ in the sense that: the matrices of $\mathcal{M}[S, \, G, \, \mathscr{L}]$ result by substituting elements from S for the variables X_1, X_2, ..., X_n of $Stan_{\mathfrak{G}}\big(G, \, \mathscr{L}\big)$. Also, if A in $\mathcal{M}[S, \, G, \, \mathscr{L}]$ is specified by a function σ from G to S according to $\big[A\big]_{r,s} = \sigma(g_r^{-1}g_s)$, for $1 \le r, \, s \le n$, then the components in the first row of A are $\sigma(g_1)$, $\sigma(g_2)$, ..., $\sigma(g_n)$ and A results when each X_k in $Stan_{\mathfrak{G}}\big(G, \, \mathscr{L}\big)$ is replaced with $\sigma(g_k)$, for $k = 1, \, 2, \, \ldots, \, n$.

3.2. Examples of standard group-pattern matrices

A group G is said to be *cyclic* when each element in G is expressible as a power of some single element in G. Since any two cyclic groups of the same order are isomorphic, it is convenient to use $\mathcal{C}_n$ to represent a cyclic group of order n.

EXAMPLE 3.2. For $G = \mathcal{C}_1 = \{e\}$, the only list is $\mathscr{L} = \big(e\big)$ and we have

$$Stan_{\mathfrak{G}}\big(\mathcal{C}_1, \, \mathscr{L}\big) = \big[X_1\big].$$

Thus, there is just one standard group-pattern matrix for $\mathcal{C}_1$.

For $G - \mathcal{C}_2 = \big\{e, \, \alpha\big\}$ with $\alpha^2 = e$, the only list is $\mathscr{L} = \big(e, \, \alpha\big)$ and we have

$$Stan_{\mathfrak{G}}\big(\mathcal{C}_2, \, \mathscr{L}\big) = \begin{bmatrix} X_1 & X_2 \\ X_2 & X_1 \end{bmatrix}.$$

Thus, there is just one standard group-pattern matrix for $\mathcal{C}_2$.

For $\mathcal{C}_3 = \{1,\, \omega,\, \omega^2\}$, there are two lists: $\mathscr{L}_1 = (1,\, \omega,\, \omega^2)$ and $\mathscr{L}_2 = (1,\, \omega^2,\, \omega)$. We use each of the corresponding multiplication tables

$\cdot$	1	ω	ω^2
1	1	ω	ω^2
ω^2	ω^2	1	ω
ω	ω	ω^2	1

$\cdot$	1	ω^2	ω
1	1	ω^2	ω
ω	ω	1	ω^2
ω^2	ω^2	ω	1

to conclude that

$$Stan_{\mathfrak{G}}(G,\, \mathscr{L}_1) = Stan_{\mathfrak{G}}(G,\, \mathscr{L}_2) = \begin{bmatrix} X_1 & X_2 & X_3 \\ X_3 & X_1 & X_2 \\ X_2 & X_3 & X_1 \end{bmatrix}.$$

Thus, there is just one standard group-pattern matrix for $\mathcal{C}_3$.

EXAMPLE 3.3. For $\mathcal{C}_4 = \{1,\, i,\, -1,\, -i\}$ with $i^2 = -1$, there are six lists

$$\mathscr{L}_1 = (1,\, i,\, -1,\, -i), \qquad \mathscr{L}_2 = (1,\, -i,\, -1,\, i),$$
$$\mathscr{L}_3 = (1,\, i,\, -i,\, -1), \qquad \mathscr{L}_4 = (1,\, -i,\, i,\, -1),$$
$$\mathscr{L}_5 = (1,\, -1,\, i,\, -i), \qquad \mathscr{L}_6 = (1,\, -1,\, -i,\, i).$$

The lists $\mathscr{L}_1$ and $\mathscr{L}_2$ have the respective multiplication tables

(3.3)

$\cdot$	1	i	-1	$-i$
1	1	i	-1	$-i$
$-i$	$-i$	1	i	-1
-1	-1	$-i$	1	i
i	i	-1	$-i$	1

$\cdot$	1	$-i$	-1	i
1	1	$-i$	-1	i
i	i	1	$-i$	-1
-1	-1	i	1	$-i$
$-i$	$-i$	-1	i	1

and they specify the same standard group-pattern matrix

(3.4)
$$Stan_{\mathfrak{G}}(\mathcal{C}_4,\, \mathscr{L}_1) = Stan_{\mathfrak{G}}(\mathcal{C}_4,\, \mathscr{L}_2) = \begin{bmatrix} X_1 & X_2 & X_3 & X_4 \\ X_4 & X_1 & X_2 & X_3 \\ X_3 & X_4 & X_1 & X_2 \\ X_2 & X_3 & X_4 & X_1 \end{bmatrix}.$$

The lists $\mathscr{L}_3$ and $\mathscr{L}_4$ have the respective multiplication tables

(3.5)

$\cdot$	1	i	$-i$	-1
1	1	i	$-i$	-1
$-i$	$-i$	1	-1	i
i	i	-1	1	$-i$
-1	-1	$-i$	i	1

$\cdot$	1	$-i$	i	-1
1	1	$-i$	i	-1
i	i	1	-1	$-i$
$-i$	$-i$	-1	1	i
-1	-1	i	$-i$	1

and they specify the same standard group-pattern matrix

(3.6)
$$Stan_{\mathfrak{G}}(\mathcal{C}_4,\, \mathscr{L}_3) = Stan_{\mathfrak{G}}(\mathcal{C}_4,\, \mathscr{L}_4) = \begin{bmatrix} X_1 & X_2 & X_3 & X_4 \\ X_3 & X_1 & X_4 & X_2 \\ X_2 & X_4 & X_1 & X_3 \\ X_4 & X_3 & X_2 & X_1 \end{bmatrix}.$$

The lists $\mathscr{L}_5$ and $\mathscr{L}_6$ have the respective multiplication tables

(3.7)

$\cdot$	1	-1	i	$-i$
1	1	-1	i	$-i$
-1	-1	1	$-i$	i
$-i$	$-i$	i	1	-1
i	i	$-i$	-1	1

$\cdot$	1	-1	$-i$	i
1	1	-1	$-i$	i
-1	-1	1	i	$-i$
i	i	$-i$	1	-1
$-i$	$-i$	i	-1	1

and they specify the same standard group-pattern matrix

(3.8)
$$Stan_{\mathfrak{G}}(\mathcal{C}_4, \mathscr{L}_5) = Stan_{\mathfrak{G}}(\mathcal{C}_4, \mathscr{L}_6) = \begin{bmatrix} X_1 & X_2 & X_3 & X_4 \\ X_2 & X_1 & X_4 & X_3 \\ X_4 & X_3 & X_1 & X_2 \\ X_3 & X_4 & X_2 & X_1 \end{bmatrix}.$$

Consequently, there are three standard group-pattern matrices for $\mathcal{C}_4$.

EXAMPLE 3.4. The notation $\mathcal{C}_2 \times \mathcal{C}_2$ for the noncyclic group of order 4 is natural when one interprets its four elements as given by

$$e = (1, 1), \quad \alpha = (1, -1), \quad \beta = (-1, 1), \quad \gamma = (-1, -1),$$

where products are computed by multiplying corresponding components. Thus, for example, we have $\alpha\beta = (1, -1)(-1, 1) = \big((1)(-1), (-1)(1)\big) = (-1, -1) = \gamma$. See Section 3.5. A different interpretation for $\mathcal{C}_2 \times \mathcal{C}_2$ was given in Section 1.3.

The $3! = 6$ lists for the elements of $\mathcal{C}_2 \times \mathcal{C}_2$ are

$$\mathscr{L}_1 = (e, \alpha, \beta, \gamma), \quad \mathscr{L}_2 = (e, \beta, \alpha,, \gamma), \quad \mathscr{L}_3 = (e, \gamma, \alpha, \beta),$$
$$\mathscr{L}_4 = (e, \alpha, \gamma, \beta), \quad \mathscr{L}_5 = (e, \beta, \gamma, \alpha), \quad \mathscr{L}_6 = (e, \gamma, \beta, \alpha).$$

These lists specify the corresponding multiplication tables

$\cdot$	e	α	β	γ
e	e	α	β	γ
α	α	e	γ	β
β	β	γ	e	α
γ	γ	β	α	e

$\cdot$	e	β	α	γ
e	e	β	α	γ
β	β	e	γ	α
α	α	γ	e	β
γ	γ	α	β	e

$\cdot$	e	γ	α	β
e	e	γ	α	β
γ	γ	e	β	α
α	α	β	e	γ
β	β	α	γ	e

$\cdot$	e	α	γ	β
e	e	α	γ	β
α	α	e	β	γ
γ	γ	β	e	α
β	β	γ	α	e

$\cdot$	e	β	γ	α
e	e	β	γ	α
β	β	e	α	γ
γ	γ	α	e	β
α	α	γ	β	e

$\cdot$	e	γ	β	α
e	e	γ	β	α
γ	γ	e	α	β
β	β	α	e	γ
α	α	β	γ	e

and each of these six tables yields

$$Stan_{\mathfrak{G}}(\mathcal{C}_2 \times \mathcal{C}_2, \mathscr{L}_k) = \begin{bmatrix} X_1 & X_2 & X_3 & X_4 \\ X_2 & X_1 & X_4 & X_3 \\ X_3 & X_4 & X_1 & X_2 \\ X_4 & X_3 & X_2 & X_1 \end{bmatrix}, \quad \text{for } k = 1, 2, \ldots, 6.$$

Consequently, there is just one standard group-pattern matrix for $\mathcal{C}_2 \times \mathcal{C}_2$.

3.3. The number of standard group-pattern matrices for G

Let $\mathfrak{N}_{\mathfrak{G}}(G)$ be the number of standard group-pattern matrices that exist for a finite group G. The preceding section shows that

$$\mathfrak{N}_{\mathfrak{G}}(\mathcal{C}_1) = 1, \mathfrak{N}_{\mathfrak{G}}(\mathcal{C}_2) = 1, \;\; \mathfrak{N}_{\mathfrak{G}}(\mathcal{C}_3) = 1, \;\; \mathfrak{N}_{\mathfrak{G}}(\mathcal{C}_4) = 3, \;\; \mathfrak{N}_{\mathfrak{G}}(\mathcal{C}_2 \times \mathcal{C}_2) = 1.$$

Let G be a group of order n. The multiplication tables for G of the form (2.2) on page 11 are uniquely specified by the lists of the n elements in G. With the first element of each list restricted to be the identity element e, there are $(n-1)!$ multiplication tables. However, as Example 3.3 demonstrates, some multiplication tables may be redundant. For example, only one of the two tables in each of (3.3), (3.5), and (3.7) is needed. A comparison of the two tables in each of (3.3), (3.5), and (3.7) shows that either one is obtained from the other by replacing i everywhere with $-i$. That corresponds to an automorphism of $\mathcal{C}_4$ in the following sense.

DEFINITION 3.5. An **automorphism** of a group G is a one-to-one function χ from G onto G that satisfies

$$(3.9) \qquad\qquad \chi(xy) = \chi(x)\chi(y), \quad \text{for each } x, y \text{ in } G.$$

The symbolism $Aut(G)$ designates the set of automorphisms for G.

PROPOSITION 3.6. *The set $Aut(G)$ forms a group with composition of functions as its multiplication.*

PROOF. The identity function id on G satisfies $id(xy) = xy = id(x)\,id(y)$, for any x, y in G, and therefore belongs to $Aut(G)$. For ϕ, ψ in $Aut(G)$ and any x, y in G, we use (3.9) to obtain

$$(\phi \circ \psi)(xy) = \phi\big(\psi(xy)\big) = \phi\big(\psi(x)\psi(y)\big) = \phi\big(\psi(x)\big)\phi\big(\psi(y)\big) = (\phi \circ \psi)(x)\,(\phi \circ \psi)(y)$$

and conclude that the product $\phi \circ \psi$ belongs to $Aut(G)$. Let ϕ^{-1} be the inverse function for ϕ in $Aut(G)$. That yields $\phi^{-1} \circ \phi = \phi \circ \phi^{-1} = id$. For x, y in G, we set $u = \phi^{-1}(x)$ and $v = \phi^{-1}(y)$ in order to deduce that $x = \phi(u)$, $y = \phi(v)$, and

$$\phi^{-1}(xy) = \phi^{-1}\big(\phi(u)\,\phi(v)\big) = \phi^{-1}\big(\phi(uv)\big) = uv = \phi^{-1}(x)\,\phi^{-1}(y).$$

Thus, ϕ^{-1} belongs to $Aut(G)$ and $Aut(G)$ is a group. This completes the proof. $\square$

Just as $|G|$ designates the order of a group G, the notation $|Aut(G)|$ is suitable to represent the number of automorphisms of G.

THEOREM 3.7. *For a group G of order n that has m as the number of its automorphisms, the number $\mathfrak{N}_{\mathfrak{G}}(G)$ of standard group-pattern matrices for G is*

$$(3.10) \qquad\qquad \mathfrak{N}_{\mathfrak{G}}(G) = \frac{(n-1)!}{m} = \frac{(n-1)!}{|Aut(G)|} = \frac{(|G|-1)!}{|Aut(G)|}.$$

PROOF. Let $\mathscr{L}_1$ and $\mathscr{L}_2$ denote any two lists for the n elements of G in which their first components equal e and let $\mathscr{L}_1$ be represented by $\mathscr{L}_1 = (g_1, g_2, \ldots, g_n)$ with $g_1 = e$. Then, there is a one-to-one function ϕ from G onto G such that $\mathscr{L}_2$ is given by $\mathscr{L}_2 = \big(\phi(g_1), \phi(g_2), \ldots, \phi(g_n)\big)$ and $\phi(g_1) = e$. We use (3.2) and (3.1) to see that the standard group-pattern matrix specified by $\mathscr{L}_1$ is given by

$$(3.11) \quad \big[Stan_{\mathfrak{G}}(G, \mathscr{L}_1)\big]_{r,s} = f_{\mathscr{L}_1}\big(g_r^{-1} g_s\big), \quad \text{for } r, s = 1, 2, \ldots, n,$$

in terms of the function $f_{\mathscr{L}_1}$ from G onto $\{X_1, X_2, \ldots, X_n\}$ having

$$(3.12) \qquad f_{\mathscr{L}_1}(g_k) = X_k, \quad \text{for } k = 1, 2, \ldots, n.$$

Also, (3.2) and (3.1) shows that the standard group-pattern matrix specified by $\mathscr{L}_2$ is given by

$$(3.13) \quad \left[Stan_{\mathfrak{G}}(G, \mathscr{L}_2)\right]_{r,s} = f_{\mathscr{L}_2}\left((\phi(g_r))^{-1}(\phi(g_s))\right), \quad \text{for } r, s = 1, 2, \ldots, n,$$

in terms of the function $f_{\mathscr{L}_2}$ from G onto $\{X_1, X_2, \ldots, X_n\}$ having

$$(3.14) \qquad f_{\mathscr{L}_2}(\phi(g_k)) = X_k, \quad \text{for } k = 1, 2, \ldots, n.$$

In particular, (3.12) and (3.14) yield

$$(3.15) \qquad f_{\mathscr{L}_1} = f_{\mathscr{L}_2} \circ \phi.$$

We proceed to consider two situations marked by (i) and (ii).

(i) Suppose that $\mathscr{L}_1$ and $\mathscr{L}_2$ specify the same standard group-pattern matrix. Then, we use $Stan_{\mathfrak{G}}(G, \mathscr{L}_1) = Stan_{\mathfrak{G}}(G, \mathscr{L}_2)$ to see that (3.11) and (3.13) yield

$$(3.16) \qquad f_{\mathscr{L}_1}(g_r^{-1}g_s) = f_{\mathscr{L}_2}\left((\phi(g_r))^{-1}(\phi(g_s))\right), \quad \text{for } r, s = 1, 2, \ldots, n.$$

We employ (3.15) to rewrite (3.16) as

$$f_{\mathscr{L}_2}\left(\phi(g_r^{-1}g_s)\right) = f_{\mathscr{L}_2}\left((\phi(g_r))^{-1}(\phi(g_s))\right), \quad \text{for } r, s = 1, 2, \ldots, n.$$

Since $f_{\mathscr{L}_2}$ is one-to-one, we find that

$$(3.17) \qquad \phi(g_r^{-1}g_s) = (\phi(g_r))^{-1}(\phi(g_s)), \quad \text{for } r, s = 1, 2, \ldots, n.$$

Setting $s = 1$ in (3.17), we use $g_1 = e$ and $\phi(e) = e$ to obttain $\phi(g_r^{-1}) = (\phi(g_r))^{-1}$, for $r = 1, 2, \ldots, n$. Thus, we see that (3.17) yields

$$(3.18) \qquad \phi(g_r^{-1}g_s) = \phi(g_r^{-1})\phi(g_s), \quad \text{for } r, s = 1, 2, \ldots, n.$$

For any x and y in G, there are unique integers r_0, s_0 satisfying $1 \le r_0, s_0 \le n$ such that $g_{r_0}^{-1} = x$ and $g_{s_0} = y$. We combine this with (3.18) to deduce that

$$\phi(xy) = \phi(g_{r_0}^{-1}g_{s_0}) = \phi(g_{r_0}^{-1})\phi(g_{s_0}) = \phi(x)\phi(y).$$

Thus, in view of Definition 3.9, ϕ is an automorphism of G.

(ii) Suppose that ϕ is an automorphism of G. We first use (3.9) to verify that $\phi(e) = \phi(ee) = \phi(e)\phi(e)$ and $\phi(e) = e$. Thus, starting with a typical list $\mathscr{L}_1 = (g_1, g_2, \ldots, g_n)$ having $g_1 = e$, we set

$$\mathscr{L}_2 - (\psi(y_1), \phi(g_2), \ldots, \phi(g_n)), \quad \text{with } \phi(g_1) = e,$$

and repeat the context of (3.11)–(3.15) for $\mathscr{L}_1$ and $\mathscr{L}_2$. However, we now have $(\phi(g_r))^{-1}\phi(g_r) = e = \phi(g_r^{-1}g_r) = \phi(g_r^{-1})\phi(g_r)$ and

$$(3.19) \qquad (\phi(g_r))^{-1} = \phi(g_r^{-1}), \quad \text{for } r = 1, 2, \ldots, n.$$

A rewriting of (3.15) yields

(3.20) $$f_{\mathscr{L}_2} = f_{\mathscr{L}_1} \circ \phi^{-1}.$$

We use (3.13), (3.20), (3.19), ϕ as an automorphism, and (3.11) to deduce

$$\begin{aligned}
\left[Stan_{\mathfrak{G}}(G, \mathscr{L}_2)\right]_{r,s} &= f_{\mathscr{L}_2}\left(\left(\phi(g_r)\right)^{-1}\left(\phi(g_s)\right)\right) \\
&= \left(f_{\mathscr{L}_1} \circ \phi^{-1}\right)\left(\phi(g_r^{-1})\,\phi(g_s)\right) \\
&= \left(f_{\mathscr{L}_1} \circ \phi^{-1}\right)\left(\phi(g_r^{-1}g_s)\right) \\
&= f_{\mathscr{L}_1}\left(g_r^{-1}g_s\right) \\
&= \left[Stan_{\mathfrak{G}}(G, \mathscr{L}_1)\right]_{r,s}, \quad \text{for } r, s = 1, 2, \ldots, n.
\end{aligned}$$

Thus, we have $Stan_{\mathfrak{G}}(G, \mathscr{L}_2) = Stan_{\mathfrak{G}}(G, \mathscr{L}_1)$.

The arguments of (i) and (ii) show that: for the two lists $\mathscr{L}_1 = (g_1, g_2, \ldots, g_n)$ and $\mathscr{L}_2$ of a group G, in order for $\mathscr{L}_1$ and $\mathscr{L}_2$ to specify the same standard group-pattern matrix it is necessary and sufficient that there is an automorphism ϕ of G such that $\mathscr{L}_2 = \left(\phi(g_1), \phi(g_2), \ldots, \phi(g_n)\right)$.

Let $\mathfrak{L}$ denote the set of lists for G of the kind considered here where the first element of each list is the identity element e of G. Then, the number of elements in $\mathfrak{L}$ is $(n-1)!$ because that is the number of ways of placing the remaining $n-1$ elements of G into the remaining $n-1$ positions. For each automorphism ϕ of G which necessarily yields $\phi(e) = e$, there is a function F_ϕ from $\mathfrak{L}$ to $\mathfrak{L}$ defined by

$$F_\phi\left((g_1, g_2, \ldots, g_n)\right) = \left(\phi(g_1), \phi(g_2), \ldots, \phi(g_n)\right),$$
$$\text{for each } (g_1, g_2, \ldots, g_n) \text{ in } \mathfrak{L}.$$

Clearly, F_ϕ is one-to-one and onto.

For any two elements $\mathscr{L}_1$ and $\mathscr{L}_2$ of $\mathfrak{L}$, we write $\mathscr{L}_1 \sim \mathscr{L}_2$ if and only if there is an automorphism ϕ of G such that $\mathscr{L}_2 = F_\phi(\mathscr{L}_1)$. In other words, we have $\mathscr{L}_1 \sim \mathscr{L}_2$ if only if $\mathscr{L}_1$ and $\mathscr{L}_2$ specify the same standard group-pattern matrix. Thus, $\sim$ is an equivalence relation for $\mathfrak{L}$ and it partitions $\mathfrak{L}$ into equivalence classes such that the number of these equivalence classes is equal to the number $\mathfrak{N}_{\mathfrak{G}}(G)$ of distinct standard group-pattern matrices for G. Let $\phi_1, \phi_2, \ldots, \phi_m$ denote the automorphisms of G. For any $\mathscr{L}$ in $\mathfrak{L}$, the elements equivalent to $\mathscr{L}$ are

$$F_{\phi_1}(\mathscr{L}), F_{\phi_2}(\mathscr{L}), \ldots, F_{\phi_m}(\mathscr{L}).$$

Consequently, each of the $\mathfrak{N}_{\mathfrak{G}}(G)$ equivalence classes contains m elements. Since the number $(n-1)!$ of elements in $\mathfrak{L}$ is equal to the product of the number $\mathfrak{N}_{\mathfrak{G}}(G)$ of equivalence classes and the number m of elements in each equivalence class, we obtain (3.10). This completes the proof.   □

EXAMPLE 3.8. For the group $\mathcal{C}_4 = \{1, i, -1, -i\}$ in Example 3.3 on page 28, we found that $\mathfrak{N}_{\mathfrak{G}}(\mathcal{C}_4) = 3$. To obtain this by using (3.10), we note that: if ϕ is an automorphism of $\mathcal{C}_4$, then either $\phi(i) = i$ or $\phi(i) = -i$. Thus, there are two automorphisms for $\mathcal{C}_4$ and, with $m = 2$, (3.10) yields $\mathfrak{N}_{\mathfrak{G}}(\mathcal{C}_4) = (4-1)!/2 = 3$.

For the group $\mathcal{C}_2 \times \mathcal{C}_2 = \{e, \alpha, \beta, \gamma\}$ in Example 3.4 on page 29, we deduced that $\mathfrak{N}_{\mathfrak{G}}(\mathcal{C}_2 \times \mathcal{C}_2) = 1$. To see that (3.10) yields the same result, we note that: if ϕ is any automorphism of $\mathcal{C}_2 \times \mathcal{C}_2$, then we must have $\phi(e) = e$ while $\phi(\alpha)$, $\phi(\beta)$, $\phi(\gamma)$ can be any permutation of α, β, γ. Consequently, the group $\mathcal{C}_2 \times \mathcal{C}_2$ has $m = 3! = 6$ automorphisms and (3.10) therefore yields $\mathfrak{N}_{\mathfrak{G}}(\mathcal{C}_2 \times \mathcal{C}_2) = (4-1)!/3! = 1$.

3.4. Program to obtain all of the automorphisms for G

3.4.1. Motivation that is provided by the proof of Theorem 3.7.

Let $\mathscr{L}_0 = \big(g_1, g_2, \ldots, g_n\big)$ with $g_1 = e$ be a list of the elements in a group G having order n and set $Z_0 = Stan_{\mathfrak{G}}(G, \mathscr{L}_0)$, where $Stan_{\mathfrak{G}}(G, \mathscr{L}_0)$ is the standard group-pattern matrix that is specified by G and $\mathscr{L}_0$. As π ranges through the $(n-1)!$ permutations of $\{1, 2, \ldots, n\}$ having $\pi(1) = 1$, we see that; the corresponding functions ϕ_π from G to G defined by $\phi_\pi(g_k) = g_{\pi(k)}$, for $1 \le k \le n$, include all of the one-to-one functions ϕ from G onto G having $\phi(e) = e$; and the corresponding lists $\mathscr{L}_\pi = \big(g_{\pi(1)}, g_{\pi(2)}, \ldots, g_{\pi(n)}\big)$ include all those for G whose first element is e. The proof of Theorem 3.7 shows that ϕ_π is an automorphism of G if and only if $Stan_{\mathfrak{G}}(G, \mathscr{L}_\pi) = Z_0$, where $Stan_{\mathfrak{G}}(G, \mathscr{L}_\pi)$ is the standard group-pattern matrix specified by G and $\mathscr{L}_\pi$.

3.4.2. Use of one standard group-pattern matrix to obtain the others.

To employ computer algebra based on the preceding observations, we need to know how each standard group-pattern matrix $\mathcal{S}_\pi = Stan_{\mathfrak{G}}(G, \mathscr{L}_\pi)$ can be obtained from Z_0 through application of π. With $\mathfrak{R} = \mathbb{Q}\big[X_1, X_2, \ldots, X_n\big]$ as the polynomial ring in $X_1, X_2, \ldots, X_n$ over the field $\mathbb{Q}$ of rational numbers, there is a function σ from G to $\mathfrak{R}$ such that $\sigma(g_k) = X_k$, for $1 \le k \le n$, and

$$(3.21) \qquad \big[Z_0\big]_{r,s} = \sigma\big(g_r^{-1} g_s\big), \quad \text{for } r, s = 1, 2, \ldots, n.$$

There is also a function τ from G to $\mathfrak{R}$ such that $\tau(g_{\pi(k)}) = X_k$, for $1 \le k \le n$, and

$$(3.22) \qquad \big[\mathcal{S}_\pi\big]_{r,s} = \tau\big(g_{\pi(r)}^{-1} g_{\pi(s)}\big), \quad \text{for } r, s = 1, 2, \ldots, n.$$

To visualize how $\mathcal{S}_\pi$ is obtainable from Z_0, we observe how the multiplication table

$$(3.23)$$

$\cdot$	g_1	g_2	$\cdots$	g_n
g_1^{-1}	$g_1^{-1}g_1$	$g_1^{-1}g_2$	$\cdots$	$g_1^{-1}g_n$
g_2^{-1}	$g_2^{-1}g_1$	$g_2^{-1}g_2$	$\cdots$	$g_2^{-1}g_n$
$\vdots$	$\vdots$	$\vdots$	$\ddots$	$\vdots$
g_n^{-1}	$g_n^{-1}g_1$	$g_n^{-1}g_2$	$\cdots$	$g_n^{-1}g_n$

for G and $\mathscr{L}_0$ can be modified to obtain the multiplication table

$$(3.24)$$

$\cdot$	$g_{\pi(1)}$	$g_{\pi(2)}$	$\cdots$	$g_{\pi(n)}$
$g_{\pi(1)}^{-1}$	$g_{\pi(1)}^{-1}g_{\pi(1)}$	$g_{\pi(1)}^{-1}g_{\pi(2)}$	$\cdots$	$g_{\pi(1)}^{-1}g_{\pi(n)}$
$g_{\pi(2)}^{-1}$	$g_{\pi(2)}^{-1}g_{\pi(1)}$	$g_{\pi(2)}^{-1}g_{\pi(2)}$	$\cdots$	$g_{\pi(2)}^{-1}g_{\pi(n)}$
$\vdots$	$\vdots$	$\vdots$	$\ddots$	$\vdots$
$g_{\pi(n)}^{-1}$	$g_{\pi(n)}^{-1}g_{\pi(1)}$	$g_{\pi(n)}^{-1}g_{\pi(2)}$	$\cdots$	$g_{\pi(n)}^{-1}g_{\pi(n)}$

for G and $\mathscr{L}_\pi$. Namely, we see that (3.24) is obtained from (3.23) by applying the same permutation π, with $\pi(1) = 1$, to both the rows and the columns of (3.23).

Therefore, we use (3.21) to see that the $n \times n$ matrix B defined by

$$(3.25) \qquad [B]_{r,s} = \sigma\big(g_{\pi(r)}^{-1} g_{\pi(s)}\big), \quad \text{for } r,\, s = 1,\, 2,\, \ldots,\, n,$$

is related to Z_0 by means of $B = P^T Z_0 P$, where P is the $n \times n$ permutation matrix with $[P]_{1,1} = 1$ such that

$$(3.26) \qquad [\pi(1),\, \pi(2),\, \ldots,\, \pi(n)] = [1,\, 2,\, \ldots,\, n] P.$$

In view of (3.21), (3.22), and (3.25), we obtain $\mathcal{S}_\pi$ from B by replacing each $X_{\pi(k)}$ in B with X_k, for $1 \le k \le n$. A computer can obtain S_π from B by setting $b_k = [B]_{1,k}$, for $1 \le k \le n$, and then replacing each b_k in B with X_k, for $1 \le k \le n$.

The program of Subsection 3.4.3 considers each $n \times n$ permutation matrix P having $[P]_{1,1} = 1$ and checks whether the corresponding $B = P^T Z_0 P$ yields an S_π for which $S_\pi = Z_0$. When this condition is satisfied, the program uses (3.26) to represent the corresponding automorphism by $\big(\pi(1),\, \pi(2),\, \ldots,\, \pi(n)\big)$.

If A is an injective group-pattern matrix for G and $\mathcal{L}_0$, then the program can deduce Z_0 from A by setting $a_k = [A]_{1,k}$, for $1 \le k \le n$ and then replacing each a_k in A with X_k, for $1 \le k \le n$.

3.4.3. Implementation. In regard to a version of *Mathematica* such as [48], we enter in a *Mathematica* notebook the input

```
automorphisms[A_] :=
    Module[ {n, id, list, P, counter, a, rule1,
            Z0, L0, B, rule2, b},
    n = Dimensions[A][[1]];
    id = IdentityMatrix[n];
    list = Permutations[id];
    Do[P[i] = list[[i]], {i, 1, (n - 1)!}];
    Clear[counter];  counter = 0;
    Do[a[k] = A[[1, k]], {k, 1, n}];
    rule1 = Table[a[k] -> X[k], {k, 1, n}];
    Z0 = A /. rule1;
    L0 = Table[k, {k, 1, n}];
    Do[( B[j] = Transpose[P[j]].Z0.P[j];
        Do[b[k] = B[j][[1, k]], {k, 1, n}];
        rule2 = Table[b[k] -> X[k], {k, 1, n}];
        B[j] = B[j] /. rule2;
        If[TrueQ[B[j] == Z0],  (lastCount = counter + 1;
            counter = counter + 1;
            Print["L[", counter, "] = " , L0.P[j]] ), {} ];
        Clear[B]      ), {j, 1, (n - 1)!}];
    Print["Number m of automorphisms is ", lastCount];
    Print["Number of standard matrices is ",
        (n - 1)!/lastCount ] ]
```

and evaluate it. Then, to find the automorphisms of a group G in relation to a list $\mathcal{L}_0$ for G, we evaluate a *Mathematica* representation A of an injective group-pattern matrix A for G with $\mathcal{L}_0$ and then evaluate automorphisms[A].

It is advisable to check that the input for A is correct. Helpful for that purpose is the program of page 19 to evaluate injectiveGroupPatternQ[A].

EXAMPLE 3.9. For the quaternion group Q_8 of (2.21) on page 21 and its list $\mathscr{L}_0 = (g_1, g_2, g_3, g_4, g_5, g_6, g_7, g_8)$ given by $g_1 = 1$, $g_2 = i$, $g_3 = -1$, $g_4 = -i$, $g_5 = j$, $g_6 = k$, $g_7 = -j$, $g_8 = -k$, we use (2.22) to see that the matrix

$$A = \begin{bmatrix} 1 & 2 & 3 & 4 & 5 & 6 & 7 & 8 \\ 4 & 1 & 2 & 3 & 8 & 5 & 6 & 7 \\ 3 & 4 & 1 & 2 & 7 & 8 & 5 & 6 \\ 2 & 3 & 4 & 1 & 6 & 7 & 8 & 5 \\ 7 & 6 & 5 & 8 & 1 & 4 & 3 & 2 \\ 8 & 7 & 6 & 5 & 2 & 1 & 4 & 3 \\ 5 & 8 & 7 & 6 & 3 & 2 & 1 & 4 \\ 6 & 5 & 8 & 7 & 4 & 3 & 2 & 1 \end{bmatrix}.$$

is an injective group-pattern matrix for Q_8 and $\mathscr{L}_0$. After evaluating the input for page 34 and the input A for A, we find that evaluation of `automorphisms[A]` yields twenty-four automorphisms for Q_8 specified with respect to $\mathscr{L}_0$ by

```
L[1]  = [1,2,3,4,5,6,7,8],
L[2]  = [1,2,3,4,8,5,6,7],
L[3]  = [1,2,3,4,7,8,5,6],
L[4]  = [1,2,3,4,6,7,8,5],
L[5]  = [1,4,3,2,5,8,7,6],
L[6]  = [1,4,3,2,6,5,8,7],
L[7]  = [1,4,3,2,7,6,5,8],
L[8]  = [1,4,3,2,8,7,6,5],
L[9]  = [1,5,3,7,2,8,4,6],
L[10] = [1,7,3,5,2,6,4,8],
L[11] = [1,6,3,8,2,5,4,7],
L[12] = [1,8,3,6,2,7,4,5],
L[13] = [1,5,3,7,6,2,8,4],
L[14] = [1,7,3,5,8,2,6,4],
L[15] = [1,8,3,6,5,2,7,4],
L[16] = [1,6,3,8,7,2,5,4],
L[17] = [1,5,3,7,4,6,2,8],
L[18] = [1,7,3,5,4,8,2,6],
L[19] = [1,8,3,6,4,5,2,7],
L[20] = [1,6,3,8,4,7,2,5],
L[21] = [1,5,3,7,8,4,6,2],
L[22] = [1,7,3,5,6,4,8,2],
L[23] = [1,6,3,8,5,4,7,2],
L[24] = [1,8,3,6,7,4,5,2]
```

followed by the two assertions

```
Number of automorphisms for G is 24
Number of standard matrices for G is 210
```

as the remaining output. The computation requires less than three seconds.

For instance, the automorphism represented by L[6] is the function ϕ_6 from Q_8 onto Q_8 having $\phi_6(g_1) = g_1$, $\phi_6(g_2) = g_4$, $\phi_6(g_3) = g_3$, $\phi_6(g_4) = g_2$, $\phi_6(g_5) = g_6$, $\phi_6(g_6) = g_5$, $\phi_6(g_7) = g_8$, $\phi_6(g_8) = g_7$. To download the *Mathematica* notebook P-035.nb used for the preceding computations, apply the directions of Appendix H.

3.5. Direct product of groups

DEFINITION 3.10. For $k = 1, 2, \ldots, d$, let G_k denote a group of order n_k and let G denote the set of d-tuples

$$(w_1, w_2, \ldots, w_d), \quad \text{where, for } k = 1, 2, \ldots, d, w_k \text{ is an element of } G_k.$$

A product uv in G is defined for any u, v in G with respect to the representations

$$u = (u_1, u_2, \ldots, u_d) \quad \text{and} \quad v = (v_1, v_2, \ldots, v_d), \quad \text{having } u_k, v_k \text{ in } G_k,$$

by the rule that

$$(3.27) \quad uv = (u_1 v_1, u_2 v_2, \ldots, u_d v_d), \quad \text{where, for } k = 1, 2, \ldots, d, u_k v_k \text{ is in } G_k.$$

For this context, G is the **direct product** of G_1, G_2, $\ldots$, G_d and this is expressed with the notation $G = G_1 \times G_2 \times \cdots \times G_d$. Clearly, G has $n = n_1 n_2 \cdots n_d$ elements.

PROPOSITION 3.11. *The direct product* $G = G_1 \times G_2 \times \cdots \times G_d$ *is a group.*

PROOF. For $u = (u_1, u_2, \ldots, u_d)$, $v = (v_1, v_2, \ldots, v_d)$, $w = (w_1, w_2, \ldots, w_d)$ in G, we use (3.27) with the associativity $(u_k v_k) w_k = u_k (v_k w_k)$ of multiplication in G_k, for $k = 1, 2, \ldots, d$, to conclude that associativity $(uv) w = u (v w)$ is valid in G. In terms of the unit element e_k of G_k, for $k = 1, 2, \ldots, d$, the unit element for G is naturally given by $e = (e_1, e_2, \ldots, e_d)$. For $k = 1, 2, \ldots, d$, the inverse of u_k in G_k is denoted by u_k^{-1}. Thus, the element $(u_1^{-1}, u_2^{-1}, \ldots, u_d^{-1})$ in G is the inverse of u. This shows that G is a group and completes the proof. $\square$

EXAMPLE 3.12. In terms of the cyclic group $\mathcal{C}_3 = \{e, \rho, \rho^2\}$, having $\rho^3 = e$, we introduce the list $\mathscr{L}_0 = (g_1, g_2, g_3, g_4, g_5, g_6, g_7, g_8, g_9)$ for $\mathcal{C}_3 \times \mathcal{C}_3$, where

$$g_1 = (e, e), \qquad g_2 = (\rho, e), \qquad g_3 = (\rho^2, e),$$
$$g_4 = (e, \rho), \qquad g_5 = (\rho, \rho), \qquad g_6 = (\rho^2, \rho),$$
$$g_7 = (e, \rho^2), \qquad g_8 = (\rho, \rho^2), \qquad g_9 = (\rho^2, \rho^2).$$

The list $\mathscr{L}_0$ specifies the multiplication table

(3.28)

$\cdot$	g_1	g_2	g_3	g_4	g_5	g_6	g_7	g_8	g_9
g_1	g_1	g_2	g_3	g_4	g_5	g_6	g_7	g_8	g_9
g_3	g_3	g_1	g_2	g_6	g_4	g_5	g_9	g_7	g_8
g_2	g_2	g_3	g_1	g_5	g_6	g_4	g_8	g_9	g_7
g_7	g_7	g_8	g_9	g_1	g_2	g_3	g_4	g_5	g_6
g_9	g_9	g_7	g_8	g_3	g_1	g_2	g_6	g_4	g_5
g_8	g_8	g_9	g_7	g_2	g_3	g_1	g_5	g_6	g_4
g_4	g_4	g_5	g_6	g_7	g_8	g_9	g_1	g_2	g_3
g_6	g_6	g_4	g_5	g_9	g_7	g_8	g_3	g_1	g_2
g_5	g_5	g_6	g_4	g_8	g_9	g_7	g_2	g_3	g_1

for $\mathcal{C}_3 \times \mathcal{C}_3$. A comparison of (2.20) on page 20 with (3.28) shows that the group $\mathfrak{P}$ in Example 2.15 has the structure of $\mathcal{C}_3 \times \mathcal{C}_3$. Therefore, an injective group-pattern matrix for $\mathcal{C}_3 \times \mathcal{C}_3$ is provided by A in (2.19) on page 20. For the representation $\mathbf{A}$ of that matrix in a *Mathematica* notebook, the evaluations of `automorphisms[A_]` as well as `A` and `automorphisms[A]` yield forty-eight automorphisms for $\mathcal{C}_3 \times \mathcal{C}_3$ of which the first seven are given by

$$
\begin{aligned}
\text{L[1]} &= \text{[1,2,3,4,5,6,7,8,9]}, \\
\text{L[2]} &= \text{[1,2,3,6,4,5,8,9,7]}, \\
\text{L[3]} &= \text{[1,2,3,5,6,4,9,7,8]}, \\
\text{L[4]} &= \text{[1,2,3,7,8,9,4,5,6]}, \\
\text{L[5]} &= \text{[1,2,3,8,9,7,6,4,5]}, \\
\text{L[6]} &= \text{[1,2,3,9,7,8,5,6,4]}, \\
\text{L[7]} &= \text{[1,3,2,4,6,5,7,9,8]}
\end{aligned}
$$

followed by forty-one more and statements about $m = 48$ with $\mathfrak{N}_{\mathfrak{G}}(\mathcal{C}_3 \times \mathcal{C}_3) = 840$. Consequently, the group $Aut(\mathcal{C}_3 \times \mathcal{C}_3)$ of automorphisms for $\mathcal{C}_3 \times \mathcal{C}_3$ has order 48. For all of the automorphisms, download the notebook P-037.nb from

$$\texttt{http://homepages.uc.edu/~chalklr/group-pattern.htm}$$

as described in Appendix H. This computation required less than 24 seconds.

OBSERVATION 3.13. Multiplication for $Aut(\mathcal{C}_3 \times \mathcal{C}_3)$ is composition of functions read right-to-left as is customary. Thus, we have $L[2] \circ L[4] = L[5]$. If the elements of $Aut(\mathcal{C}_3 \times \mathcal{C}_3)$ are viewed as permutation symbols read left-to-right, then

$$
L[2] \cdot L[4] = \begin{pmatrix} 1\,2\,3\,4\,5\,6\,7\,8\,9 \\ 1\,2\,3\,6\,4\,5\,8\,9\,7 \end{pmatrix} \begin{pmatrix} 1\,2\,3\,4\,5\,6\,7\,8\,9 \\ 1\,2\,3\,7\,8\,9\,4\,5\,6 \end{pmatrix} = \begin{pmatrix} 1\,2\,3\,4\,5\,6\,7\,8\,9 \\ 1\,2\,3\,9\,7\,8\,5\,6\,4 \end{pmatrix} = L[6].
$$

This motivates our inclusion of the following well-known result.

PROPOSITION 3.14. *Let G be a group with its multiplication denoted by $\circ$ and let $*$ be the multiplication defined for G by $x * y = y \circ x$, for each x, y in G. Then, G with $*$ forms a group that is isomorphic to the group G with $\circ$.*

PROOF. Let f be the function from G to G defined by $f(x) = x^{-1}$, for each x in G, where x^{-1} is the inverse of x in G with respect to $\circ$. Then, f is one-to-one, onto, and satisfies

$$f(x \circ y) = (x \circ y)^{-1} = y^{-1} \circ x^{-1} = f(y) \circ f(x) = f(x) * f(y), \quad \text{for each } x,\ y \text{ in } G.$$

These properties of f shows that: the multiplication $*$ is associative; the identity element e for $\circ$ is an identity element for $*$; and, for each x in G, an inverse for x with $\circ$ is an inverse for x with $*$. Thus, G with $*$ is a group and f is a group isomorphism of G with $\circ$ onto G with $*$. This completes the proof. $\square$

Definition 2.2 on page 12 provides context for the next result.

PROPOSITION 3.15. *Let G be a group of order n, let S denote a set that has at least two elements, and let $\mathscr{L}_1, \mathscr{L}_2, \ldots, \mathscr{L}_{(n-1)!}$ denote the $(n-1)!$ lists for G having e as their first element. Then, the number of distinct sets among*

$$(3.29) \qquad \mathcal{M}[S,\ G,\ \mathscr{L}_1],\ \mathcal{M}[S,\ G,\ \mathscr{L}_2],\ \ldots,\ \mathcal{M}[S,\ G,\ \mathscr{L}_{(n-1)!}]$$

is equal to the number $\mathfrak{N}_{\mathfrak{G}}(G)$ of standard group-pattern matrices for G.

PROOF. For $n \leq 3$, we have $\mathfrak{N}_{\mathfrak{G}}(G) = 1$ and the assertion is clearly valid.

For $n \geq 4$, suppose that $\mathfrak{S}_1$, $\mathfrak{S}_2$ are unequal standard group-pattern matrices for G. Let $\mathfrak{M}_1$, $\mathfrak{M}_2$ be the sets of (3.29) that they specify. The first rows of $\mathfrak{S}_1$ and $\mathfrak{S}_2$ are equal and each of the components in their principal diagonals are equal to X_1. Thus, there are integers r, s such that $2 \leq r \leq n$, $1 \leq s \leq n$, $r \neq s$, and $\left[\mathfrak{S}_1\right]_{r,s} \neq \left[\mathfrak{S}_2\right]_{r,s}$. We set $X_k = \left[\mathfrak{S}_1\right]_{r,s}$ and $X_\ell = \left[\mathfrak{S}_2\right]_{r,s}$, where $2 \leq k,\, \ell \leq n$ and $k \neq \ell$. In view of

$$\left[\mathfrak{S}_1\right]_{1,k} = X_k = \left[\mathfrak{S}_1\right]_{r,s} \quad \text{and} \quad \left[\mathfrak{S}_2\right]_{1,k} = X_k \neq X_\ell = \left[\mathfrak{S}_2\right]_{r,s},$$

we observe that: for each matrix A in $\mathfrak{M}_1$, the $(1, k)$-component of A and the (r, s)-component of A are equal. However, there are matrices in $\mathfrak{M}_2$ for which the $(1, k)$-component and the (r, s)-component are not equal. This yields $\mathfrak{M}_1 \neq \mathfrak{M}_2$ and shows that the number of distinct sets in (3.29) is equal to the number $\mathfrak{N}_{\mathfrak{G}}(G)$ of distinct standard group-pattern matrices for G. This completes the proof. $\square$

3.6. Computer algebra for the standard group-pattern matrices

We can use a version of *Mathematica* such as [48] to exhibit all of the standard group-pattern matrices for any group G of order $n \leq 10$ whenever the computer has at least 16 GB of random access memory. For $n = 11$, $n = 12$, $\ldots$, the memory demands increase rapidly. We begin by entering the input statement

```
standard[A_] := Module[
    {n, id, list, P, i, j, k, a, Z1, B, b, s, Stan},
    n = Dimensions[A][[1]];
    id = IdentityMatrix[n];
    list = Permutations[id];
    Do[ P[i] = list[[i]], {i, 1, (n-1)!}];
    Clear[counter];  counter = 0;
    Do[ a[k] = A[[1, k]], {k, 1, n}];
    rule1 = Table[ a[k] -> X[k], {k, 1, n}];
    Z0 = A /. rule1;
    Do[ ( B[j] = P[j].Z0.Transpose[P[j]];
        Do[ b[k] = B[j][[1, k]], {k, 1, n}];
        rule2 = Table[b[k] -> X[k], {k, 1, n}];
        B[j] = B[j] /. rule2;
        Do[ If[ TrueQ[ B[j] == Stan[k] ],
                s[j, k] = 1, s[j, k] = 0],
                        {k, 1, counter}];
        If[TrueQ[ Sum[ s[j, p], {p, 1, counter}] == 0],
                ( Print["Stan[", counter + 1, "]= ",
                            B[j] // MatrixForm];
        Stan[counter + 1] = B[j];
        lastCount = counter + 1;
        counter = counter + 1),    {}   ];
        Clear[B, s]), {j, 1, (n-1)!}];
    Print["Number of standard matrices = ", lastCount];
    Print["Number of automorphisms = ", (n - 1)!/lastCount] ]
```

in a *Mathematica* notebook and evaluating it. Then, to exhibit all of the standard group-pattern matrices for G, we input a *Mathematica* representation A of any injective group-pattern matrix A for G and then evaluate A as well as standard[A].

The program also yields $\mathfrak{N}_{\mathfrak{G}}(G)$ and $m = |Aut(G)|$. However, to merely obtain m and $\mathfrak{N}_{\mathfrak{G}}(G)$, the program of page 34 is much more efficient.

OBSERVATION 3.16. For the line of the preceding input that appears as

```
Do[ P[i] = list[[i]], {i, 1, (n-1)!}];
```

and for the same line in the input of the computer program on page 34, there is a natural question. However, with *Mathematica* employed as the system of computer algebra, the first $(n-1)!$ elements in its list of $n \times n$ permutation matrices are those having their $(1, 1)$-component equal to 1. They are precisely the ones needed.

OBSERVATION 3.17. As a check that the input A for the later evaluation of standard[A] is a representation of an injective group-pattern matrix, we can first enter and evaluate the input statement of page 19. Then, we enter and evaluate each of A and injectiveGroupPatternQ[A].

EXAMPLE 3.18. For the cyclic group $G = C_5$ of order 5, the matrix A with $n = 5$ in (2.4) on page 12 shows that a standard group-pattern matrix for C_5 is given by

$$(3.30) \qquad A_0 = \begin{bmatrix} X_1 & X_2 & X_3 & X_4 & X_5 \\ X_5 & X_1 & X_2 & X_3 & X_4 \\ X_4 & X_5 & X_1 & X_2 & X_3 \\ X_3 & X_4 & X_5 & X_1 & X_2 \\ X_2 & X_3 & X_4 & X_5 & X_1 \end{bmatrix}.$$

After evaluating a *Mathematica* representation A0 for A_0 in (3.30), we evaluate standard[A0] and find that the output yields the six matrices

$$Stan[1] = \begin{bmatrix} X_1 & X_2 & X_3 & X_4 & X_5 \\ X_5 & X_1 & X_2 & X_3 & X_4 \\ X_4 & X_5 & X_1 & X_2 & X_3 \\ X_3 & X_4 & X_5 & X_1 & X_2 \\ X_2 & X_3 & X_4 & X_5 & X_1 \end{bmatrix}, \qquad Stan[2] = \begin{bmatrix} X_1 & X_2 & X_3 & X_4 & X_5 \\ X_4 & X_1 & X_2 & X_5 & X_3 \\ X_5 & X_4 & X_1 & X_3 & X_2 \\ X_2 & X_3 & X_5 & X_1 & X_4 \\ X_3 & X_5 & X_4 & X_2 & X_1 \end{bmatrix},$$

$$Stan[3] = \begin{bmatrix} X_1 & X_2 & X_3 & X_4 & X_5 \\ X_5 & X_1 & X_4 & X_2 & X_3 \\ X_4 & X_3 & X_1 & X_5 & X_2 \\ X_3 & X_5 & X_2 & X_1 & X_4 \\ X_2 & X_4 & X_5 & X_3 & X_1 \end{bmatrix}, \qquad Stan[4] = \begin{bmatrix} X_1 & X_2 & X_3 & X_4 & X_5 \\ X_4 & X_1 & X_5 & X_3 & X_2 \\ X_5 & X_3 & X_1 & X_2 & X_4 \\ X_2 & X_5 & X_4 & X_1 & X_3 \\ X_3 & X_4 & X_2 & X_5 & X_1 \end{bmatrix},$$

$$Stan[5] = \begin{bmatrix} X_1 & X_2 & X_3 & X_4 & X_5 \\ X_3 & X_1 & X_5 & X_2 & X_4 \\ X_2 & X_4 & X_1 & X_5 & X_3 \\ X_5 & X_3 & X_4 & X_1 & X_2 \\ X_4 & X_5 & X_2 & X_3 & X_1 \end{bmatrix}, \qquad Stan[6] = \begin{bmatrix} X_1 & X_2 & X_3 & X_4 & X_5 \\ X_3 & X_1 & X_4 & X_5 & X_2 \\ X_2 & X_5 & X_1 & X_3 & X_4 \\ X_5 & X_4 & X_2 & X_1 & X_3 \\ X_4 & X_3 & X_5 & X_2 & X_1 \end{bmatrix}$$

as the standard group-pattern matrices for C_5. It also outputs $\mathfrak{N}_{\mathfrak{G}}(G) = 6$ and $(5 - 1)!/\mathfrak{N}_{\mathfrak{G}}(G) = 4$ as the number of automorphisms for C_5.

EXERCISES

1. For the dihedral group $\mathcal{D}_3$ consisting of the six symmetries for an equilateral triangle in Example 1.5 where $g_1 = e$, $g_2 = \alpha$, $g = \alpha^2$, $g_4 = \beta_1$, $g_5 = \beta_2$, and $g_6 = \beta_3$, find the six automorphisms and print each of the 20 standard group-pattern matrices.

2. For the dihedral group $\mathcal{D}_4$ of the eight symmetries for a square in Example 1.6, find the automorphisms and find $\mathfrak{N}_\mathfrak{G}(\mathcal{D}_4)$.

3. For the group G that has Z_1 in (2.36) on page 25 as an injective group-pattern matrix, find $|Aut(G)|$ and $\mathfrak{N}_\mathfrak{G}(G)$.

4. For the direct product $\mathcal{C}_2 \times \mathcal{C}_2 \times \mathcal{C}_2$ as a group of order 8, use A in (1.13) on page 7 to find that $|Aut(\mathcal{C}_2 \times \mathcal{C}_2 \times \mathcal{C}_2)| = 168$ and $\mathfrak{N}_\mathfrak{G}(\mathcal{C}_2 \times \mathcal{C}_2 \times \mathcal{C}_2) = 30$.

5. For the direct product $\mathcal{C}_4 \times \mathcal{C}_2$ as a group of order 8, use the matrix A in (1.16) on page 8 to obtain $|Aut(\mathcal{C}_4 \times \mathcal{C}_2)| = 8$ and $\mathfrak{N}_\mathfrak{G}(\mathcal{C}_4 \times \mathcal{C}_2) = 630$. Also, with respect to the list of the eight elements $g_1 = e$, $g_2 = \alpha$, $g_3 = \alpha^2$, $g_4 = \alpha^3$, $g_5 = m$, $g_6 = m\alpha$, $g_7 = m\alpha^2$, and $g_8 = m\alpha^3$ that specify the multiplication table (1.15) for (1.16), find the eight automorphisms of $\mathcal{C}_4 \times \mathcal{C}_2$.

6. Among the basic results about groups, we know that: if α is an element of a group G of order n, then: (i) $\alpha^n = e$; (ii) there is a least positive integer d such that $\alpha^d = e$; and, (iii) for any integer k, the integer d divides k if and only if $\alpha^k = e$. The integer d for α is called the *period* of α (or the *order* of α). A group of order n is said to be *cyclic* when it contains an element of period n. An element of period n in a cyclic group $\mathcal{C}_n$ of order n is called a *generator* of $\mathcal{C}_n$.

For a generator α of $\mathcal{C}_n$ and an integer k that satisfies $1 \leq k \leq n$, prove that: α^k is a generator of $\mathcal{C}_n$ if and only if k is relatively prime to n.

7. Among the basic results of number theory, we know that: (i) the Euler totient function φ is defined on the set of positive integers; (ii) $\varphi(1) = 1$ and, for $n \geq 2$, if $p_1, p_2, \ldots, p_r$ are the distinct prime integers that divide n, then

$$(3.31) \qquad \varphi(n) = n \left(1 - \frac{1}{p_1}\right)\left(1 - \frac{1}{p_2}\right) \cdots \left(1 - \frac{1}{p_r}\right);$$

(iii) $\varphi(n)$ equals the number of positive integers $\leq n$ that are relatively prime to n.

Observe that the number of generators in a cyclic group $\mathcal{C}_n$ of order n is equal to $\varphi(n)$.

8. Prove that if α is a generator for a cyclic group $\mathcal{C}_n$ and ϕ is an automorphism of $\mathcal{C}_n$, then $\phi(\alpha)$ is a generator for $\mathcal{C}_n$.

9. Prove that if α and β are any two generators for $\mathcal{C}_n$, then there is a unique automorphism ϕ of $\mathcal{C}_n$ such that $\phi(\alpha) = \beta$.

10. Prove the following result. If χ is an automorphism of $\mathcal{C}_n$ and α is a generator for $\mathcal{C}_n$, then there is an integer b such that: $1 \leq b \leq n$, b is relatively prime to n, and $\chi(\alpha^k) = \alpha^{bk}$, for $0 \leq k \leq n - 1$.

11. Prove that $|Aut(\mathcal{C}_n)| = \varphi(n)$ and $\mathfrak{N}_\mathfrak{G}(\mathcal{C}_n) = (n-1)!/\varphi(n)$.

CHAPTER 4

Properties for $\mathcal{M}[\mathfrak{F},\, G,\, \mathscr{L}]$ Based on Definition 2.2

Let $\mathscr{L} = (g_1,\, g_2,\, \ldots,\, g_n)$ be a list of the n elements in a group G of order n.

4.1. Closure under addition, multiplication, and scalar multiplication

THEOREM 4.1. *Let $\mathfrak{R}$ be a ring. Let matrices A, B in $\mathcal{M}[\mathfrak{R},\, G,\, \mathscr{L}]$ be given by*

$$(4.1) \qquad [A]_{r,s} = \sigma\big(g_r^{-1}g_s\big) \quad and \quad [B]_{r,s} = \tau\big(g_r^{-1}g_s\big), \quad for\ r,\, s = 1,\, 2,\, \ldots,\, n,$$

where σ and τ are functions from G to $\mathfrak{R}$. Then, the matrix sum $A + B$ and the matrix product AB are elements of $\mathcal{M}[\mathfrak{R},\, G,\, \mathscr{L}]$ for which (4.1) yields

$$(4.2) \qquad [A+B]_{r,s} = (\sigma + \tau)\big(g_r^{-1}g_s\big) \quad and \quad [AB]_{r,s} = (\sigma * \tau)\big(g_r^{-1}g_s\big),$$

where $\sigma + \tau$ and $\sigma * \tau$ are functions from G to $\mathfrak{R}$ defined, for each x in G, through

$$(4.3) \qquad (\sigma + \tau)(x) = \sigma(x) + \tau(x) \quad and \quad (\sigma * \tau)(x) = \sum_{h \in G} \sigma(h)\, \tau\big(h^{-1}x\big).$$

Also, the scalar product cA, for c in $\mathfrak{R}$, and the transpose A^T are in $\mathcal{M}[\mathfrak{R},\, G,\, \mathscr{L}]$.

PROOF. For $1 \le r,\, s \le n$, we use (4.1) and (4.3) to deduce the first formula of (4.2) as $[A+B]_{r,s} = [A]_{r,s} + [B]_{r,s} = \sigma\big(g_r^{-1}g_s\big) + \tau\big(g_r^{-1}g_s\big) = (\sigma + \tau)\big(g_r^{-1}g_s\big)$. Thus, $A + B$ is a group-pattern matrix in $\mathcal{M}[\mathfrak{R},\, G,\, \mathscr{L}]$. Another application gives

$$(4.4) \qquad [AB]_{r,s} = \sum_{k=1}^{n} [A]_{r,k}\, [B]_{k,s} = \sum_{k=1}^{n} \sigma\big(g_r^{-1}g_k\big)\, \tau\big(g_k^{-1}g_s\big)$$

$$= \sum_{k=1}^{n} \sigma\big(g_r^{-1}g_k\big)\, \tau\Big(\big(g_r^{-1}g_k\big)^{-1}\big(g_r^{-1}g_s\big)\Big).$$

For each fixed r, as g_k ranges through the elements of G, we see that $g_r^{-1}g_k$ ranges through the elements of G. Therefore (4.4) and (4.3) give

$$[AB]_{r,s} = \sum_{h \in G} \sigma(h)\, \tau\Big(h^{-1}\big(g_r^{-1}g_s\big)\Big) = (\sigma * \tau)\big(g_r^{-1}g_s\big), \quad for\ r,\, s = 1,\, 2,\, \ldots,\, n.$$

This shows that that AB is a group-pattern matrix in $\mathcal{M}[\mathfrak{R},\, G,\, \mathscr{L}]$.

For c in $\mathfrak{R}$ and A in $\mathcal{M}[\mathfrak{R},\, G,\, \mathscr{L}]$ given by (4.1), let $c\sigma$ denote the function from G to $\mathfrak{R}$ defined by $(c\sigma)(x) = c\,\sigma(x)$, for each x in G, Then we have

$$[cA]_{r,s} = c\,[A]_{r,s} = c\,\sigma\big(g_r^{-1}g_s\big) = (c\sigma)\big(g_r^{-1}g_s\big) \quad for\ r,\, s = 1,\, 2,\, \ldots,\, n.$$

Thus, cA is a group-pattern matrix in $\mathcal{M}[\mathfrak{R},\, G,\, \mathscr{L}]$.

We define υ from G to $\mathfrak{R}$ by $\upsilon(x) = \sigma\big(x^{-1}\big)$, for each x in G, and check that

$$[A^T]_{r,s} = [A]_{s,r} = \sigma\big(g_s^{-1}g_r\big) = \sigma\Big(\big(g_r^{-1}g_s\big)^{-1}\Big) = \upsilon\big(g_r^{-1}g_s\big), \quad for\ r,\, s = 1,\, 2,\, \ldots,\, n.$$

Consequently, A^T is in $\mathcal{M}[\mathfrak{R},\, G,\, \mathscr{L}]$. This completes the proof. $\qquad\square$

4.2. Some algebraic properties of $\mathcal{M}[\mathfrak{F},\, G,\, \mathcal{L}]$

Henceforth, we assume that $\mathfrak{R}$ for $\mathcal{M}[\mathfrak{R},\, G,\, \mathcal{L}]$ is a field $\mathfrak{F}$. We recall from Section 2.7 on page 22 that $n \times n$ permutation matrices $P_1,\, P_2,\, \ldots,\, P_n$ belonging to $\mathcal{M}[\mathfrak{F},\, G,\, \mathcal{L}]$ are defined by

$$(4.5) \qquad \left[P_k\right]_{r,s} = \widehat{g_k}\big(g_r^{-1} g_s\big), \quad \text{for } 1 \leq k,\, r,\, s \leq n,$$

where, for $1 \leq k \leq n$, $\widehat{g_k}$ is the function from G to $\mathfrak{F}$ having

$$(4.6) \qquad \widehat{g_k}(x) = \begin{cases} 1, & \text{if } x = g_k, \\ 0, & \text{if } x \neq g_k, \end{cases} \quad \text{for each } x \text{ in } G.$$

Theorem 2.16 on page 22 shows that the set $\mathfrak{P} = \big\{P_1,\, P_2,\, \ldots,\, P_n\big\}$ forms a group with respect to matrix multiplication and that the function F from G to $\mathfrak{P}$ given by

$$(4.7) \qquad F(g_k) = P_k, \quad \text{for } 1 \leq k \leq n,$$

is a group isomorphism of G onto $\mathfrak{P}$. Moreover, we have $F(e) = I_n$ where I_n is the $n \times n$ identity matrix.

THEOREM 4.2. *With matrix addition and multiplication, $\mathcal{M}[\mathfrak{F},\, G,\, \mathcal{L}]$ is a ring that has I_n as an identity element for multiplication. Also, this ring is commutative if and only if G is abelian. Moreover, with matrix addition and scalar multiplication by elements of $\mathfrak{F}$, $\mathcal{M}[\mathfrak{F},\, G,\, \mathcal{L}]$ is a vector space over $\mathfrak{F}$ that has a basis given by*

$$(4.8) \qquad \mathfrak{P} = \big\{P_1,\, P_2,\, \ldots,\, P_n\big\} = \big\{F(g_1),\, F(g_2),\, \ldots,\, F(g_n)\big\},$$

where, for $1 \leq i,\, j,\, k \leq n$, the relation $P_i P_j = P_k$ is valid if and only if $g_i g_j = g_k$.

PROOF. Theorem 4.1 shows that sums and products of elements in $\mathcal{M}[\mathfrak{F},\, G,\, \mathcal{L}]$ belong to $\mathcal{M}[\mathfrak{F},\, G,\, \mathcal{L}]$. The $n \times n$ zero matrix belongs to $\mathcal{M}[\mathfrak{F},\, G,\, \mathcal{L}]$ because it is specified by the function from G to $\mathfrak{F}$ defined by $x \mapsto 0$, for each x in G. Suppose that A belongs to $\mathcal{M}[\mathfrak{F},\, G,\, \mathcal{L}]$ and is specified according to (2.3) by a function σ from G to $\mathfrak{F}$. Then $-A$ belongs to $\mathcal{M}[\mathfrak{F},\, G,\, \mathcal{L}]$ because it is specified by the function $-\sigma$ from G to $\mathfrak{F}$ defined by $(-\sigma)(x) = -\sigma(x)$, for each x in G. Matrix addition is associative and commutative. Multiplication of matrices is associative as well as distributive over addition from either side. Thus, $\mathcal{M}[\mathfrak{F},\, G,\, \mathcal{L}]$ is a ring.

For $1 \leq r,\, s \leq n$, we observe that $\big[F(e)\big]_{r,s} = \widehat{e}(g_r^{-1} g_s) = \big[I_n\big]_{r,s}$. This yields $F(e) = I_n$. Thus, the $n \times n$ identity matrix I_n is an element of $\mathcal{M}[\mathfrak{F},\, G,\, \mathcal{L}]$.

For c in $\mathfrak{F}$ and A in $\mathcal{M}[\mathfrak{F},\, G,\, \mathcal{L}]$, Theorem 4.1 shows that the scalar product cA is in $\mathcal{M}[\mathfrak{F},\, G,\, \mathcal{L}]$. The usual properties of addition and scalar multiplication for matrices show that $\mathcal{M}[\mathfrak{F},\, G,\, \mathcal{L}]$ is a vector space over $\mathfrak{F}$.

For A in $\mathcal{M}[\mathfrak{F},\, G,\, \mathcal{L}]$, there is a function σ from G to $\mathfrak{F}$ such that

$$(4.9) \qquad \big[A\big]_{r,s} = \sigma\big(g_r^{-1} g_s\big), \quad \text{for } r,\, s = 1,\, 2,\, \ldots,\, n.$$

We set $a_k = \sigma(g_k)$, for $k = 1,\, 2,\, \ldots,\, n$, and introduce the function τ from G to $\mathfrak{F}$ defined in terms of $\mathcal{L}$ and (4.6) by $\tau = a_1 \widehat{g_1} + a_2 \widehat{g_2} + \cdots + a_n \widehat{g_n}$. Since we have

$$\tau(g_k) = a_1 \widehat{g_1}(g_k) + a_2 \widehat{g_2}(g_k) + \cdots + a_n \widehat{g_n}(g_k) = a_k \widehat{g_k}(g_k) = a_k = \sigma(g_k),$$
$$\text{for } k = 1,\, 2,\, \ldots,\, n,$$

we see that $\tau = \sigma$. We use this with (4.9), (4.5), and (4.7) to obtain

$$[A]_{r,s} = \tau(g_r^{-1}g_s) = \sum_{k=1}^n a_k\,\widehat{g}_k(g_r^{-1}g_s) = \sum_{k=1}^n a_k\,[P_k]_{r,s} = \left[\sum_{k=1}^n a_k F(g_k)\right]_{r,s},$$

for $r,\,s = 1,\,2,\,\ldots,\,n$.

This yields

$$(4.10) \qquad A = a_1 F(g_1) + a_2 F(g_2) + \cdots + a_n F(g_n).$$

Thus, the linear combinations of $F(g_1)$, $F(g_2)$, ..., $F(g_n)$ span $\mathcal{M}[\mathfrak{F}, G, \mathscr{L}]$.

Let c_1, c_2, ..., c_n be elements of $\mathfrak{F}$ such that

$$c_1 F(g_1) + c_2 F(g_2) + \cdots + c_n F(g_n) = \mathcal{O},$$

where $\mathcal{O}$ is the $n \times n$ zero matrix. Then, for $r,\,s = 1,\,2,\,\ldots,\,n$, we have

$$(4.11) \qquad 0 = [\mathcal{O}]_{r,s} = \left[\sum_{k=1}^n c_k F(g_k)\right]_{r,s} = \sum_{k=1}^n c_k\,[F(g_k)]_{r,s}$$
$$= c_1\,\widehat{g}_1(g_r^{-1}g_s) + c_2\,\widehat{g}_2(g_r^{-1}g_s) + \cdots + c_n\,\widehat{g}_n(g_r^{-1}g_s).$$

For any integer k satisfying $1 \le k \le n$, there are corresponding integers r and s such that $g_r^{-1}g_s = g_k$ and (4.11) therefore yields $0 = c_k\,\widehat{g}_k(g_k) = c_k$. This shows that (4.10) is unique and (4.8) is a basis for $\mathcal{M}[\mathfrak{F}, G, \mathscr{L}]$ as a vector space over $\mathfrak{F}$.

For $1 \le i,\,j,\,k \le n$, we have $F(g_i\,g_j) = F(g_k)$ if and only if $g_i\,g_j = g_k$. In view of $P_i P_j = F(g_i)F(g_j) = F(g_i\,g_j)$ and $P_k = F(g_k)$, we see that the relation $P_i P_j = P_k$ is valid if and only if $g_i\,g_j = g_k$.

Suppose that G is abelian. Then, for $1 \le i,\,j \le n$, we have $g_i\,g_j = g_j\,g_i$ and

$$(4.12) \qquad F(g_i)\,F(g_j) = F(g_i\,g_j) = F(g_j\,g_i) = F(g_j)\,F(g_i).$$

For A and B in $\mathcal{M}[\mathfrak{F}, G, \mathscr{L}]$ given by (4.10) and

$$(4.13) \qquad B = b_1 F(g_1) + b_2 F(g_2) + \cdots + b_n F(g_n),$$

we employ (4.10), (4.13), and (4.12) to deduce that

$$AB = \sum_{1 \le i,\,j \le n} a_i\,b_j\,F(g_i)\,F(g_j) = \sum_{1 \le j,\,i \le n} b_j\,a_i\,F(g_j)\,F(g_i) = BA.$$

Thus, the ring $\mathcal{M}[\mathfrak{F}, G, \mathscr{L}]$ is commutative when G is abelian.

Suppose G is not abelian. Then, we have $h\,g \ne g\,h$, for some g, h in G, and therefore $F(h)F(g) = F(h\,g) \ne F(g\,h) = F(g)\,F(h)$, for $F(g)$, $F(h)$ in $\mathcal{M}[\mathfrak{F}, G, \mathscr{L}]$. Consequently, the ring $\mathcal{M}[\mathfrak{F}, G, \mathscr{L}]$ is not commutative when G is not abelian. This completes the proof of Theorem 4.2. $\qquad\square$

OBSERVATION 4.3. For any $n \times n$ matrices A, B over $\mathfrak{F}$ and element c in $\mathfrak{F}$, matrix multiplication and scalar multiplication yield

$$(4.14) \qquad c(AB) = (cA)B = A(cB).$$

In view of Theorem 4.2 and (4.14), we note that $\mathcal{M}[\mathfrak{F}, G, \mathscr{L}]$ *is an algebra over* $\mathfrak{F}$.

4.3. The group ring and group algebra for G over $\mathfrak{F}$

Each element in $\mathcal{M}[\mathfrak{F}, G, \mathscr{L}]$ is uniquely specified by a function from G to $\mathfrak{F}$. The one-to-one correspondence between $\mathcal{M}[\mathfrak{F}, G, \mathscr{L}]$ and the set $\mathfrak{F}[G]$ of functions from G to $\mathfrak{F}$ serves as motivation to define addition $+$, multiplication $*$, and scalar multiplication for $\mathfrak{F}[G]$ in terms of σ, τ in $\mathfrak{F}[G]$ and c in $\mathfrak{F}$ by

$$(4.15) \qquad (\sigma + \tau)(x) = \sigma(x) + \tau(x), \quad \text{for each } x \text{ in } G,$$

$$(4.16) \qquad (\sigma * \tau)(x) = \sum_{h \in G} \sigma(h)\, \tau(h^{-1}x), \quad \text{for each } x \in G.$$

and

$$(4.17) \qquad (c\sigma)(x) = c\,\sigma(x), \quad \text{for each } x \in G.$$

In regard to the list $\mathscr{L} = (g_1, g_2, \ldots, g_n)$ for G, let Φ be the function from $\mathcal{M}[\mathfrak{F}, G, \mathscr{L}]$ to $\mathfrak{F}[G]$ defined for each X in $\mathcal{M}[\mathfrak{F}, G, \mathscr{L}]$ by $\Phi(X) = f$ where f is the function in $\mathfrak{F}[G]$ such that $f(g_r^{-1}g_s) = [X]_{r,s}$, for $1 \le r, s \le n$. As an example, (4.5) and (4.6) show that $\Phi(P_k) = \widehat{g_k}$, for $1 \le k \le n$.

THEOREM 4.4. *With (4.15), (4.16), and (4.17), the set $\mathfrak{F}[G]$ is a ring as well as an algebra over $\mathfrak{F}$. A basis for $\mathfrak{F}[G]$ is given by*

$$(4.18) \qquad \mathfrak{B} = \{\widehat{g_1}, \widehat{g_2}, \ldots, \widehat{g_n}\}, \quad \text{where } \widehat{g_i} * \widehat{g_j} = \widehat{g_i g_j}, \text{ for } 1 \le i, j \le n,$$

and a group isomorphism of G onto $\mathfrak{B}$ is defined by $g_k \mapsto \widehat{g_k}$, for $1 \le k \le n$. Moreover, the ring $\mathfrak{F}[G]$ has $\widehat{e}$ as an identity element for multiplication; $\mathfrak{F}[G]$ is commutative if and only if G is abelian; and Φ is a ring isomorphism as well as an algebra isomorphism of $\mathcal{M}[\mathfrak{F}, G, \mathscr{L}]$ onto $\mathfrak{F}[G]$.

PROOF. Each f in $\mathfrak{F}[G]$ specifies a matrix X in $\mathcal{M}[\mathfrak{F}, G, \mathscr{L}]$ having $\Phi(X) = f$. Thus, Φ is onto. For A, B in $\mathcal{M}[\mathfrak{F}, G, \mathscr{L}]$ and $A \ne B$, there are integers r, s satisfying $1 \le r, s \le n$ such that

$$\Phi(A)(g_r^{-1}g_s) = [A]_{r,s} \ne [B]_{r,s} = \Phi(B)(g_r^{-1}g_s).$$

This yields $\Phi(A) \ne \Phi(B)$ and shows that Φ is one-to-one.

For any A, B in $\mathcal{M}[\mathfrak{F}, G, \mathscr{L}]$, we set $\sigma = \Phi(A)$ and $\tau = \Phi(B)$. This gives

$$[A+B]_{r,s} = [A]_{r,s} + [B]_{r,s} = \sigma(g_r^{-1}g_s) + \tau(g_r^{-1}g_s) = (\sigma + \tau)(g_r^{-1}g_s),$$

for $1 \le r, s \le n$, and therefore

$$(4.19) \qquad \Phi(A + B) = \sigma + \tau = \Phi(A) + \Phi(B).$$

We use (4.16) to deduce

$$[AB]_{r,s} = \sum_{k=1}^{n} [A]_{r,k} [B]_{k,s} = \sum_{k=1}^{n} \sigma(g_r^{-1}g_k)\, \tau(g_k^{-1}g_s)$$

$$= \sum_{k=1}^{n} \sigma(g_r^{-1}g_k)\, \tau\big((g_r^{-1}g_k)^{-1} g_r^{-1}g_s\big) = \sum_{h \in G} \sigma(h)\, \tau\big(h^{-1}(g_r^{-1}g_s)\big)$$

$$= (\sigma * \tau)(g_r^{-1}g_s), \quad \text{for } r, s = 1, 2, \ldots, n,$$

and conclude that

$$(4.20) \qquad \Phi(AB) = \sigma * \tau = \Phi(A) * \Phi(B).$$

For c in $\mathfrak{F}$, we find that

$$[cA]_{r,s} = c[A]_{r,s} = c\,\sigma\big(g_r^{-1}g_s\big) = (c\,\sigma)\big(g_r^{-1}g_s\big), \quad \text{for } r,\, s = 1,\, 2,\, \ldots,\, n,$$

and consequently

$$(4.21) \qquad\qquad \Phi(cA) = c\,\sigma = c\Phi(A).$$

In view of (4.19), (4.20), and (4.21), the structure of $\mathfrak{F}[G]$ with respect to (4.15), (4.16), and (4.17) is such that: Φ is a ring isomorphism of $\mathcal{M}[\mathfrak{F},\, G,\, \mathscr{L}]$ onto $\mathfrak{F}[G]$ and Φ is a vector-space isomorphism of $\mathcal{M}[\mathfrak{F},\, G,\, \mathscr{L}]$ onto $\mathfrak{F}[G]$.

We use (4.5) on page 42 to deduce $\Phi(P_k) = \widehat{g}_k$, for $1 \leq k \leq n$. Consequently, the vector-space isomorphism Φ of $\mathcal{M}[\mathfrak{F},\, G,\, \mathscr{L}]$ onto $\mathfrak{F}[G]$ transforms the basis $\mathfrak{P}$ of $\mathcal{M}[\mathfrak{F},\, G,\, \mathscr{L}]$ into $\mathfrak{B}$ as a basis for $\mathfrak{F}[G]$. The restriction of Φ to $\mathfrak{P}$ is a one-to-one function from $\mathfrak{P}$ onto $\mathfrak{B}$ with $\Phi(P_iP_j) = \Phi(P_i) * \Phi(P_j)$, for each $P_i,\, P_j$ in $\mathfrak{P}$. This shows that $\mathfrak{B}$ forms a group, with $*$, that is isomorphic to $\mathfrak{P}$.

The one-to-one function $\widehat{F}$ from G onto $\mathfrak{B}$ defined by $\widehat{F}(g_k) = \widehat{g}_k$, for $1 \leq k \leq n$, and F of (4.7) yield $\widehat{F}(g_k) = \widehat{g}_k = \Phi(P_k) = \Phi\big(F(g_k)\big) = (\Phi \circ F)(g_k)$, for $1 \leq k \leq n$, and $\widehat{F} = \Phi \circ F$. Thus, for $1 \leq i,\, j \leq n$, we have

$$(4.22) \qquad \widehat{F}(g_i g_j) = \Phi\big(F(g_j g_j)\big) = \Phi\big(F(g_i)\,F(g_j)\big)$$

$$= \Phi(P_iP_j) = \Phi(P_i) * \Phi(P_j) = \widehat{g}_i * \widehat{g}_j = \widehat{F}(g_i) * \widehat{F}(g_j)$$

and see that $\widehat{F}$ is a group isomorphism of G onto $\mathfrak{B}$. In particular, (4.22) yields

$$\widehat{g_i g_j} = \widehat{F}(g_i g_j) = \widehat{F}(g_i) * \widehat{F}(g_j) = \widehat{g}_i * \widehat{g}_j, \quad \text{for } 1 \leq i,\, j \leq n.$$

For the $n \times n$ identity matrix I_n in $\mathcal{M}[\mathfrak{F},\, G,\, \mathscr{L}]$, we observe that

$$[I_n]_{r,s} = \begin{cases} 1, & \text{if } r = s, \\ 0, & \text{if } r \neq s, \end{cases} = \widehat{e}\big(g_r^{-1}g_s\big), \quad \text{for } r,\, s = 1,\, 2,\, \ldots,\, n,$$

and $\Phi(I_n) = \widehat{e}$. Thus, $\widehat{e}$ is an identity element for multiplication in $\mathfrak{F}[G]$.

Theorem 4.2 shows that multiplication in $\mathcal{M}[\mathfrak{F},\, G,\, \mathscr{L}]$ is commutative if and only if G is abelian. By using the ring isomorphism Φ, we see that multiplication in $\mathfrak{F}[G]$ is commutative if and only if G is abelian.

Each c in $\mathfrak{F}$ and A, B in $\mathcal{M}[\mathfrak{F},\, G,\, \mathscr{L}]$ satisfy the relation (4.14). After setting $\sigma = \Phi(A)$ and $\tau = \Phi(B)$, we apply Φ to (4.14) and obtain

$$c\,(\sigma * \tau) = (c\,\sigma) * \tau = \sigma * (c\,\tau).$$

Since each σ and τ in $\mathfrak{F}[G]$ is expressible as $\sigma = \Phi(A)$ and $\tau = \Phi(B)$ for some A and B in $\mathcal{M}[\mathfrak{F},\, G,\, \mathscr{L}]$, we conclude that $\mathfrak{F}[G]$ forms an algebra and Φ is an algebra isomorphism of $\mathcal{M}[\mathfrak{F},\, G,\, \mathscr{L}]$ onto $\mathfrak{F}[G]$. This completes the proof. $\qquad \square$

OBSERVATION 4.5. The *group ring* or *group algebra* $\mathfrak{F}[G]$ over $\mathfrak{F}$ is a basic concept and does not need an ordering for the group G such as that provided by the list $\mathscr{L}$. For instance, properties of $\mathfrak{F}[G]$ in [**31**, page 177] are efficiently deduced directly from (4.15), (4.16), and (4.17). We indicate another traditional approach in Subsection 4.3.1.

The proof of Theorem 4.4 shows how the properties of $\mathcal{M}[\mathfrak{F},\, G,\, \mathscr{L}]$ transfer directly to give an algebraic structure for the set $\mathfrak{F}[G]$ of functions from G to $\mathfrak{F}$. That viewpoint may be novel here.

The basis $\{\widehat{g_1}, \widehat{g_2}, \ldots, \widehat{g_n}\}$ for $\mathfrak{F}[G]$ as a vector space over $\mathfrak{F}$ enables each σ and τ in $\mathfrak{F}[G]$ to have a unique representation of the form

$$\sigma = \sum_{k=1}^{n} a_k \widehat{g_k} \quad \text{and} \quad \tau = \sum_{k=1}^{n} b_k \widehat{g_k}, \quad \text{with } a_1, a_2, \ldots, a_n, b_1, b_2, \ldots, b_n \text{ in } \mathfrak{F}.$$

Then, with c in $\mathfrak{F}$, we see that (4.15)–(4.17) are expressible as

$$\sigma + \tau = \sum_{k=1}^{n} \left(a_k + b_k\right) \widehat{g_k},$$

$$\sigma * \tau = \sum_{i=1}^{n} \sum_{j=1}^{n} \left(a_i b_j\right) \widehat{g_i} * \widehat{g_j} = \sum_{i=1}^{n} \sum_{j=1}^{n} \left(a_i b_j\right) \widehat{g_i g_j} = \sum_{k=1}^{n} \left(\sum_{\substack{1 \le i,j \le n \\ g_i g_j = g_k}} \left(a_i b_j\right)\right) \widehat{g_k},$$

$$c\sigma = \sum_{k=1}^{n} \left(c a_k\right) \widehat{g_k}.$$

However, this notation is also based on an ordering for the elements of G.

4.3.1. $\mathfrak{F}[G]$ as a set of formal sums. When typical elements σ, τ in $\mathfrak{F}[G]$ are initially represented in the form

$$\sigma = \sum_{x \in G} a_x \cdot x \quad \text{and} \quad \tau = \sum_{x \in G} b_x \cdot x, \quad \text{with } a_x, b_x \text{ in } \mathfrak{F} \text{ for each } x \text{ in } G,$$

the structure of $\mathfrak{F}[G]$ can be viewed with respect to the definitions

$$\left(\sum_{x \in G} a_x \cdot x\right) + \left(\sum_{x \in G} b_x \cdot x\right) = \sum_{x \in G} (a_x + b_x) \cdot x,$$

$$\left(\sum_{x \in G} a_x \cdot x\right) * \left(\sum_{y \in G} b_y \cdot y\right) = \sum_{(x,y) \in G \times G} (a_x b_y) \cdot (xy),$$

and

$$c\left(\sum_{x \in G} a_x \cdot x\right) = \sum_{x \in G} (c\, a_x) \cdot x, \quad \text{with } c \text{ in } \mathfrak{F},$$

rather than in terms of (4.15), (4.16), and (4.17). For instance, see [**41**, page 173].

OBSERVATION 4.6. Questions about inverses for elements in $\mathcal{M}[\mathfrak{F}, G, \mathscr{L}]$ and $\mathfrak{F}[G]$ will be answered in Section 4.5 after we first provide motivation.

For the 3×3 circulant matrix A and its cofactor matrix A^c given by

$$A = \begin{bmatrix} a_1 & a_2 & a_3 \\ a_3 & a_1 & a_2 \\ a_2 & a_3 & a_1 \end{bmatrix} \quad \text{and} \quad A^c = \begin{bmatrix} b_1 & b_2 & b_3 \\ b_3 & b_1 & b_2 \\ b_2 & b_3 & b_1 \end{bmatrix},$$

where $b_1 = a_1^2 - a_2 a_3$, $b_2 = a_2^2 - a_1 a_3$, $b_3 = a_3^2 - a_1 a_2$,

we see that A^c is also a circulant matrix. Additional examples of a similar nature lead to the following question. Is it true that, for any group-pattern matrix A, the cofactor matrix A^c of A is a group-pattern matrix of the same kind? We shall prove that the answer is yes when the components of the matrices belong to a field $\mathbb{F}$ that is either $\mathbb{C}$ or any proper field extension of $\mathbb{C}$. Our argument depends on a largely overlooked explicit construction developed in [**10**] of 1981 that we present in full detail as Theorem 13.3 on page 125.

4.4. Cofactor matrices for group-pattern matrices

To exhibit the cofactor matrices of various given $n \times n$ matrices, we introduce and evaluate the following three *Mathematica* input commands

```
Cofactor[m_List?MatrixQ, {i_Integer, j_Integer}] :=
    (-1)^(i + j)Det[Drop[Transpose[Drop[Transpose[m], {j}]], {i}]]

MinorMatrix[m_List?MatrixQ] := Map[Reverse, Minors[m], {0, 1}]

CofactorMatrix[m_List?MatrixQ] :=
    MapIndexed[#1 (-1)^(Plus @@ #2) &, MinorMatrix[m], {2}]
```

in three separate cells of a *Mathematica* notebook for a version such as [**48**].

EXAMPLE 4.7. We set $a_1 = 1$, $a_2 = 5$, $a_3 = 2$, $a_4 = 4$, $a_5 = 6$, $a_6 = 3$ in (1.7) of page 5 to see that the matrix

(4.23)
$$A = \begin{bmatrix} 1 & 5 & 2 & 4 & 6 & 3 \\ 2 & 1 & 5 & 6 & 3 & 4 \\ 5 & 2 & 1 & 3 & 4 & 6 \\ 4 & 6 & 3 & 1 & 5 & 2 \\ 6 & 3 & 4 & 2 & 1 & 5 \\ 3 & 4 & 6 & 5 & 2 & 1 \end{bmatrix}$$

is a group-pattern matrix. Let A denote the representation of A in a *Mathematica* notebook. We evaluate A and then evaluate `CofactorMatrix[A]`. The output shows that the cofactor matrix A^c for A of (4.23) is given by

$$A^c = \begin{bmatrix} 1146 & -1374 & 516 & -366 & 894 & -996 \\ 516 & 1146 & -1374 & 894 & -996 & -366 \\ -1374 & 516 & 1146 & -996 & -366 & 894 \\ -366 & 894 & -996 & 1146 & -1374 & 516 \\ 894 & -996 & -366 & 516 & 1146 & -1374 \\ -996 & -366 & 894 & -1374 & 516 & 1146 \end{bmatrix}.$$

It is a group-pattern matrix of the same kind as A. The same is true for $(A^c)^c$.

THEOREM 4.8. *Let $\mathscr{L} = (g_1, g_2, \ldots, g_n)$ be a list of the elements in a group G of order n, let $\mathbb{F}$ be $\mathbb{C}$ or any proper field extension of $\mathbb{C}$, and let A be an element of $\mathcal{M}[\mathbb{F}, G, \mathscr{L}]$. Then, the cofactor matrix A^c of A belongs to $\mathcal{M}[\mathbb{F}, G, \mathscr{L}]$. Also, the matrix A is nonsingular and has a multiplicative inverse A^{-1} in $\mathcal{M}[\mathbb{F}, G, \mathscr{L}]$ if and only if $\det(A) \neq 0$.*

PROOF. In view of Proposition 2.7 on page 13, we may assume that g_1 is the identity element e of G.

Suppose that A belongs to $\mathcal{M}[\mathbb{F}, G, \mathscr{L}]$. Let σ denote the function from G to $\mathbb{F}$ such that

$$[A]_{r,s} = \sigma(g_r^{-1}g_s), \quad \text{for } r, s = 1, 2, \ldots, n.$$

Let $\mathbb{F}_1$ be a field extension of $\mathbb{F}$ in which there are variables $x_1, x_2, \ldots, x_n$ that are algebraically independent over $\mathbb{F}$. Let τ be the function from G to $\mathbb{F}_1$ such that $\tau(g_k) = x_k$, for $k = 1, 2, \ldots, n$; and, let X be the group-pattern matrix in $\mathcal{M}[\mathbb{F}_1, G, \mathscr{L}]$ having

$$(4.24) \qquad \left[X\right]_{r,s} = \tau\left(g_r^{-1}g_s\right), \quad \text{for } r, s = 1, 2, \ldots, n.$$

In the ring $\mathfrak{M}$ of $n \times n$ matrices having components in $\mathbb{F}_1$, we introduce X^c as the cofactor matrix of the matrix X. Since A is obtained from X by replacing x_k in X with $\sigma(g_k)$, for $k = 1, 2, \ldots, n$, we see that A^c is obtained from X^c by replacing x_k in X^c with $\sigma(g_k)$, for $k = 1, 2, \ldots, n$. Thus, to verify that A^c is an element of $\mathcal{M}[\mathbb{F}, G, \mathscr{L}]$, it is sufficient to establish that X^c is an element of $\mathcal{M}[\mathbb{F}_1, G, \mathscr{L}]$.

Theorem 13.3 on page 125 shows that there is an $n \times n$ nonsingular matrix M having components in $\mathbb{C}$ and there is a set $\mathfrak{B}$ of $n \times n$ block-diagonal matrices having components in $\mathbb{F}_1$ such that:

$$(4.25) \qquad \begin{cases} \text{(i)} \quad \text{the inverse of each nonsingular matrix in } \mathfrak{B} \text{ belongs to } \mathfrak{B}; \\ \text{(ii)} \quad \text{for any } n \times n \text{ matrix } \mathcal{X} \text{ having components in } \mathbb{F}_1, \\ \qquad \mathcal{X} \text{ is an element of } \mathcal{M}[\mathbb{F}_1, G, \mathscr{L}] \text{ if and only if} \\ \qquad M^{-1}\mathcal{X}M \text{ is an element of } \mathfrak{B}. \end{cases}$$

For the matrix X in $\mathcal{M}[\mathbb{F}_1, G, \mathscr{L}]$, we employ (4.24) and $\tau(e) = \tau(g_1) = x_1$ to obtain $\left[X\right]_{k,k} = x_1$, for $k = 1, 2, \ldots, n$, as well as $\left[X\right]_{r,s} \neq x_1$ for $1 \leq r, s \leq n$ and $r \neq s$. Thus, the term $(x_1)^n$ in the expansion of $\det(X)$ is the only nonzero term that is divisible by $(x_1)^n$. Hence, we have $\det(X) \neq 0$ and see that the inverse X^{-1} of X exists as an element of $\mathfrak{M}$. We introduce

$$(4.26) \qquad B = M^{-1}XM$$

and find that $\det(B) = \det(M^{-1}\mathcal{X}M) = \det(M^{-1})\det(X)\det(M) = \det(X) \neq 0$. With this, (4.25)-(ii) and (4.25)-(i) show that B and B^{-1} are elements of $\mathfrak{B}$. This enables us to rewrite (4.26) as

$$(4.27) \qquad M^{-1}X^{-1}M = \left(M^{-1}XM\right)^{-1} = B^{-1}.$$

Now, we use (4.27) and (4.25)-(ii) to verify that X^{-1} is an element of $\mathcal{M}[\mathbb{F}_1, G, \mathscr{L}]$. Due to the relation $(X^c)^T = \det(X)X^{-1}$, we conclude that $(X^c)^T$ is an element of $\mathcal{M}[\mathbb{F}_1, G, \mathscr{L}]$. Theorem 4.1 on page 41 shows that X^c, as the transpose of $(X^c)^T$, is an element of $\mathcal{M}[\mathbb{F}_1, G, \mathscr{L}]$. Consequently, A^c is an element of $\mathcal{M}[\mathbb{F}, G, \mathscr{L}]$.

Of course, if A is nonsingular, then $\det(A) \neq 0$.

Suppose that $\det(A) \neq 0$. Then, A is nonsingular and A^{-1} exists in $\mathfrak{M}$. Since A^c is an element of $\mathcal{M}[\mathbb{F}, G, \mathscr{L}]$ and $\left(A^c\right)^T$ is therefore an element of $\mathcal{M}[\mathbb{F}, G, \mathscr{L}]$, we use the identity

$$(4.28) \qquad A^{-1} = \left(1/\det(A)\right)(A^c)^T$$

to conclude that A^{-1} is an element of $\mathcal{M}[\mathbb{F}, G, \mathscr{L}]$. This completes the proof. $\square$

EXAMPLE 4.9. The field extension $\mathbb{F}$ of $\mathbb{C}$ may contain variables $a_1, a_2, \ldots, a_n$ that are algebraically independent over $\mathbb{C}$. If such variables are the respective components of the first row for an $n \times n$ matrix A that belongs to $\mathcal{M}[\mathbb{F}, G, \mathscr{L}]$, then the determinant of A has the nonzero term $(a_1)^n$ and the inverse of A belongs to $\mathcal{M}[\mathbb{F}, G, \mathscr{L}]$.

4.5. Applications of the isomorphism for $\mathbb{F}[G]$ and $\mathcal{M}[\mathbb{F}, G, \mathscr{L}]$

PROPOSITION 4.10. *Let $\mathscr{L} = (g_1, g_2, \ldots, g_n)$ be a list of the elements in a group G of order n, let $\mathbb{F}$ be $\mathbb{C}$ or any proper field extension of $\mathbb{C}$, and let Ψ denote the function from $\mathbb{F}[G]$ onto $\mathcal{M}[\mathbb{F}, G, \mathscr{L}]$ defined, for each σ in $\mathbb{F}[G]$, by*

$$(4.29) \qquad \Psi(\sigma) = A, \quad \text{where} \quad [A]_{r,s} = \sigma(g_r^{-1} g_s), \quad \text{for } r, s = 1, 2, \ldots, n.$$

Then, σ in $\mathbb{F}[G]$ has a two-sided inverse in $\mathbb{F}[G]$ if and only if $\det(\Psi(\sigma)) \neq 0$; and, when $\det(\Psi(\sigma)) \neq 0$, the inverse σ^{-1} of σ is obtained from $A = \Psi(\sigma)$ by computing the inverse A^{-1} of A in $\mathcal{M}[\mathbb{F}, G, \mathscr{L}]$ and then writing $\sigma^{-1} = \Psi^{-1}(A^{-1})$.

PROOF. We note that Ψ, as the inverse for Φ of Theorem 4.4, is an isomorphism of $\mathbb{F}[G]$ onto $\mathcal{M}[\mathbb{F}, G, \mathscr{L}]$. Thus, σ in $\mathbb{F}[G]$ has a multiplicative inverse in $\mathbb{F}[G]$ if and only if the matrix $A = \Psi(\sigma)$ in $\mathcal{M}[\mathbb{F}, G, \mathscr{L}]$ has a multiplicative inverse in $\mathcal{M}[\mathbb{F}, G, \mathscr{L}]$. Theorem 4.8 shows that occurs if and only if $\det(A) \neq 0$.

Suppose $A = \Psi(\sigma)$ and $\det(A) \neq 0$. Then, by Theorem 4.8, A^{-1} exists in $\mathcal{M}[\mathbb{F}, G, \mathscr{L}]$. Thus, there is function τ from G to $\mathbb{F}$ such that $\Psi(\tau) = A^{-1}$. Since $\Phi = \Psi^{-1}$ is an isomorphism of $\mathcal{M}[\mathbb{F}, G, \mathscr{L}]$ onto $\mathbb{F}[G]$, we obtain

$$\sigma * \tau = \Phi(A) * \Phi(A^{-1}) = \Phi(AA^{-1}) = \Phi(I_n) = \widehat{e}$$

and

$$\tau * \sigma = \Phi(A^{-1}) * \Phi(A) = \Phi(A^{-1}A) = \Phi(I_n) = \widehat{e}.$$

Thus, σ has the inverse $\sigma^{-1} = \Phi(A^{-1}) = \Psi^{-1}(A^{-1})$. This completes the proof. $\square$

PROBLEM 4.11. *A multiplication table for the group $C_2 \times C_2 = \{e, u, v, w\}$ is given in (1.2) on page 3. For σ_1 and σ_2 defined in $\mathbb{C}[C_2 \times C_2]$ via (4.6) by*

$$(4.30) \qquad \sigma_1 = \widehat{e} + 2\widehat{u} + 3\widehat{v} + 4\widehat{w} \quad \text{and} \quad \sigma_2 = 2\widehat{e} + 3\widehat{u} + 5\widehat{v} + 7\widehat{w},$$

establish that σ_1 does not have an inverse and find an inverse for σ_2.

SOLUTION. We set $\mathscr{L} = (g_1, g_2, g_3, g_4)$, where $g_1 = e$, $g_2 = u$, $g_3 = v$, $g_4 = w$. The corresponding group-pattern table

$\cdot$	$g_1 = e$	$g_2 = u$	$g_3 = v$	$g_4 = w$
$g_1^{-1} = e$	e	u	v	w
$g_2^{-1} = u$	u	e	w	v
$g_3^{-1} = v$	v	w	e	u
$g_4^{-1} = w$	w	v	u	e

(4.31)

shows that an isomorphism Ψ of $\mathbb{C}[C_2 \times C_2]$ onto $\mathcal{M}[\mathbb{C}, C_2 \times C_2, \mathscr{L}]$ is defined by

$$(4.32) \qquad \Psi(\sigma) = \begin{bmatrix} \sigma(e) & \sigma(u) & \sigma(v) & \sigma(w) \\ \sigma(u) & \sigma(e) & \sigma(w) & \sigma(v) \\ \sigma(v) & \sigma(w) & \sigma(e) & \sigma(u) \\ \sigma(w) & \sigma(v) & \sigma(u) & \sigma(e) \end{bmatrix}, \quad \text{for any } \sigma \text{ in } \mathbb{C}[C_2 \times C_2].$$

Since σ_1 in (4.30) yields $\sigma_1(e) = 1$, $\sigma_1(u) = 2$, $\sigma_1(v) = 3$, $\sigma_1(w) = 4$, we find that

$$\Psi(\sigma_1) = \begin{bmatrix} 1 & 2 & 3 & 4 \\ 2 & 1 & 4 & 3 \\ 3 & 4 & 1 & 2 \\ 4 & 3 & 2 & 1 \end{bmatrix} \quad \text{and} \quad \det\big(\Psi(\sigma_1)\big) = 0.$$

Thus, $\Psi(\sigma_1)$ does not have an inverse and therefore σ_1 does not have an inverse.

For σ_2 in (4.30), we use (4.32) to obtain

$$\Psi(\sigma_2) = \begin{bmatrix} 2 & 3 & 5 & 7 \\ 3 & 2 & 7 & 5 \\ 5 & 7 & 2 & 3 \\ 7 & 5 & 3 & 2 \end{bmatrix}$$

and

$$\big(\Psi(\sigma_2)\big)^{-1} = \frac{1}{357} \begin{bmatrix} 52 & -67 & -101 & 137 \\ -67 & 52 & 137 & -101 \\ -101 & 137 & 52 & -67 \\ 137 & -101 & -67 & 52 \end{bmatrix} = \Psi(\tau_2)$$

where

(4.33) $\tau_2 = (52/357)\,\widehat{e} + (-67/357)\,\widehat{u} + (-101/357)\,\widehat{v} + (137/357)\,\widehat{w}.$

Thus, σ_2^{-1} exists and is given by $\sigma_2^{-1} = \tau_2$. This completes the solution. □

PROBLEM 4.12. *For the group* $\mathcal{D}_3 = \big\{e, \alpha, \alpha^2, \beta_1, \beta_2, \beta_3\big\}$ *of symmetries of an equilateral triangle considered in* Section 1.5, *find an inverse for*

(4.34) $\sigma_3 = \widehat{e} + \widehat{\alpha} + \widehat{\alpha^2} + \widehat{\beta_1}, \quad in \ \mathbb{C}[\mathcal{D}_3].$

SOLUTION. We set $\mathscr{L} = \big(g_1, g_2, g_3, g_4, g_5, g_6\big)$, where $g_1 = e$, $g_2 = \alpha$, $g_3 = \alpha^2$, $g_4 = \beta_1$, $g_5 = \beta_2$, $g_6 = \beta_3$. Then, the multiplication table assumes the form

	$g_1 = e$	$g_2 = \alpha$	$g_3 = \alpha^2$	$g_4 = \beta_1$	$g_5 = \beta_2$	$g_6 = \beta_3$
$g_1^{-1} = e$	e	α	α^2	β_1	β_2	β_3
$g_2^{-1} = \alpha^2$	α^2	e	α	β_2	β_3	β_1
$g_3^{-1} = \alpha$	α	α^2	e	β_3	β_1	β_2
$g_4^{-1} = \beta_1$	β_1	β_2	β_3	e	α	α^2
$g_5^{-1} = \beta_2$	β_2	β_3	β_1	α^2	e	α
$g_6^{-1} = \beta_3$	β_3	β_1	β_2	α	α^2	e

(4.35)

and shows that an isomorphism of $\mathbb{C}[\mathcal{D}_3]$ onto $\mathcal{M}[\mathbb{C}, \mathcal{D}_3, \mathscr{L}]$ is defined by

(4.36) $$\Psi(\sigma) = \begin{bmatrix} \sigma(e) & \sigma(\alpha) & \sigma(\alpha^2) & \sigma(\beta_1) & \sigma(\beta_2) & \sigma(\beta_3) \\ \sigma(\alpha^2) & \sigma(e) & \sigma(\alpha) & \sigma(\beta_2) & \sigma(\beta_3) & \sigma(\beta_1) \\ \sigma(\alpha) & \sigma(\alpha^2) & \sigma(e) & \sigma(\beta_3) & \sigma(\beta_1) & \sigma(\beta_2) \\ \sigma(\beta_1) & \sigma(\beta_2) & \sigma(\beta_3) & \sigma(e) & \sigma(\alpha) & \sigma(\alpha^2) \\ \sigma(\beta_2) & \sigma(\beta_3) & \sigma(\beta_1) & \sigma(\alpha^2) & \sigma(e) & \sigma(\alpha) \\ \sigma(\beta_3) & \sigma(\beta_1) & \sigma((\beta_2) & \sigma(\alpha) & \sigma(\alpha^2) & \sigma(e) \end{bmatrix}.$$

We use (4.36) and σ_3 in (4.34) to obtain

$$\Psi(\sigma_3) = \begin{bmatrix} 1 & 1 & 1 & 1 & 0 & 0 \\ 1 & 1 & 1 & 0 & 0 & 1 \\ 1 & 1 & 1 & 0 & 1 & 0 \\ 1 & 0 & 0 & 1 & 1 & 1 \\ 0 & 0 & 1 & 1 & 1 & 1 \\ 0 & 1 & 0 & 1 & 1 & 1 \end{bmatrix}$$

and

$$\left(\Psi(\sigma_3)\right)^{-1} = \frac{1}{8} \begin{bmatrix} 1 & 1 & 1 & 5 & -3 & -3 \\ 1 & 1 & 1 & -3 & -3 & 5 \\ 1 & 1 & 1 & -3 & 5 & -3 \\ 5 & -3 & -3 & 1 & 1 & 1 \\ -3 & -3 & 5 & 1 & 1 & 1 \\ -3 & 5 & -3 & 1 & 1 & 1 \end{bmatrix} = \Psi(\tau_3),$$

where

(4.37) $\qquad \tau_3 = (1/8)\widehat{e} + (1/8)\widehat{\alpha} + (1/8)\widehat{\alpha^2} + (5/8)\widehat{\beta_1} + (-3/8)\widehat{\beta_2} + (-3/8)\widehat{\beta_3}.$

Thus, we have $\sigma_3^{-1} = \tau_3$. This completes the solution. $\qquad\qquad\square$

EXERCISES

1. For σ_2 in (4.30) and τ_2 in (4.33), use (4.18) to check that $\sigma_2 * \tau_2 = \widehat{e}$.

2. For σ_3 in (4.34) and τ_3 in (4.37), use (4.18) to check that $\sigma_3 * \tau_3 = \widehat{e}$.

3. For the cyclic group $\mathcal{C}_4 = \{e, \rho, \rho^2, \rho^3\}$ of order 4 with $\rho^4 = e = \rho^0$, find an inverse in $\mathbb{C}[\mathcal{C}_4]$ for the element $\sigma_4 = \widehat{e} + 2\widehat{\rho} + 3\widehat{\rho^2} + 4\widehat{\rho^3}$ of $\mathbb{C}[\mathcal{C}_4]$.

4. For $\sigma_4 = \widehat{e} + 2\widehat{\rho} + 3\widehat{\rho^2} + 4\widehat{\rho^3}$ and $\tau_4 = (-9/40)\widehat{e} + (11/40)\widehat{\rho} + (1/40)\widehat{\rho^2} + (1/40)\widehat{\rho^3}$ in the group ring $\mathbb{C}[\mathcal{C}_4]$, use (4.18) to compute the product $\sigma_4 * \tau_4$.

5. The group-pattern matrix A_2 at the bottom of page 3 has the interpretation of $\mathcal{C}_6$ as $\mathcal{C}_3 \times \mathcal{C}_2$. It shows that the matrix

(4.38) $\qquad A = \begin{bmatrix} 1 & 2 & 3 & 4 & 5 & 7 \\ 3 & 1 & 2 & 7 & 4 & 5 \\ 2 & 3 & 1 & 5 & 7 & 4 \\ 4 & 5 & 7 & 1 & 2 & 3 \\ 7 & 4 & 5 & 3 & 1 & 2 \\ 5 & 7 & 4 & 2 & 3 & 1 \end{bmatrix}$

is an injective group-pattern matrix of that same kind. Use the technique of page 47 to compute A^c and $(A^c)^c$. Check that A^c, $(A^c)^c$, and A^T are group-pattern matrices of the same kind as A_2.

6. Prove that: for σ and τ in $\mathbb{F}[G]$, if τ is a one-sided inverse for σ, then τ is an inverse for σ.

7. Conjecture: The statement of Theorem 4.8 remains valid when $\mathbb{F}$ is replaced by any field $\mathfrak{F}$.

CHAPTER 5

Historical Perspective, Part 1

Thomas Muir observed in [**36**, Vol. 2, pages 401–404] that the determinants for matrices of the type

$$(5.1) \qquad \begin{bmatrix} a_1 \end{bmatrix}, \qquad \begin{bmatrix} a_1 & a_2 \\ a_2 & a_1 \end{bmatrix}, \qquad \begin{bmatrix} a_1 & a_2 & a_3 \\ a_3 & a_1 & a_2 \\ a_2 & a_3 & a_1 \end{bmatrix}, \qquad \begin{bmatrix} a_1 & a_2 & a_3 & a_4 \\ a_4 & a_1 & a_2 & a_3 \\ a_3 & a_4 & a_1 & a_2 \\ a_2 & a_3 & a_4 & a_1 \end{bmatrix}, \quad \cdots$$

and for matrices of the type

$$(5.2) \qquad \begin{bmatrix} a_1 \end{bmatrix}, \qquad \begin{bmatrix} a_1 & a_2 \\ a_2 & a_1 \end{bmatrix}, \qquad \begin{bmatrix} a_1 & a_2 & a_3 \\ a_2 & a_3 & a_1 \\ a_3 & a_1 & a_2 \end{bmatrix}, \qquad \begin{bmatrix} a_1 & a_2 & a_3 & a_4 \\ a_2 & a_3 & a_4 & a_1 \\ a_3 & a_4 & a_1 & a_2 \\ a_4 & a_1 & a_2 & a_3 \end{bmatrix}, \quad \cdots$$

were both considered by Eugène Charles Catalan in [**4**] of 1846. The matrices of (5.1) are currently termed *circulant matrices*. We use *leftward-circulant matrices* for the ones in (5.2).

5.1. Factorization for the determinant of a leftward-circulant matrix

An $n \times n$ leftward-circulant matrix B has the form

$$(5.3) \qquad B = \begin{bmatrix} a_1 & a_2 & a_3 & \cdots & a_{n-1} & a_n \\ a_2 & a_3 & a_4 & \cdots & a_n & a_1 \\ a_3 & a_4 & a_5 & \cdots & a_1 & a_2 \\ \vdots & \vdots & \vdots & \vdots & \vdots & \vdots \\ a_{n-1} & a_n & a_1 & \cdots & a_{n-3} & a_{n-2} \\ a_n & a_1 & a_2 & \cdots & a_{n-2} & a_{n-1} \end{bmatrix}.$$

In [**47**] of 1853, William H. Spottiswode studied the determinant of a matrix like B in (5.3) and, apart from a $\pm$ sign, he expressed it as a product of linear combinations of its components. For details, see [**36**, Vol. 2, page 405].

To evaluate $\det(B)$, Luigi Cremona employed in [**16**] of 1856 a primitive nth root ρ of unity in $\mathbb{C}$ such as $\rho = \cos\left(\frac{2\pi}{n}\right) + i \sin\left(\frac{2\pi}{n}\right)$, with $i^2 = -1$, in order to introduce the determinant of the matrix

$$(5.4) \qquad M = \begin{bmatrix} 1 & 1 & 1 & \cdots & 1 & 1 \\ 1 & \rho^{1 \cdot 1} & \rho^{1 \cdot 2} & \cdots & \rho^{1 \cdot (n-2)} & \rho^{1 \cdot (n-1)} \\ 1 & \rho^{2 \cdot 1} & \rho^{2 \cdot 2} & \cdots & \rho^{2 \cdot (n-2)} & \rho^{2 \cdot (n-1)} \\ \vdots & \vdots & \vdots & \vdots & \vdots & \vdots \\ 1 & \rho^{(n-2) \cdot 1} & \rho^{(n-2) \cdot 2} & \cdots & \rho^{(n-2)(n-2)} & \rho^{(n-2)(n-1)} \\ 1 & \rho^{(n-1) \cdot 1} & \rho^{(n-1) \cdot 2} & \cdots & \rho^{(n-1)(n-2)} & \rho^{(n-1)(n-1)} \end{bmatrix}.$$

53

After checking that $\det(M) \neq 0$, Luigi Cremona introduced

$$(5.5) \qquad \theta_k = a_1 + \rho^{(k-1)\cdot 1} a_2 + \rho^{(k-1)\cdot 2} a_3 + \cdots + \rho^{(k-1)(n-1)} a_n,$$

$$\text{for } k = 1, 2, \ldots, n,$$

in [**16**] and then proceeded to compute $\det(B) \det(M) = \det(BM)$. We find that

$$\det(B) \det(M) = \det(BM)$$

$$= \begin{vmatrix} \theta_1 & \theta_2 & \theta_3 & \cdots & \theta_{n-1} & \theta_n \\ \theta_1 & \rho^{1\cdot(n-1)}\theta_2 & \rho^{2(n-1)}\theta_3 & \cdots & \rho^{(n-2)(n-1)}\theta_{n-1} & \rho^{(n-1)(n-1)}\theta_n \\ \theta_1 & \rho^{1\cdot(n-2)}\theta_2 & \rho^{2(n-2)}\theta_3 & \cdots & \rho^{(n-2)(n-2)}\theta_{n-1} & \rho^{(n-1)(n-2)}\theta_n \\ \vdots & \vdots & \vdots & \ddots & \vdots & \vdots \\ \theta_1 & \rho^{1\cdot 2}\theta_2 & \rho^{2\cdot 2}\theta_3 & \cdots & \rho^{(n-2)\cdot 2}\theta_{n-1} & \rho^{(n-1)\cdot 2}\theta_n \\ \theta_1 & \rho^{1\cdot 1}\theta_2 & \rho^{2\cdot 1}\theta_3 & \cdots & \rho^{(n-2)\cdot 1}\theta_{n-1} & \rho^{(n-1)\cdot 1}\theta_n \end{vmatrix}$$

$$= \theta_1 \theta_2 \cdots \theta_n \begin{vmatrix} 1 & 1 & 1 & \cdots & 1 & 1 \\ 1 & \rho^{1\cdot(n-1)} & \rho^{2(n-1)} & \cdots & \rho^{(n-2)(n-1)} & \rho^{(n-1)(n-1)} \\ 1 & \rho^{1\cdot(n-2)} & \rho^{2(n-2)} & \cdots & \rho^{(n-2)(n-2)} & \rho^{(n-1)(n-2)} \\ \vdots & \vdots & \vdots & \ddots & \vdots & \vdots \\ 1 & \rho^{1\cdot 2} & \rho^{2\cdot 2} & \cdots & \rho^{(n-2)\cdot 2} & \rho^{(n-1)\cdot 2} \\ 1 & \rho^{1\cdot 1} & \rho^{2\cdot 1} & \cdots & \rho^{(n-2)\cdot 1} & \rho^{(n-1)\cdot 1} \end{vmatrix}$$

$$= \theta_1 \theta_2 \cdots \theta_n \, (-1)^{(n-1)(n-2)/2} \det(M),$$

and thereby conclude that B in (5.3) has its determinant given by

$$(5.6) \qquad \det(B) = (-1)^{(n-1)(n-2)/2} \theta_1 \theta_2 \cdots \theta_n.$$

The expression $(n-1)(n-2)/2$ appears incorrectly in [**36**, Vol. 2, page 408] as $n(n-1)/2$. For corroboration about that, compare (6.26) and (6.27) on page 70.

Thomas Muir mentions in [**36**, Vol. 2, page 408] that the introduction of ρ for the preceding argument may be attributed to Francisco Brioschi.

5.2. Factorization for the determinant of a circulant matrix

An $n \times n$ circulant matrix has the form

$$(5.7) \qquad A = \begin{bmatrix} a_1 & a_2 & a_3 & \cdots & a_{n-1} & a_n \\ a_n & a_1 & a_2 & \cdots & a_{n-2} & a_{n-1} \\ a_{n-1} & a_n & a_1 & \cdots & a_{n-3} & a_{n-2} \\ \vdots & \vdots & \vdots & \vdots & \vdots & \vdots \\ a_3 & a_4 & b_5 & \cdots & a_1 & a_2 \\ a_2 & a_3 & a_4 & \cdots & a_n & a_1 \end{bmatrix}.$$

The matrix M of (5.4) is well suited for the task of evaluating $\det(A)$ because it yields the convenient formula

$$(5.8) \qquad AM = M \begin{bmatrix} \theta_1 & 0 & 0 & \cdots & 0 & 0 \\ 0 & \theta_2 & 0 & \cdots & 0 & 0 \\ 0 & 0 & \theta_3 & \cdots & 0 & 0 \\ \vdots & \vdots & \vdots & \ddots & \vdots & \vdots \\ 0 & 0 & 0 & \cdots & \theta_{n-1} & 0 \\ 0 & 0 & 0 & \cdots & 0 & \theta_n \end{bmatrix}$$

that involves multiplication of matrices where each θ_k is given by (5.5) and

$$(5.9) \qquad \theta_k = \sum_{\nu=1}^{n} \rho^{(k-1)(\nu-1)} a_\nu, \quad \text{for } k = 1, 2, \ldots, n.$$

Since (5.8) yields

$$\det(A) \det(M) = \det(AM) = \det(M)\, \theta_1\, \theta_2 \cdots \theta_n,$$

we see that the determinant of A in (5.7) is given by

$$(5.10) \qquad \det(A) = \prod_{k=1}^{n} \left[\sum_{\nu=1}^{n} \rho^{(k-1)(\nu-1)} a_\nu \right].$$

The identity (5.8) and its use in deducing (5.10) is reexamined in Observation 6.5 on page 66.

5.3. Procedure of Anton Puchta
for discovering factorizations

The matrix A_1 and the factorization of its determinant given by

$$(5.11) \qquad A_1 = \begin{bmatrix} a & b \\ b & a \end{bmatrix} \quad \text{and} \quad \det(A_1) = (a+b)(a-b)$$

suggested to Anton Puchta in [**42**] of 1877 that he introduce the matrix

$$(5.12) \qquad T_1 = \begin{bmatrix} 1 & 1 \\ 1 & -1 \end{bmatrix}$$

as a convenient way to specify that factorization.

For the matrix

$$(5.13) \qquad A_2 = \left[\begin{array}{cc|cc} a & b & c & d \\ b & a & d & c \\ \hline c & d & a & b \\ d & c & b & a \end{array} \right],$$

there is a corresponding matrix suggested with respect to T_1 in (5.12) by

$$\left[\begin{array}{c|c} 1 \cdot T_1 & 1 \cdot T_1 \\ \hline 1 \cdot T_1 & (-1) \cdot T_1 \end{array} \right]$$

that enabled Puchta to introduce in [**42**] of 1877 the matrix

$$(5.14) \qquad T_2 = \left[\begin{array}{cc|cc} 1 & 1 & 1 & 1 \\ 1 & -1 & 1 & -1 \\ \hline 1 & 1 & -1 & -1 \\ 1 & -1 & -1 & 1 \end{array} \right]$$

as an effortless way to present in [**42**] the factorization

$$(5.15) \quad \det(A_2) = (a+b+c+d)(a-b+c-d)(a+b-c-d)(a-b-c+d).$$

For the matrix

$$(5.16) \qquad A_3 = \left[\begin{array}{cccc|cccc} a & b & c & d & e & f & g & h \\ b & a & d & c & f & e & h & g \\ c & d & a & b & g & h & e & f \\ d & c & b & a & h & g & f & e \\ \hline e & f & g & h & a & b & c & d \\ f & e & h & g & b & a & d & c \\ g & h & e & f & c & d & a & b \\ h & g & f & e & d & c & b & a \end{array}\right],$$

the use of T_2 in (5.14) to form the expression

$$\left[\begin{array}{c|c} 1 \cdot T_2 & 1 \cdot T_2 \\ \hline 1 \cdot T_2 & (-1) \cdot T_2 \end{array}\right]$$

enabled Anton Puchta to introduce in [**42**] of 1877 the matrix

$$(5.17) \qquad T_3 = \left[\begin{array}{cccc|cccc} 1 & 1 & 1 & 1 & 1 & 1 & 1 & 1 \\ 1 & -1 & 1 & -1 & 1 & -1 & 1 & -1 \\ 1 & 1 & -1 & -1 & 1 & 1 & -1 & -1 \\ 1 & -1 & -1 & 1 & 1 & -1 & -1 & 1 \\ \hline 1 & 1 & 1 & 1 & -1 & -1 & -1 & -1 \\ 1 & -1 & 1 & -1 & -1 & 1 & -1 & 1 \\ 1 & 1 & -1 & -1 & -1 & -1 & 1 & 1 \\ 1 & -1 & -1 & 1 & -1 & 1 & 1 & -1 \end{array}\right]$$

as an effortless way to write in [**42**] the factorization

$$(5.18) \quad \det(A_3) = (a+b+c+d+e+f+g+h)(a-b+c-d+e-f+g-h)$$
$$\times (a+b-c-d+e+f-g-h)(a-b-c+d+e-f-g+h)$$
$$\times (a+b+c+d-e-f-g-h)(a-b+c-d-e+f-g+h)$$
$$\times (a+b-c-d-e-f+g+h)(a-b-c+d-e+f+g-h).$$

A continuation of this procedure yields

$$(5.19) \quad T_4 = \left[\begin{array}{cccccccc|cccccccc} 1 & 1 & 1 & 1 & 1 & 1 & 1 & 1 & 1 & 1 & 1 & 1 & 1 & 1 & 1 & 1 \\ 1 & -1 & 1 & -1 & 1 & -1 & 1 & -1 & 1 & -1 & 1 & -1 & 1 & -1 & 1 & -1 \\ 1 & 1 & -1 & -1 & 1 & 1 & -1 & -1 & 1 & 1 & -1 & -1 & 1 & 1 & -1 & -1 \\ 1 & -1 & -1 & 1 & 1 & -1 & -1 & 1 & 1 & -1 & -1 & 1 & 1 & -1 & -1 & 1 \\ 1 & 1 & 1 & 1 & -1 & -1 & -1 & -1 & 1 & 1 & 1 & 1 & -1 & -1 & -1 & -1 \\ 1 & -1 & 1 & -1 & -1 & 1 & -1 & 1 & 1 & -1 & 1 & -1 & -1 & 1 & -1 & 1 \\ 1 & 1 & -1 & -1 & -1 & -1 & 1 & 1 & 1 & 1 & -1 & -1 & -1 & -1 & 1 & 1 \\ 1 & -1 & -1 & 1 & -1 & 1 & 1 & -1 & 1 & -1 & -1 & 1 & -1 & 1 & 1 & -1 \\ \hline 1 & 1 & 1 & 1 & 1 & 1 & 1 & 1 & -1 & -1 & -1 & -1 & -1 & -1 & -1 & -1 \\ 1 & -1 & 1 & -1 & 1 & -1 & 1 & -1 & -1 & 1 & -1 & 1 & -1 & 1 & -1 & 1 \\ 1 & 1 & -1 & -1 & 1 & 1 & -1 & -1 & -1 & -1 & 1 & 1 & -1 & -1 & 1 & 1 \\ 1 & -1 & -1 & 1 & 1 & -1 & -1 & 1 & -1 & 1 & 1 & -1 & -1 & 1 & 1 & -1 \\ 1 & 1 & 1 & 1 & -1 & -1 & -1 & -1 & -1 & -1 & -1 & -1 & 1 & 1 & 1 & 1 \\ 1 & -1 & 1 & -1 & -1 & 1 & -1 & 1 & -1 & 1 & -1 & 1 & 1 & -1 & 1 & -1 \\ 1 & 1 & -1 & -1 & -1 & -1 & 1 & 1 & -1 & -1 & 1 & 1 & 1 & 1 & -1 & -1 \\ 1 & -1 & -1 & 1 & -1 & 1 & 1 & -1 & -1 & 1 & 1 & -1 & 1 & -1 & -1 & 1 \end{array}\right]$$

to specify the factors for the determinant of the matrix Z_2 in (2.37) on page 25.

5.4. Verifications by M. Noether
for formulations of A. Puchta

In [**36**, Vol. 3, pages 388–389] of 1920 and [**36**, Vol. 4, pages 385–387] of 1923, Thomas Muir pointed out that, while the technique of Anton Puchta for formulating factorizations of determinants for special matrices was clearly presented, there was insufficient explanation as to why it produced correct results. Thus, until the later research of Richard Dedekind and Georg Frobenius is considered, a general context is postponed to first show how Max Noether in [**39, 40**] of 1880 verified particular results of Anton Puchta and did so with a useful notation for the factorizations.

Starting directly with the matrices

$$A_1 = \begin{bmatrix} a & b \\ b & a \end{bmatrix}, \quad A_2 = \begin{bmatrix} a & b & c & d \\ b & a & d & c \\ c & d & a & b \\ d & c & b & a \end{bmatrix}, \quad A_3 = \begin{bmatrix} a & b & c & d & e & f & g & h \\ b & a & d & c & f & e & h & g \\ c & d & a & b & g & h & e & f \\ d & c & b & a & h & g & f & e \\ e & f & g & h & a & b & c & d \\ f & e & h & g & b & a & d & c \\ g & h & e & f & c & d & a & b \\ h & g & f & e & d & c & b & a \end{bmatrix}$$

from (5.11), (5.13), (5.16), Max Noether established in [**39**] of 1880 the formulas

$$(5.20) \quad \det(A_1) = \prod_{\epsilon_1 \in \{1,-1\}} (a + \epsilon_1 b) = (a+b)(a-b),$$

$$(5.21) \quad \det(A_2) = \prod_{\epsilon_1, \epsilon_2 \in \{1,-1\}} (a + \epsilon_1 b + \epsilon_2 c + \epsilon_1 \epsilon_2 d)$$
$$= (a+b+c+d)(a+b-c-d)(a-b+c-d)(a-b-c+d),$$

$$(5.22) \quad \det(A_3) = \prod_{\epsilon_1, \epsilon_2, \epsilon_3 \in \{1,-1\}} (a + \epsilon_1 b + \epsilon_2 c + \epsilon_1 \epsilon_2 d + \epsilon_3 e + \epsilon_1 \epsilon_3 f + \epsilon_2 \epsilon_3 g + \epsilon_1 \epsilon_2 \epsilon_3 h).$$

Of course, (5.20) is obvious.

To verify (5.21), let ϵ_1, ϵ_2 be elements of $\{1, -1\}$ and introduce

$$E_2 = \begin{bmatrix} 1 & 0 & 0 & 0 \\ 0 & \epsilon_1 & 0 & 0 \\ 0 & 0 & \epsilon_2 & 0 \\ 0 & 0 & 0 & \epsilon_1 \epsilon_2 \end{bmatrix}$$

as well as

$$F_2 = E_2 A_2 E_2 = \begin{bmatrix} a & \epsilon_1 b & \epsilon_2 c & \epsilon_1 \epsilon_2 d \\ \epsilon_1 b & a & \epsilon_1 \epsilon_2 d & \epsilon_2 c \\ \epsilon_2 c & \epsilon_1 \epsilon_2 d & a & \epsilon_1 b \\ \epsilon_1 \epsilon_2 d & \epsilon_2 c & \epsilon_1 b & a \end{bmatrix}.$$

With $\epsilon_1^2 = 1$ and $\epsilon_2^2 = 1$, this yields $\det(E_2) = 1$ and

$$\det(A_2) = \det(E_2) \det(A_2) \det(E_2) = \det(E_2 A_2 E_2) = \det(F_2).$$

By adding to the first row of F_2 each of its other rows, we see that both $\det(F_2)$ and $\det(A_2)$ have the four factors $(a + \epsilon_1 b + \epsilon_2 c + \epsilon_1 \epsilon_2 d)$, for ϵ_1, ϵ_2 in $\{1, -1\}$. Moreover, the coefficient of a^4 in $\det(A_2)$ is 1. Thus, (5.21) is valid.

To verify (5.22), let ϵ_1, ϵ_2, ϵ_3 be elements of $\{1, -1\}$ and introduce

$$E_3 = \begin{bmatrix} 1 & 0 & 0 & 0 & 0 & 0 & 0 & 0 \\ 0 & \epsilon_1 & 0 & 0 & 0 & 0 & 0 & 0 \\ 0 & 0 & \epsilon_2 & 0 & 0 & 0 & 0 & 0 \\ 0 & 0 & 0 & \epsilon_1\epsilon_2 & 0 & 0 & 0 & 0 \\ 0 & 0 & 0 & 0 & \epsilon_3 & 0 & 0 & 0 \\ 0 & 0 & 0 & 0 & 0 & \epsilon_1\epsilon_3 & 0 & 0 \\ 0 & 0 & 0 & 0 & 0 & 0 & \epsilon_2\epsilon_3 & 0 \\ 0 & 0 & 0 & 0 & 0 & 0 & 0 & \epsilon_1\epsilon_2\epsilon_3 \end{bmatrix}$$

as well as $F_3 = E_3 A_3 E_3$. This yields $\det(E_3) = 1$,

$$F_3 = \begin{bmatrix} a & \epsilon_1 b & \epsilon_2 c & \epsilon_1\epsilon_2 d & \epsilon_3 e & \epsilon_1\epsilon_3 f & \epsilon_2\epsilon_3 g & \epsilon_1\epsilon_2\epsilon_3 h \\ \epsilon_1 b & a & \epsilon_1\epsilon_2 d & \epsilon_2 c & \epsilon_1\epsilon_3 f & \epsilon_3 e & \epsilon_1\epsilon_2\epsilon_3 h & \epsilon_2\epsilon_3 g \\ \epsilon_2 c & \epsilon_1\epsilon_2 d & a & \epsilon_1 b & \epsilon_2\epsilon_3 g & \epsilon_1\epsilon_2\epsilon_3 h & \epsilon_3 e & \epsilon_1\epsilon_3 f \\ \epsilon_1\epsilon_2 d & \epsilon_2 c & \epsilon_1 b & a & \epsilon_1\epsilon_2\epsilon_3 h & \epsilon_2\epsilon_3 g & \epsilon_1\epsilon_3 f & \epsilon_3 e \\ \epsilon_3 e & \epsilon_1\epsilon_3 f & \epsilon_2\epsilon_3 g & \epsilon_1\epsilon_2\epsilon_3 h & a & \epsilon_1 b & \epsilon_2 c & \epsilon_1\epsilon_2 d \\ \epsilon_1\epsilon_3 f & \epsilon_3 e & \epsilon_1\epsilon_2\epsilon_3 h & \epsilon_2\epsilon_3 g & \epsilon_1 b & a & \epsilon_1\epsilon_2 d & \epsilon_2 c \\ \epsilon_2\epsilon_3 g & \epsilon_1\epsilon_2\epsilon_3 h & \epsilon_3 e & \epsilon_1\epsilon_3 f & \epsilon_2 c & \epsilon_1\epsilon_2 d & a & \epsilon_1 b \\ \epsilon_1\epsilon_2\epsilon_3 h & \epsilon_2\epsilon_3 g & \epsilon_1\epsilon_3 f & \epsilon_3 e & \epsilon_1\epsilon_2 d & \epsilon_2 c & \epsilon_1 b & a \end{bmatrix},$$

and $\det(A_3) = \det(E_3) \det(A_3) \det(E_3) = \det\left(E_3 A_3 E_3\right) = \det F_3$. By adding to the first row of F_3 each of the other rows, we see that both $\det(F_3)$ and $\det(A_3)$ have the eight distinct factors

$$\left(a + \epsilon_1 b + \epsilon_2 c + \epsilon_1\epsilon_2 d + \epsilon_3 e + \epsilon_1\epsilon_3 f + \epsilon_2\epsilon_3 g + \epsilon_1\epsilon_2\epsilon_3 h\right), \quad \text{for } \epsilon_1, \epsilon_2, \epsilon_3 \text{ in } \{1, -1\}.$$

Moreover, the coefficient of a^8 in $det(A_3)$ is 1. Thus, (5.22) is valid.

OBSERVATION 5.1. Here, to easily see that (5.22) yields (5.18), we write directly below the preceding formula the factors for $\det(A)$ that it yields. We obtain

$$\begin{aligned} \det(A_3) = {}& \left(a + b + c + d + e + f + g + h\right)\left(a - b + c - d + e - f + g - h\right) \\ & \times \left(a + b - c - d + e + f - g - h\right)\left(a - b - c + d + e - f - g + h\right) \\ & \times \left(a + b + c + d - e - f - g - h\right)\left(a - b + c - d - e + f - g + h\right) \\ & \times \left(a + b - c - d - e - f + g + h\right)\left(a - b - c + d - e + f + g - h\right). \end{aligned}$$

5.5. Factorization procedure of Anton Puchta
for an example in [43]

Anton Puchta noted in [43] of 1881 that the determinant of the matrix

$$(5.23) \qquad A_5 = \left[\begin{array}{ccc|ccc|ccc} a & b & c & d & e & f & g & h & i \\ c & a & b & f & d & e & i & g & h \\ b & c & a & e & f & d & h & i & g \\ \hline g & h & i & a & b & c & d & e & f \\ i & g & h & c & a & b & f & d & e \\ h & i & g & b & c & a & e & f & d \\ \hline d & e & f & g & h & i & a & b & c \\ f & d & e & i & g & h & c & a & b \\ e & f & d & h & i & g & b & c & a \end{array}\right]$$

is expressible as a product of nine factors specified by the nine rows of the matrix

$$(5.24) \qquad T_5 = \begin{bmatrix} 1 & 1 & 1 & 1 & 1 & 1 & 1 & 1 & 1 \\ 1 & \omega & \omega^2 & 1 & \omega & \omega^2 & 1 & \omega & \omega^2 \\ 1 & \omega^2 & \omega & 1 & \omega^2 & \omega & 1 & \omega^2 & \omega \\ 1 & 1 & 1 & \omega & \omega & \omega & \omega^2 & \omega^2 & \omega^2 \\ 1 & \omega & \omega^2 & \omega & \omega^2 & 1 & \omega^2 & 1 & \omega \\ 1 & \omega^2 & \omega & \omega & 1 & \omega^2 & \omega^2 & \omega & 1 \\ 1 & 1 & 1 & \omega^2 & \omega^2 & \omega^2 & \omega & \omega & \omega \\ 1 & \omega & \omega^2 & \omega^2 & 1 & \omega & \omega & \omega^2 & 1 \\ 1 & \omega^2 & \omega & \omega^2 & \omega & 1 & \omega & 1 & \omega^2 \end{bmatrix}.$$

For example, the fifth row of T_5 in (5.24) specifies the factor

$$\left(a + \omega b + \omega^2 c + \omega d + \omega^2 e + f + \omega^2 g + h + \omega i \right)$$

of $\det(A_5)$. Puchta's method to obtain (5.24) was based on the arrangement of the 3×3 circulant submatrices of A_5 in (5.23) and the factorization

$$(5.25) \qquad \begin{vmatrix} a & b & c \\ c & a & b \\ b & c & a \end{vmatrix} = \left(a + b + c \right)\left(a + \omega b + \omega^2 c \right)\left(a + \omega^2 b + \omega c \right),$$

where $\omega^2 + \omega + 1 = 0$ and therefore $\omega^3 = 1$. Since the matrix

$$L = \begin{bmatrix} 1 & 1 & 1 \\ 1 & \omega & \omega^2 \\ 1 & \omega^2 & \omega \end{bmatrix}$$

specifies the factorization (5.25), the corresponding expression

$$\begin{bmatrix} 1 \cdot L & 1 \cdot L & 1 \cdot L \\ 1 \cdot L & \omega \cdot L & \omega^2 \cdot L \\ 1 \cdot L & \omega^2 \cdot L & \omega \cdot L \end{bmatrix}$$

provides the rule for forming the matrix T_5 in (5.24).

5.6. Verification for the preceding example of A Puchta

Here, we show how the method of M. Noether for (5.21)–(5.22) can be modified to derive a factorization for the determinant of the matrix A_5 in (5.23) on page 58. Let ϵ_1, ϵ_2 denote elements of $\{1, \omega, \omega^2\}$ where ω is a primitive cube root of unity and therefore satisfies $\omega^2 + \omega + 1 = 0$ as well as $\omega^3 = 1$. We introduce

$$E_5 = \begin{bmatrix} 1 & 0 & 0 & 0 & 0 & 0 & 0 & 0 & 0 \\ 0 & \epsilon_1 & 0 & 0 & 0 & 0 & 0 & 0 & 0 \\ 0 & 0 & \epsilon_1^2 & 0 & 0 & 0 & 0 & 0 & 0 \\ 0 & 0 & 0 & \epsilon_2 & 0 & 0 & 0 & 0 & 0 \\ 0 & 0 & 0 & 0 & \epsilon_1\epsilon_2 & 0 & 0 & 0 & 0 \\ 0 & 0 & 0 & 0 & 0 & \epsilon_1^2\epsilon_2 & 0 & 0 & 0 \\ 0 & 0 & 0 & 0 & 0 & 0 & \epsilon_2^2 & 0 & 0 \\ 0 & 0 & 0 & 0 & 0 & 0 & 0 & \epsilon_1\epsilon_2^2 & 0 \\ 0 & 0 & 0 & 0 & 0 & 0 & 0 & 0 & \epsilon_1^2\epsilon_2^2 \end{bmatrix}.$$

In view of $\epsilon_1^3 = 1$ and $\epsilon_2^3 = 1$, we observe that E_5 is nonsingular,

$$E_5^{-1} = \begin{bmatrix} 1 & 0 & 0 & 0 & 0 & 0 & 0 & 0 & 0 \\ 0 & \epsilon_1^2 & 0 & 0 & 0 & 0 & 0 & 0 & 0 \\ 0 & 0 & \epsilon_1 & 0 & 0 & 0 & 0 & 0 & 0 \\ 0 & 0 & 0 & \epsilon_2^2 & 0 & 0 & 0 & 0 & 0 \\ 0 & 0 & 0 & 0 & \epsilon_1^2 \epsilon_2^2 & 0 & 0 & 0 & 0 \\ 0 & 0 & 0 & 0 & 0 & \epsilon_1 \epsilon_2^2 & 0 & 0 & 0 \\ 0 & 0 & 0 & 0 & 0 & 0 & \epsilon_2 & 0 & 0 \\ 0 & 0 & 0 & 0 & 0 & 0 & 0 & \epsilon_1^2 \epsilon_2 & 0 \\ 0 & 0 & 0 & 0 & 0 & 0 & 0 & 0 & \epsilon_1 \epsilon_2 \end{bmatrix},$$

and $\det(E_5) = \det(E_5^{-1}) = 1$. For the matrix $F_5 = E_5^{-1} A_5 E_5$, we deduce

$$(5.26) \qquad \det(A_5) = \det(E_5^{-1}) \det(A_5) \det(E_5) = \det\left(E_5^{-1} A_5 E_5\right) = \det(F_5)$$

and

$$F_5 = \begin{bmatrix} a & \epsilon_1 b & \epsilon_1^2 c & \epsilon_2 d & \epsilon_1 \epsilon_2 e & \epsilon_1^2 \epsilon_2 f & \epsilon_2^2 g & \epsilon_1 \epsilon_2^2 h & \epsilon_1^2 \epsilon_2^2 i \\ \epsilon_1^2 c & a & \epsilon_1 b & \epsilon_1^2 \epsilon_2 f & \epsilon_2 d & \epsilon_1 \epsilon_2 e & \epsilon_1^2 \epsilon_2^2 i & \epsilon_2^2 g & \epsilon_1 \epsilon_2^2 h \\ \epsilon_1 b & \epsilon_1^2 c & a & \epsilon_1 \epsilon_2 e & \epsilon_1^2 \epsilon_2 f & \epsilon_2 d & \epsilon_1 \epsilon_2^2 h & \epsilon_1^2 \epsilon_2^2 i & \epsilon_2^2 g \\ \epsilon_2^2 g & \epsilon_1 \epsilon_2^2 h & \epsilon_1^2 \epsilon_2^2 i & a & \epsilon_1 b & \epsilon_1^2 c & \epsilon_2 d & \epsilon_1 \epsilon_2 e & \epsilon_1^2 \epsilon_2 f \\ \epsilon_1^2 \epsilon_2^2 i & \epsilon_2^2 g & \epsilon_1 \epsilon_2^2 h & \epsilon_1^2 c & a & \epsilon_1 b & \epsilon_1^2 \epsilon_2 f & \epsilon_2 d & \epsilon_1 \epsilon_2 e \\ \epsilon_1 \epsilon_2^2 h & \epsilon_1^2 \epsilon_2^2 i & \epsilon_2^2 g & \epsilon_1 b & \epsilon_1^2 c & a & \epsilon_1 \epsilon_2 e & \epsilon_1^2 \epsilon_2 f & \epsilon_2 d \\ \epsilon_2 d & \epsilon_1 \epsilon_2 e & \epsilon_1^2 \epsilon_2 f & \epsilon_2^2 g & \epsilon_1 \epsilon_2^2 h & \epsilon_1^2 \epsilon_2^2 i & a & \epsilon_1 b & \epsilon_1^2 c \\ \epsilon_1^2 \epsilon_2 f & \epsilon_2 d & \epsilon_1 \epsilon_2 e & \epsilon_1^2 \epsilon_2^2 i & \epsilon_2^2 g & \epsilon_1 \epsilon_2^2 h & \epsilon_1^2 c & a & \epsilon_1 b \\ \epsilon_1 \epsilon_2 e & \epsilon_1^2 \epsilon_2 f & \epsilon_2 d & \epsilon_1 \epsilon_2^2 h & \epsilon_1^2 \epsilon_2^2 i & \epsilon_2^2 g & \epsilon_1 b & \epsilon_1^2 c & a \end{bmatrix}.$$

By adding to the first row of F_5 each of the other rows, we find that both $\det(F_5)$ and $\det(A_5)$ have as factors each of the nine distinct expressions

$$\left(a + \epsilon_1 b + \epsilon_1^2 c + \epsilon_2 d + \epsilon_1 \epsilon_2 e + \epsilon_1^2 \epsilon_2 f + \epsilon_2^2 g + \epsilon_1 \epsilon_2^2 h + \epsilon_1^2 \epsilon_2^2 i \right), \quad \text{for } \epsilon_1, \epsilon_2 \text{ in } \{1, \omega, \omega^2\}.$$

Also, the coefficient of a^9 in A_5 is 1. Consequently, the determinant of the matrix

$$A_5 = \left[\begin{array}{ccc|ccc|ccc} a & b & c & d & e & f & g & h & i \\ c & a & b & f & d & e & i & g & h \\ b & c & a & e & f & d & h & i & g \\ \hline g & h & i & a & b & c & d & e & f \\ i & g & h & c & a & b & f & d & e \\ h & i & g & b & c & a & e & f & d \\ \hline d & e & f & g & h & i & a & b & c \\ f & d & e & i & g & h & c & a & b \\ e & f & d & h & i & g & b & c & a \end{array} \right]$$

is given by

$$\det(A_5) = \prod_{\epsilon_1, \epsilon_2 \in \{1, \omega, \omega^2\}} \left(a + \epsilon_1 b + \epsilon_1^2 c + \epsilon_2 d + \epsilon_1 \epsilon_2 e + \epsilon_1^2 \epsilon_2 f + \epsilon_2^2 g + \epsilon_1 \epsilon_2^2 h + \epsilon_1^2 \epsilon_2^2 i \right).$$

5.7. Observations

A check of the computations in this chapter based on computer algebra is presented in a *Mathematica* notebook that can be downloaded by first visiting the web page

`http://homepages.uc.edu/~chalklr/group-pattern.htm`

and then clicking on particular pages for Chapter 5. See Appendix H.

The research of Luigi Cremona, Francisco Brioschi, Anton Puchta, Max Noether and others about factorizations for the determinants of various special matrices suggests the existence of a general context about such matters.

We include in Chapter 21 two remarkable letters that Richard Dedekind wrote to Georg Frobenius. The letter on page 193 dated March 25 of 1896 explains that, in February of 1886, Dedekind studied group-determinants like the object labeled H in that letter. It shows that, when the finite group is abelian, Dedekind knew how to express H as a product of linear combinations of its components. That result includes the factorizations of this chapter and numerous other ones.

In Chapters 6 through 9, emphasis is placed on diagonalizations of circulant matrices and their natural generalizations to other abelian-group-pattern matrices. Dedekind's result about H will be established in that context.

We continue this explanation in Chapter 10 by indicating how Georg Frobenius was directly influenced by the letter from Dedekind dated April 6 of 1896 that describes other unpublished research done by Dedekind in February of 1886.

EXERCISES

1. Use a system of computer algebra to check that the Puchta matrix T_2 in (5.14) on page 55 is nonsingular and that, for A_2 in (5.13), the matrix product $T_2^{-1} A_2 T_2$ is a diagonal matrix whose diagonal components are the factors of $\det(A_2)$ in (5.15).

2. Use a system of computer algebra to check that the Puchta matrix T_3 in (5.17) on page 56 is nonsingular and that, for A_3 in (5.16), the matrix product $T_3^{-1} A_3 T_3$ is a diagonal matrix whose diagonal elements are the factors of $\det(A_3)$ in (5.18).

3. Use a system of computer algebra to check that the Puchta matrix T_4 in (5.19) is nonsingular and that, for Z_2 in (2.37) on page 25, the matrix product $T_4^{-1} Z_2 T_4$ is a diagonal matrix.

4. Use a system of computer algebra to check that the Puchta matrix T_5 in (5.24) on page 59 is nonsingular and that, for A_5 in (5.23) on page 58, the matrix product $T_5^{-1} A_5 T_5$ is a diagonal matrix.

CHAPTER 6

Viewpoints about Circulant Matrices

For a cyclic group G of order n generated by h, let $\mathscr{L} = (g_1, g_2, \ldots, g_n)$ be the list of elements in G given by $g_k = h^{k-1}$, for $1 \leq k \leq n$, where $h^n = h^0 = e$. An $n \times n$ matrix A with components in a set S is a ***circulant matrix*** if and only if it is a group-pattern matrix for G and $\mathscr{L}$; i.e., it is an element of $\mathcal{M}[S, G, \mathscr{L}]$. This occurs if and only if there is a function σ from G to S such that

$$(6.1) \qquad \left[A\right]_{r,s} = \sigma\!\left(g_r^{-1} g_s\right), \quad \text{for } r,\, s = 1,\, 2 \ldots,\, n.$$

In this chapter, let S be a field $\mathcal{F}$ of the following kind.

Henceforth, for a given positive integer n, $\mathcal{F}$ denotes a field that contains a primitive nth root ρ of unity having period n. This means that the polynomial $X^n - 1$ has n distinct roots in $\mathcal{F}$ given by $\rho, \rho^2, \ldots, \rho^n$. Thus, $\mathcal{F}$ can be any field extension $\mathbb{F}$ of $\mathbb{C}$. There are numerous other possibilities. For instance, when $n = 3$, $\mathcal{F}$ can be the field of Section C.4 on page 214.

6.1. Properties of the field $\mathcal{F}$

The polynomial $f(X) = X^n - 1$ factors as

$$f(X) = \left(X - \rho\right)\left(X - \rho^2\right) \cdots \left(X - \rho^n\right)$$
$$= (X - \rho) g(X), \quad \text{with } g(\rho) \neq 0.$$

Thus, we have $f'(X) = g(X) + (X - \rho) g'(x)$, $f'(\rho) = g(\rho) \neq 0$, $n X^{n-1} = f'(X) \neq 0$, and $n \cdot 1 = \overbrace{1 + 1 + \cdots + 1}^{n} \neq 0$. Thus, $1/(n \cdot 1)$ is in $\mathcal{F}$ and we shall write it as $1/n$.

PROPOSITION 6.1. *A field $\mathfrak{F}$ contains a primitive nth root of unity having period n if and only if the polynomial $f(X) = X^n - 1$ has n distinct roots in $\mathfrak{F}$.*

PROOF. (i). If ζ is a primitive nth root of unity in $\mathfrak{F}$ having period n, then $\zeta, \zeta^2, \ldots, \zeta^n$ are n distinct roots of $f(X)$ in $\mathfrak{F}$.

(ii). Suppose that $\alpha_1, \alpha_2, \ldots, \alpha_n$ are n distinct roots of $f(X)$ in $\mathfrak{F}$. Then, with respect to multiplication, the set U of these n roots forms a group of order n that is a finite subgroup of the multiplicative group $\mathfrak{F}^*$ of nonzero elements in $\mathfrak{F}$. The result of [**31**, page 194, Theorem 4.9] establishes that each finite subgroup of the multiplicative group of any field is cyclic. Thus, U is a cyclic group of order n. Let ζ be a generator of U. Then, each of $\alpha_1, \alpha_2, \ldots, \alpha_n$ is expressible as ζ^k for some k that satisfies $1 \leq k \leq n$. Hence, ζ is a primitive nth root of unity in $\mathfrak{F}$ that has period n. This completes the proof. $\qquad \square$

6.2. The matrix M that corresponds to (5.4) for [16] of 1856

In terms of ρ, an $n \times n$ matrix M is defined by

$$(6.2) \qquad [M]_{r,s} = \rho^{(r-1)(s-1)}, \quad \text{for } r, s = 1, 2, \ldots, n.$$

The n rows of M specify n distinct functions $\chi_1, \chi_2, \ldots, \chi_n$ from G to $\mathcal{F}^*$ having

$$(6.3) \qquad \chi_r(g_s) = \rho^{(r-1)(s-1)}, \quad \text{for } r = 1, 2, \ldots, n \text{ and } s = 1, 2, \ldots, n.$$

In particular, the element $g_1 = e$ in G yields $\chi_r(e) = 1$, for $1 \le r \le n$.

THEOREM 6.2. *The functions $\chi_1, \chi_2, \ldots, \chi_n$ from G to $\mathcal{F}$ satisfy*

$$(6.4) \qquad \chi_r(xy) = \chi_r(x)\chi_r(y), \quad \text{for } x, y \text{ in } G \text{ and } 1 \le r \le n,$$

$$(6.5) \qquad \chi_r(g_s) = \chi_s(g_r), \quad \text{for } r, s = 1, 2, \ldots, n,$$

$$(6.6) \qquad \chi_r(g_s^{-1}) = (\chi_r(g_s))^{-1} = (\chi_s(g_r))^{-1} = \chi_s(g_r^{-1}), \quad \text{for } r, s = 1, 2, \ldots, n,$$

and, for x in G,

$$(6.7) \qquad \sum_{k=1}^{n} \chi_k(x) = \begin{cases} n, & \text{if } x = e, \\ 0, & \text{if } x \ne e. \end{cases}$$

PROOF. To verify (6.4) for x, y in G, let i, j denote the unique integers that satisfy $1 \le i, j \le n$ and yield $x = g_i = h^{i-1}$ as well as $y = g_j = h^{j-1}$.
(i) Suppose that $i + j - 1 \le n$. Then, we use (6.3) to check that

$$\chi_r(xy) = \chi_r(g_i g_j) = \chi_r(h^{i+j-2}) = \chi_r(g_{i+j-1}) = \rho^{(r-1)(i+j-2)}$$

$$= \rho^{(r-1)(i-1)}\rho^{(r-1)(j-1)} = \chi_r(x)\chi_r(y).$$

(ii) Suppose that $i + j - 1 \ge n + 1$. Then, with $1 \le i + j - 1 - n \le n - 1$, we apply the definition of χ_r in (6.3) with the properties $h^n = e$ and $\rho^n = 1$ to obtain

$$\chi_r(xy) = \chi_r(g_i g_j) = \chi_r(h^{i+j-2}) = \chi_r(h^{i+j-2-n}) = \chi_r(g_{i+j-1-n})$$

$$= \rho^{(r-1)(i+j-2-n)} = \rho^{(r-1)(i-1)}\rho^{(r-1)(j-1)} = \chi_r(x)\chi_r(y).$$

Hence, we conclude that (6.4) is valid.

We note that (6.5) is an immediate consequence of (6.3).

For (6.6), we see that (6.4) yields $1 = \chi_r(e) = \chi_r(g_s^{-1}g_s) = \chi_r(g_s^{-1})\chi_r(g_s)$ and therefore $\chi_r(g_s^{-1}) = (\chi_r(g_s))^{-1}$, for $1 \le r, s \le n$. This and (6.5) establish (6.6).

To verify (6.7) when $x = e$, each $\chi_k(e)$ is 1 and their sum is n. When $x \ne e$, there is an integer β such that $1 \le \beta \le n - 1$, $x = g_{\beta+1} = h^\beta$, $\rho^\beta \ne 1$, and

$$\sum_{k=1}^{n} \chi_k(x) = \sum_{k=1}^{n} \chi_k(g_{\beta+1}) = \sum_{k=1}^{n} \rho^{(k-1)\beta} = 1 + \rho^\beta + (\rho^\beta)^2 + \cdots + (\rho^\beta)^{n-1}$$

$$= \frac{(\rho^\beta)^n - 1}{\rho^\beta - 1} = \frac{1 - 1}{\rho^\beta - 1} = 0.$$

This shows that (6.7) is valid when $x \ne e$ and completes the proof. $\qquad \square$

The $n \times n$ matrix M of (6.2) is given by

$$(6.8) \qquad [M]_{r,s} = \chi_r(g_s), \quad \text{for } r, s = 1, 2, \ldots, n.$$

We introduce the $n \times n$ matrix L defined by

$$(6.9) \qquad [L]_{r,s} = \chi_r(g_s^{-1})/n, \quad \text{for } r, s = 1, 2, \ldots, n.$$

PROPOSITION 6.3. *The matrix M is nonsingular and L is its inverse.*

PROOF. For $r, s = 1, 2, \ldots, n$, we apply (6.9), (6.8), (6.6), (6.4), and (6.7) to verify that

$$[LM]_{r,s} = \sum_{\mu=1}^{n} [L]_{r,\mu} [M]_{\mu,s} = \sum_{\mu=1}^{n} (1/n)\, \chi_r(g_\mu^{-1})\, \chi_\mu(g_s)$$

$$= \sum_{\mu=1}^{n} (1/n)\, \chi_\mu(g_r^{-1})\, \chi_\mu(g_s) = \sum_{\mu=1}^{n} (1/n)\, \chi_\mu(g_r^{-1} g_s) = \begin{cases} 1, & \text{if } r = s, \\ 0, & \text{if } r \neq s. \end{cases}$$

Thus, LM is the $n \times n$ identity matrix and $M^{-1} = L$. This completes the proof. $\square$

THEOREM 6.4. *Suppose that A is an $n \times n$ circulant matrix given by (6.1). Then, the matrix $D = M^{-1}AM$ is the $n \times n$ diagonal matrix having*

$$(6.10) \qquad [D]_{k,k} = \sum_{\nu=1}^{n} \rho^{(k-1)(\nu-1)}\, \sigma(g_\nu), \quad \text{for } k = 1, 2, \ldots, n.$$

Moreover, (6.10) yields the determinant of A as

$$(6.11) \qquad \det(A) = \prod_{k=1}^{n} \left[\sum_{\nu=1}^{n} \rho^{(k-1)(\nu-1)} \sigma(g_\nu) \right].$$

PROOF. Let A be given by (6.1). For $D = M^{-1}AM = LAM$, we apply (6.9), (6.1), (6.8), (6.6), (6.5) to obtain

$$[D]_{r,s} = [LAM]_{r,s} = \sum_{\mu=1}^{n} \sum_{\nu=1}^{n} [L]_{r,\mu} [A]_{\mu,\nu} [M]_{\nu,s}$$

$$= \sum_{\mu=1}^{n} \sum_{\nu=1}^{n} (1/n)\, \chi_r(g_\mu^{-1})\, \sigma(g_\mu^{-1} g_\nu)\, \chi_\nu(g_s)$$

$$= \sum_{\mu=1}^{n} (1/n)\, \chi_\mu(g_r^{-1}) \sum_{\nu=1}^{n} \sigma(g_\mu^{-1} g_\nu)\, \chi_s(g_\nu), \quad \text{for } 1 \leq r, s \leq n.$$

As ν ranges from 1 to n, we see that g_ν and $g_\mu^{-1} g_\nu$ range through G. We use (6.4), (6.5), and the preceding observation to verify that

$$\sum_{\nu=1}^{n} \sigma(g_\mu^{-1} g_\nu)\, \chi_s(g_\nu) = \sum_{\nu=1}^{n} \sigma(g_\mu^{-1} g_\nu)\, \chi_s(g_\mu^{-1} g_\nu)\, \chi_s(g_\mu) = \chi_\mu(g_s) \sum_{\nu=1}^{n} \chi_s(g_\nu)\, \sigma(g_\nu).$$

By combining the preceding displayed formulas and employing (6.7), we find that

$$[D]_{r,s} = \left[\sum_{\mu=1}^{n} \chi_\mu(g_r^{-1})\, \chi_\mu(g_s)/n \right] \left[\sum_{\nu=1}^{n} \chi_s(g_\nu)\, \sigma(g_\nu) \right]$$

$$= \left[\sum_{\mu=1}^{n} \chi_\mu(g_r^{-1} g_s)/n \right] \left[\sum_{\nu=1}^{n} \chi_s(y_\nu)\, \sigma(g_\nu) \right] = \begin{cases} \displaystyle\sum_{\nu=1}^{n} \chi_r(g_\nu)\, \sigma(g_\nu), & \text{if } r = s, \\ 0, & \text{if } r \neq s. \end{cases}$$

Thus, D is an $n \times n$ diagonal matrix for which (6.10) is valid. We use that with $\det(A) = \det(D)$ to obtain (6.11) and complete the proof. $\square$

OBSERVATION 6.5. When the circulant matrix A for Theorem 6.4 is A in (5.7) on page 54, we have $\sigma(g_k) = a_k$, for $1 \le k \le n$. Then, (6.10) and (6.11) yield (5.9) and (5.10) on page 55.

THEOREM 6.6. *Suppose D is an $n \times n$ diagonal matrix having components in $\mathcal{F}$. Then, the matrix $A = MDM^{-1}$ is the $n \times n$ circulant matrix given by (6.1) with*

$$(6.12) \qquad \sigma(x) = \sum_{k=1}^{n} (1/n)\, \chi_k(x^{-1}) \left[D\right]_{k,k}, \quad \text{for each } x \text{ in } G.$$

PROOF. We use $A = MDM^{-1}$, the property of D as a diagonal matrix, (6.8), (6.9), (6.5), (6.4), and (6.12) to deduce, for $1 \le r$, $s \le n$, that

$$\left[A\right]_{r,s} = \left[MDM^{-1}\right]_{r,s} = \sum_{\mu=1}^{n}\sum_{\nu=1}^{n} \left[M\right]_{r,\mu} \left[D\right]_{\mu,\nu} \left[M^{-1}\right]_{\nu,s} = \sum_{k=1}^{n} \left[M\right]_{r,k} \left[D\right]_{k,k} \left[L\right]_{k,s}$$

$$= \sum_{k=1}^{n} \chi_r(g_k) \left[D\right]_{k,k} \chi_k(g_s^{-1})/n = \sum_{k=1}^{n} \chi_k(g_s^{-1}) \chi_k(g_r) \left[D\right]_{k,k}/n$$

$$= \sum_{k=1}^{n} \chi_k(g_s^{-1} g_r) \left[D\right]_{k,k}/n = \sum_{k=1}^{n} (1/n)\, \chi_k\left((g_r^{-1} g_s)^{-1}\right) \left[D\right]_{k,k} = \sigma(g_r^{-1} g_s).$$

Thus, A is the circulant matrix (6.1) given by (6.12). This completes the proof. $\square$

OBSERVATION 6.7. Theorem 4.2 on page 42 shows that the set $\mathcal{M}[\mathcal{F}, G, \mathscr{L}]$ of $n \times n$ circulant matrices having components in $\mathcal{F}$ is a ring and a vector space over $\mathcal{F}$ with respect to matrix addition, multiplication, and scalar multiplication. Theorems 6.4 and 6.6 provide more information.

Let $\mathcal{D}_n[\mathcal{F}]$ denote the set of $n \times n$ diagonal matrices having components in $\mathcal{F}$. With matrix addition, multiplication, and scalar multiplication, $\mathcal{D}_n[\mathcal{F}]$ clearly forms a ring and a vector space over $\mathcal{F}$.

THEOREM 6.8. *There is a ring and vector-space isomorphism Φ of $\mathcal{M}[\mathcal{F}, G, \mathscr{L}]$ onto $\mathcal{D}_n[\mathcal{F}]$.*

PROOF. We use Theorem 6.4 to see that: for each A in $\mathcal{M}[\mathcal{F}, G, \mathscr{L}]$, the matrix $D = M^{-1}AM$ is an element of $\mathcal{D}_n[\mathcal{F}]$. Thus, a function Φ from $\mathcal{M}[\mathcal{F}, G, \mathscr{L}]$ to $\mathcal{D}_n[\mathcal{F}]$ is defined by

$$(6.13) \qquad \Phi(A) = M^{-1}AM, \quad \text{for each } A \text{ in } \mathcal{M}[\mathcal{F}, G, \mathscr{L}].$$

We apply (6.13) to deduce that Φ is one-to-one. Theorem 6.6 shows that: for any D in $\mathcal{D}_n[\mathcal{F}]$, the matrix $A = MDM^{-1}$ belongs to $\mathcal{M}[\mathcal{F}, G, \mathscr{L}]$ and yields $\Phi(A) = D$. Thus, Φ is onto.

For A_1, A_2 in $\mathcal{M}[\mathcal{F}, G, \mathscr{L}]$ and c in $\mathcal{F}$, we note that (6.13) yields

$$\Phi(A_1 + A_2) = \Phi(A_1) + \Phi(A_2), \quad \Phi(A_1 A_2) = \Phi(A_1)\,\Phi(A_2), \quad \Phi(cA_1) = c\,\Phi(A_1).$$

Thus, Φ is a ring isomorphism as well as a vector-space isomorphism of $\mathcal{M}[\mathcal{F}, G, \mathscr{L}]$ onto $\mathcal{D}_n[\mathcal{F}]$. This completes the proof. $\square$

6.3. Circulant and leftward-circulant matrices

We continue the context of the preceding section. For convenience, we repeat the circulant matrices of (5.1) as

$$(6.14) \qquad \begin{bmatrix} a_1 \end{bmatrix}, \qquad \begin{bmatrix} a_1 & a_2 \\ a_2 & a_1 \end{bmatrix}, \qquad \begin{bmatrix} a_1 & a_2 & a_3 \\ a_3 & a_1 & a_2 \\ a_2 & a_3 & a_1 \end{bmatrix}, \qquad \begin{bmatrix} a_1 & a_2 & a_3 & a_4 \\ a_4 & a_1 & a_2 & a_3 \\ a_3 & a_4 & a_1 & a_2 \\ a_2 & a_3 & a_4 & a_1 \end{bmatrix}, \qquad \cdots$$

and we repeat the leftward-circulant matrices of (5.2) as

$$(6.15) \qquad \begin{bmatrix} a_1 \end{bmatrix}, \qquad \begin{bmatrix} a_1 & a_2 \\ a_2 & a_1 \end{bmatrix}, \qquad \begin{bmatrix} a_1 & a_2 & a_3 \\ a_2 & a_3 & a_1 \\ a_3 & a_1 & a_2 \end{bmatrix}, \qquad \begin{bmatrix} a_1 & a_2 & a_3 & a_4 \\ a_2 & a_3 & a_4 & a_1 \\ a_3 & a_4 & a_1 & a_2 \\ a_4 & a_1 & a_2 & a_3 \end{bmatrix}, \qquad \cdots$$

Just as the multiplication table

(6.16)

$\cdot$	g_1	g_2	g_3	$\cdots$	g_s	$\cdots$	g_{n-1}	g_n
g_1^{-1}	h^0	h^1	h^2	$\cdots$	h^{s-1}	$\cdots$	h^{n-2}	h^{n-1}
g_2^{-1}	h^{n-1}	h^0	h^1	$\cdots$	h^{s-2}	$\cdots$	h^{n-3}	h^{n-2}
g_3^{-1}	h^{n-2}	h^{n-1}	h^0	$\cdots$	h^{s-3}	$\cdots$	h^{n-4}	h^{n-3}
$\vdots$	$\vdots$	$\vdots$	$\vdots$	$\vdots$	$\vdots$	$\vdots$	$\vdots$	$\vdots$
g_r^{-1}	h^{n-r+1}	h^{n-r+2}	h^{n-r+3}	$\cdots$	h^{s-r}	$\cdots$	h^{n-r-1}	h^{n-r}
$\vdots$	$\vdots$	$\vdots$	$\vdots$	$\vdots$	$\vdots$	$\vdots$	$\vdots$	$\vdots$
g_{n-1}^{-1}	h^2	h^3	h^4	$\cdots$	h^{s+1}	$\cdots$	h^0	h^1
g_n^{-1}	h^1	h^2	h^3	$\cdots$	h^s	$\cdots$	h^{n-1}	h^0

serves to connect the visual appearance of the circulant matrices in (6.14) with the context for (6.1), we use the multiplication table

(6.17)

$\cdot$	g_1	g_2	g_3	$\cdots$	g_s	$\cdots$	g_{n-1}	g_n
g_1	h^0	h^1	h^2	$\cdots$	h^{s-1}	$\cdots$	h^{n-2}	h^{n-1}
g_2	h^1	h^2	h^3	$\cdots$	h^s	$\cdots$	h^{n-1}	h^0
g_3	h^2	h^3	h^4	$\cdots$	h^{s+1}	$\cdots$	h^0	h^1
$\vdots$	$\vdots$	$\vdots$	$\vdots$	$\vdots$	$\vdots$	$\vdots$	$\vdots$	$\vdots$
g_r	h^{r-1}	h^r	h^{r+1}	$\cdots$	h^{r+s-2}	$\cdots$	h^{r-3}	h^{r-2}
$\vdots$	$\vdots$	$\vdots$	$\vdots$	$\vdots$	$\vdots$	$\vdots$	$\vdots$	$\vdots$
g_{n-1}	h^{n-2}	h^{n-1}	h^0	$\cdots$	h^{s-3}	$\cdots$	h^{n-4}	h^{n-3}
g_n	h^{n-1}	h^0	h^1	$\cdots$	h^{s-2}	$\cdots$	h^{n-3}	h^{n-2}

to conclude that: *an $n \times n$ matrix B having components in $\mathcal{F}$ is a leftward-circulant matrix if and only if there is a function τ from G to $\mathcal{F}$ such that the element in the rth row and sth column of B is given by*

$$(6.18) \qquad \begin{bmatrix} B \end{bmatrix}_{r,s} = \tau(g_r \, g_s), \quad \text{for } r, s = 1, 2, \ldots, n.$$

The $n \times n$ circulant matrix A of (6.14) is given by (6.23) of page 70 and the $n \times n$ leftward circulant matrix B of (6.15) is given by (6.24). We shall show that $B = NA$ and $A = NB$ in (6.25), where N is the matrix over $\mathcal{F}$ studied next.

We introduce the $n \times n$ matrix N and the function ϕ from G to $\mathcal{F}$ defined by

(6.19) $\qquad [N]_{r,s} = \begin{cases} 1, & \text{if } g_r\,g_s = e, \\ 0, & \text{if } g_r\,g_s \neq e. \end{cases} \quad \text{and} \quad \phi(x) = \begin{cases} 1, & \text{if } x = e, \\ 0, & \text{if } x \neq e, \end{cases}$

for $1 \leq r, s \leq n$ and x in G. This gives $[N]_{r,s} = \phi(g_r\,g_s)$, for $1 \leq r, s \leq n$. Thus, N is an $n \times n$ leftward circulant matrix; and, (6.17) shows that it is visualized as

(6.20) $\qquad N = \begin{bmatrix} 1 & 0 & 0 & \cdots & 0 & 0 & 0 \\ 0 & 0 & 0 & \cdots & 0 & 0 & 1 \\ 0 & 0 & 0 & \cdots & 0 & 1 & 0 \\ 0 & 0 & 0 & \cdots & 1 & 0 & 0 \\ \vdots & \vdots & \vdots & \ddots & \vdots & \vdots & \vdots \\ 0 & 0 & 1 & \cdots & 0 & 0 & 0 \\ 0 & 1 & 0 & \cdots & 0 & 0 & 0 \end{bmatrix}.$

PROPOSITION 6.9. *The matrix N is an $n \times n$ permutation matrix such that:* $N^T = N$, $N^{-1} = N$, $N = (1/n)M^2$, *and*

(6.21) $\qquad \det(N) = (-1)^{(n-1)(n-2)/2}.$

PROOF. We use (6.19) to see that each row of N and each column of N has precisely one component equal to 1 and $n-1$ components equal to 0. Thus, N is a permutation matrix. Because (6.19) yields $[N]_{s,r} = [N]_{r,s}$, for $1 \leq r, s \leq n$, the matrix N is symmetric. Since each permutation matrix has an inverse that is equal to its transpose, we observe that $N^{-1} = N^T = N$.

For $1 \leq r, s \leq n$, we see that (6.8), (6.5), (6.4), (6.7), and (6.19) yield

$$[(1/n)M^2]_{r,s} = \sum_{k=1}^{n} (1/n)\,[M]_{r,k}\,[M]_{k,s} = (1/n)\sum_{k=1}^{n} \chi_r(g_k)\,\chi_k(g_s)$$

$$= (1/n)\sum_{k=1}^{n} \chi_k(g_r)\,\chi_k(g_s) = (1/n)\sum_{k=1}^{n} \chi_k(g_r\,g_s) = \begin{cases} 1, & \text{if } g_r\,g_s = e, \\ 0, & \text{if } g_r\,g_s \neq e \end{cases} = [N]_{r,s}.$$

Hence, we have $N = (1/n)M^2$.

We use the $n \times n$ matrix N in (6.20) to obtain $\det(N) = q_{n-1}$, where $q_0 = 1$,

$$q_1 = 1, \quad q_2 = \begin{vmatrix} 0 & 1 \\ 1 & 0 \end{vmatrix}, \quad q_3 = \begin{vmatrix} 0 & 0 & 1 \\ 0 & 1 & 0 \\ 1 & 0 & 0 \end{vmatrix}, \quad q_4 = \begin{vmatrix} 0 & 0 & 0 & 1 \\ 0 & 0 & 1 & 0 \\ 0 & 1 & 0 & 0 \\ 1 & 0 & 0 & 0 \end{vmatrix}, \quad \ldots.$$

For $n \geq 2$, we expand q_n to obtain $q_n = (-1)^{n+1}q_{n-1}$. For $n \geq 3$, this yields

$$\det(N) = q_{n-1} = (-1)^n q_{n-2} = (-1)^n (-1)^{n-1} q_{n-3} = (-1)^n (-1)^{n-1} (-1)^{n-2} q_{n-4}$$

$$= (-1)^n (-1)^{n-1} \cdots (-1)^{n-k} q_{n-(k+2)} = (-1)^n (-1)^{n-1} \ldots (-1)^3 q_1$$

$$= (-1)^{n-2}(-1)^{n-3}\cdots(-1)^1 = (-1)^{1+2+\cdots+(n-2)} = (-1)^{(n-1)(n-2)/2}.$$

Thus, (6.21) is valid for $n \geq 3$. We check that it is also valid for $n = 1$ and $n = 2$. This completes the proof. $\qquad\qquad \square$

THEOREM 6.10. *Suppose that A is an $n \times n$ circulant matrix having components in $\mathcal{F}$ and suppose that B is an $n \times n$ leftward-circulant matrix having components in $\mathcal{F}$. Then, the matrices NA and AN are leftward-circulant matrices while the matrices NB and BN are circulant matrices.*

PROOF. There are functions σ and τ from G to $\mathcal{F}$ such that

$$(6.22) \qquad [A]_{r,s} = \sigma(g_r^{-1} g_s) \quad \text{and} \quad [B]_{r,s} = \tau(g_r g_s), \quad \text{for } r, s = 1, 2, \ldots, n.$$

(i) For $1 \leq r, s \leq n$, we use (6.22) and (6.19) to obtain

$$[NA]_{r,s} = \sum_{k=1}^{n} [N]_{r,k} [A]_{k,s} = \sum_{k=1}^{n} \phi(g_r g_k) \, \sigma(g_k^{-1} g_s)$$
$$= \sigma\left((g_r^{-1})^{-1} g_s\right) = \sigma(g_r g_s).$$

Thus, in view of (6.18) with $\tau = \sigma$, we find that NA is a leftward-circulant matrix.

(ii). Let ψ be the function from G to $\mathcal{F}$ defined by $\psi(x) = \sigma(x^{-1})$, for x in G. For $1 \leq r, s \leq n$, we observe that (6.22) and (6.19) yield

$$[AN]_{r,s} = \sum_{k=1}^{n} [A]_{r,k} [N]_{k,s} = \sum_{k=1}^{n} \sigma(g_r^{-1} g_k) \, \phi(g_k g_s)$$
$$= \sigma(g_r^{-1} g_s^{-1}) = \sigma(g_s^{-1} g_r^{-1}) = \sigma\left((g_r g_s)^{-1}\right)$$
$$= \psi(g_r g_s).$$

Thus, with $\tau = \psi$ in (6.18), we see that AN is a leftward-circulant matrix.

(iii). For $1 \leq r, s \leq n$, we apply (6.22) and (6.19) to deduce that

$$[NB]_{r,s} = \sum_{k=1}^{n} [N]_{r,k} [B]_{k,s} = \sum_{k=1}^{n} \phi(g_r g_k) \, \tau(g_k g_s)$$
$$= \tau(g_r^{-1} g_s).$$

Thus, with $\sigma = \tau$ in (6.1), this shows that NB is a circulant matrix.

(iv). Let υ be the function from G to $\mathcal{F}$ defined by $\upsilon(x) = \tau(x^{-1})$, for x in G. For $1 \leq r, s \leq n$, we find that (6.22) and (6.19) give

$$[BN]_{r,s} = \sum_{k=1}^{n} [B]_{r,k} [N]_{k,s} = \sum_{k=1}^{n} \tau(g_r g_k) \, \phi(g_k g_s)$$
$$= \tau(g_r g_s^{-1}) = \tau(g_s^{-1} g_r) = \tau\left((g_r^{-1} g_s)^{-1}\right)$$
$$= \upsilon(g_r^{-1} g_s).$$

By comparing this to (6.1) with $\sigma = \upsilon$, we conclude that BN is a circulant matrix. This completes the proof. $\qquad\square$

COROLLARY 6.11. *Let A_1, A_2 be $n \times n$ circulant matrices over $\mathcal{F}$ and let B_1, B_2 be $n \times n$ leftward-circulant matrices over $\mathcal{F}$. Then, the products $A_1 A_2$ and $B_1 B_2$ are circulant matrices while $A_1 B_1$ and $B_1 A_1$ are leftward-circulant matrices.*

PROOF. Since the $n \times n$ circulant matrices are the group-pattern matrices for the group G and list $\mathscr{L}$ used to define (6.16), we apply Theorem 4.1 on page 41 to see that the product $A_1 A_2$ of $n \times n$ circulant matrices is a circulant matrix.

Theorem 6.10 and the preceding observation show that each of

$$B_1 N, \quad N B_2, \quad \text{and} \quad (B_1 N)(N B_2) = B_1 (NN) B_2 = B_1 B_2$$

is a circulant matrix.

Since A_1, $B_1 N$, and $A_1(B_1 N)$ are circulant matrices, we see that the matrix

$$\big(A_1(B_1 N)\big)N = (A_1 B_1)(NN) = A_1 B_1$$

is a leftward-circulant matrix.

Because $N B_1$, A_1, and $(N B_1) A_1$ are circulant matrices, the matrix

$$N\big((N B_1) A_1\big) = (NN)(B_1 A_1) = B_1 A_1$$

is a leftward-circulant matrix. This completes the proof. $\square$

OBSERVATIONS 6.12. When σ is the function from G to $\mathcal{F}$ having $\sigma(g_k) = a_k$, for $1 \le k \le n$, the $n \times n$ circulant matrix A specified by (6.1) is visualized as

$$(6.23) \qquad A = \begin{bmatrix} a_1 & a_2 & a_3 & \cdots & a_{n-1} & a_n \\ a_n & a_1 & a_2 & \cdots & a_{n-2} & a_{n-1} \\ a_{n-1} & a_n & a_1 & \cdots & a_{n-3} & a_{n-2} \\ \vdots & \vdots & \vdots & \vdots & \vdots & \vdots \\ a_3 & a_4 & a_5 & \cdots & a_1 & a_2 \\ a_2 & a_3 & a_4 & \cdots & a_n & a_1 \end{bmatrix}.$$

When τ equals σ and is therefore the function from G to $\mathcal{F}$ defined by $\tau(g_k) = a_k$, for $1 \le k \le n$, the leftward-circulant matrix B specified by (6.18) is visualized as

$$(6.24) \qquad B = \begin{bmatrix} a_1 & a_2 & a_3 & \cdots & a_{n-1} & a_n \\ a_2 & a_3 & a_4 & \cdots & a_n & a_1 \\ a_3 & a_4 & a_5 & \cdots & a_1 & a_2 \\ \vdots & \vdots & \vdots & \vdots & \vdots & \vdots \\ a_{n-1} & a_n & a_1 & \cdots & a_{n-3} & a_{n-2} \\ a_n & a_1 & a_2 & \cdots & a_{n-2} & a_{n-1} \end{bmatrix}.$$

The matrices A and B are related by means of N in (6.19) and (6.20) through

$$(6.25) \qquad\qquad B = NA \quad \text{and} \quad A = NB.$$

A verification of (6.25) is provided by (i) and (iii) of the proof for Theorem 6.10.

In regard to (5.6) and (5.10) on pages 54–55, we use (6.25), (6.21), and (6.11) with $\sigma(g_\nu) = a_\nu$, for $1 \le \nu \le n$, to deduce

$$(6.26) \qquad \det(B) = \det(N)\det(A) = (-1)^{(n-1)(n-2)/2} \det(A),$$

where

$$(6.27) \qquad \det(A) = \prod_{k=1}^{n} \left[\sum_{\nu=1}^{n} \rho^{(k-1)(\nu-1)} a_\nu \right].$$

6.4. Additional results

6.4.1. The transpose of a circulant matrix A. Since a circulant matrix A is a group-pattern matrix with respect to the G and $\mathscr{L}$ of page 63, we use Theorem 4.1 on page 41 to see that the transpose A^T of a circulant matrix A is a circulant matrix. We apply Theorem 6.10 to deduce that: if A is an $n \times n$ circulant matrix with components in $\mathcal{F}$, then NAN is a circulant matrix.

PROPOSITION 6.13. *If A is an $n \times n$ circulant matrix over $\mathcal{F}$, then A^T is a circulant matrix and $NAN = A^T$.*

PROOF. Let A be given by (6.1). Then, for $1 \leq r, s \leq n$, we deduce that

$$(6.28) \quad \left[NAN\right]_{r,s} = \sum_{\mu=1}^{n} \sum_{\nu=1}^{n} \left[N\right]_{r,\mu} \left[A\right]_{\mu,\nu} \left[N\right]_{\nu,s} = \sum_{k=1}^{n} \left[N\right]_{r,\mu} \sigma\!\left(g_\mu^{-1} g_\nu\right) \left[N\right]_{\nu,s}.$$

For $1 \leq \mu, \nu, r, s \leq n$, we have $\left[N\right]_{r,\mu} \left[N\right]_{\nu,s} \neq 0$ if and only if $g_r g_\mu = e = g_\nu g_s$ or equivalently $g_\mu = g_r^{-1}$ and $g_\nu = g_s^{-1}$. With this, we find that (6.28) and (6.1) give

$$\left[NAN\right]_{r,s} = \sigma\!\left(g_r g_s^{-1}\right) = \sigma\!\left(g_s^{-1} g_r\right) = \left[A\right]_{s,r} = \left[A^T\right]_{r,s}, \quad \text{for } 1 \leq r, s = 1, 2, \ldots, n.$$

Thus, we obtain $NAN = A^T$. This completes the proof. $\qquad\square$

6.4.2. Symmetric formulation for Theorems 6.4 and 6.6.

THEOREM 6.14. *Suppose that A_1 and A_2 are $n \times n$ matrices over $\mathcal{F}$ that satisfy*

$$(6.29) \qquad\qquad A_1 M = M A_2.$$

Then, A_1 is a circulant matrix if and only if A_2 is a diagonal matrix; and, A_2 is a circulant matrix if and only if A_1 is a diagonal matrix.

PROOF. (i). By writing (6.29) as $M^{-1} A_1 M = A_2$, we use Theorems 6.6 and 6.4 to see that A_1 is a circulant matrix if and only if A_2 is a diagonal matrix.

(ii) Since M is a symmetric matrix, we note that (6.29) yields

$$A_2^T M = A_2^T M^T = (M A_2)^T = (A_1 M)^T = M^T A_1^T = M A_1^T.$$

Thus, by Part (i), A_2^T is a circulant matrix if and only if A_1^T is a diagonal matrix. Each of Theorem 4.1 on page 41 and Proposition 6.13 show that A_2^T is a circulant matrix if and only if A_2 is a circulant matrix. Of course, A_1^T is a diagonal matrix if and only if A_1 is a diagonal matrix. Consequently, A_2 is a circulant matrix if and only if A_1 is a diagonal matrix. This completes the proof. $\qquad\square$

OBSERVATION 6.15. While the preceding proof is correct, it is desirable to actually explain the symmetry of (6.29). For that, we note that M and L of (6.8) and (6.9) are given in terms of (6.3) by

$$\left[M\right]_{r,s} = \rho^{(r-1)(s-1)} \quad \text{and} \quad \left[L\right]_{r,s} = (1/n)\rho^{-(r-1)(s-1)}, \quad \text{for } r, s = 1, 2, \ldots, n.$$

Consequently, there is an $n \times n$ matrix function $F(\rho)$ such that $M = F(\rho)$ and $L = (1/n)\,F(1/\rho)$. Here, we make the key observation that ρ is a primitive nth root of unity in $\mathcal{F}$ having period n if and only if $\hat{\rho} = 1/\rho$ is a primitive nth root of unity in $\mathcal{F}$ having period n. Thus, each result in this chapter remains valid when we replace ρ throughout with $\hat{\rho} = 1/\rho$. For any $n \times n$ matrix A over $\mathcal{F}$, Theorems 6.4 and 6.6 show that an $n \times n$ matrix A over $\mathcal{F}$ is a circulant matrix if and only if $(1/n)\,F(1/\rho)\,A\,F(\rho)$ is a diagonal matrix; and, that is therefore true if and only if $(1/n) F(\rho) A F(1/\rho)$ is a diagonal matrix. With $(1/n) F(1/\rho) A F(\rho) = M^{-1}AM$ and $(1/n) F(\rho) A F(1/\rho) = MAM^{-1}$, the symmetry is explained.

6.5. Modifications of Theorem 6.4 and Theorem 6.6

Continuing the context for (6.1) on page 63 where the list $\mathscr{L} = (g_1, g_2, \ldots, g_n)$ for (6.1) has $g_1 = h^0 = e$, we see that: if $(a_1, a_2, \ldots, a_n)$ is the first row of an $n \times n$ circulant matrix A specified by (6.1), then the function σ for (6.1) is given by $\sigma(g_k) = a_k$, for $1 \le k \le n$. Thus, each circulant matrix is uniquely specified by the components of its first row.

An $n \times n$ diagonal matrix D is uniquely specified by its n principal diagonal components $d_k = [D]_{k,k}$, for $1 \le k \le n$. Of course, when $d_1, d_2, \ldots, d_n$ are given, they yield a unique $n \times n$ diagonal matrix D having $[D]_{k,k} = d_k$, for $1 \le k \le n$.

MODIFICATION 6.16. If an $n \times n$ circulant matrix A over $\mathcal{F}$ has $(a_1, a_2, \ldots, a_n)$ as its first row, then the matrix $D = M^{-1}AM$ is an $n \times n$ diagonal matrix and

$$(6.30) \qquad [d_1, d_2, \ldots, d_n] = [a_1, a_2, \ldots, a_n] M.$$

PROOF. For this modification of Theorem 6.4, we see that (6.10) yields

$$[d_1, d_2, \ldots, d_n]^T = M [a_1, a_2, \ldots, a_n]^T.$$

We transpose this expression and use $M^T = M$ to obtain (6.30). □

MODIFICATION 6.17. If D is an $n \times n$ diagonal matrix over $\mathcal{F}$, then the matrix $A = MDM^{-1}$ is a circulant matrix and the components of its first row are given by

$$(6.31) \qquad [a_1, a_2, \ldots, a_n] = [d_1, d_2, \ldots, d_n] M^{-1}.$$

PROOF. For this modification of Theorem 6.6, we use (6.12) to obtain

$$(6.32) \qquad a_\ell = \sigma(g_\ell) = \sum_{k=1}^{n} (1/n) \chi_k(g_\ell^{-1}) [D]_{k,k} = \sum_{k=1}^{n} d_k [M^{-1}]_{k,\ell},$$
$$\text{for } 1 \le \ell \le n.$$

Formula (6.31) is a consequence of (6.32). □

EXERCISES

1. For $n = 5$, verify that A in (6.23) and B in (6.24) satisfy (6.25).

2. Show that N of (6.19) is given by $N = nL^2$ with respect to L of (6.9).

3. Use the validity of Modification 6.16 to establish Modification 6.17.

4. Use the validity of Modification 6.17 to establish Modification 6.16.

CHAPTER 7

Diagonalizations
of Cyclic-Group-Pattern Matrices

Of course, circulant matrices are merely one type of group-pattern matrix for cyclic groups of order $n \geq 4$. To facilitate the transition from the diagonalizations of circulant matrices in the preceding chapter to the diagonalizations of group-pattern matrices for any finite abelian group in the next chapter, we provide details here for the diagonalization of any group-pattern matrices for a finite cyclic group.

7.1. Context and principal results

Let G be a cyclic group of order n, let $\mathscr{L} = (g_1, g_2, \ldots, g_n)$ be any list of the n distinct elements of G, and let $\mathcal{F}$ be a field of the kind defined in Section 6.1 that contains a primitive nth root ρ of unity having period n. Then, an $n \times n$ matrix A with components in $\mathcal{F}$ is a group-pattern matrix for G and $\mathscr{L}$ if and only if there is a function σ from G to $\mathcal{F}$ such that

(7.1) $$[A]_{r,s} = \sigma\big(g_r^{-1}g_s\big), \quad \text{for } r, s = 1, 2, \ldots, n.$$

To construct a nonsingular $n \times n$ matrix M over $\mathcal{F}$ for which $M^{-1}AM$ is a diagonal matrix over $\mathcal{F}$ if and only if A is a group-pattern matrix over $\mathcal{F}$ for G and $\mathscr{L}$, we note that: there is an element h in G and there are n distinct integers $\alpha_1, \alpha_2, \ldots, \alpha_n$ subject to $0 \leq \alpha_k \leq n-1$, for $1 \leq k \leq n$, such that

(7.2) $$g_k = h^{\alpha_k}, \quad \text{for } k = 1, 2, \ldots . n.$$

For $k = 1, 2, \ldots, n$, let χ_k be the function from G to $\mathcal{F}^*$ defined by

(7.3) $$\chi_k\big(h^\beta\big) = \rho^{\alpha_k \beta}, \quad \text{for } \beta = 0, 1, \ldots, n-1.$$

We introduce M as the $n \times n$ matrix over $\mathcal{F}$ having

(7.4) $$[M]_{r,s} = \chi_r(g_s), \quad \text{for } r, s = 1, 2, \ldots, n.$$

THEOREM 7.1. *The functions* $\chi_1, \chi_2, \ldots, \chi_n$ *from* G *to* $\mathcal{F}$ *satisfy*

(7.5) $\quad \chi_k(xy) = \chi_k(x)\chi_k(y), \quad$ *for x, y in G and $1 \leq k \leq n$,*

(7.6) $\quad \chi_r(g_s) = \chi_s(g_r), \quad$ *for $r, s = 1, 2, \ldots, n$,*

(7.7) $\quad \chi_r\big(g_s^{-1}\big) = \big(\chi_r(g_s)\big)^{-1} = \big(\chi_o(g_r)\big)^{-1} = \chi_s\big(g_r^{-1}\big), \quad$ *for $r, s = 1, 2, \ldots, n$,*

and, for x in G,

(7.8) $$\sum_{k=1}^{n} \chi_k(x) = \begin{cases} n, & \text{if } x = e, \\ 0, & \text{if } x \neq e. \end{cases}$$

73

PROOF. For integers i, j that satisfy $h^i = h^j$, we have $h^{i-j} = e$, $i - j = qn$ for some integer q, $\rho^i = \rho^j$, and $\rho^{\alpha_k i} = \rho^{\alpha_k j}$. Thus, (7.3) is valid for any integer β.

For x, y in G, there are integers u, v such that $x = h^u$ and $y = h^v$. Then, we obtain $\chi_k(xy) = \chi_k(h^{u+v}) = \rho^{\alpha_k(u+v)} = \rho^{\alpha_k u}\rho^{\alpha_k v} = \chi_k(x)\,\chi_k(y)$. Thus, (7.5) is valid.

For $1 \le r$, $s \le n$, we use (7.3) and (7.2) to deduce

$$\chi_r(g_s) = \chi_r(h^{\alpha_s}) = \rho^{\alpha_r \alpha_s} = \rho^{\alpha_s \alpha_r} = \chi_s(h^{\alpha_r}) = \chi_s(g_r).$$

Hence, (7.6) is valid.

For $1 \le r$, $s \le n$, we find that (7.5) and (7.3) yield

$$\chi_r(g_s)\,\chi_r(g_s^{-1}) = \chi_r(e) = \chi_r(h^0) = \rho^0 = 1.$$

We use this result and (7.6) to verify that (7.7) is valid.

Since $\alpha_1, \alpha_2, \ldots, \alpha_n$ is a permutation of $0, 1, \ldots, n-1$, we see that: for $x = h^\beta$,

$$(7.9) \qquad \sum_{k=1}^{n} \chi_k(x) = \sum_{k=1}^{n} \rho^{\alpha_k \beta} = \sum_{k=1}^{n} \left(\rho^\beta\right)^{k-1}.$$

If $x = e$, then $\rho^\beta = 1$ and the sum in (7.9) equals n. If $x \ne e$, then $\rho^\beta \ne 1$ and

$$\sum_{k=1}^{n} \chi_k(x) = 1 + \rho^\beta + \left(\rho^\beta\right)^2 + \cdots + \left(\rho^\beta\right)^{n-1} = \frac{\left(\rho^\beta\right)^n - 1}{\rho^\beta - 1} = \frac{1 - 1}{\rho^\beta - 1} = 0,$$

Thus, (7.8) is valid. This completes the proof. □

To supplement M in (7.4), we introduce L as the $n \times n$ matrix over $\mathcal{F}$ having

$$(7.10) \qquad \left[L\right]_{r,s} = (1/n)\,\chi_r(g_s^{-1}), \quad \text{for } r, s = 1, 2, \ldots, n.$$

PROPOSITION 7.2. *The matrix M of (7.4) is nonsingular and L is its inverse.*

PROOF. For $1 \le r$, $s \le n$, we use (7.10), (7.4), (7.7), (7.5), and (7.8) to obtain

$$[LM]_{r,s} = \sum_{\mu=1}^{n} \left[L\right]_{r,\mu} \left[M\right]_{\mu,s} = \sum_{\mu=1}^{n} (1/n)\,\chi_r(g_\mu^{-1})\,\chi_\mu(g_s)$$

$$= (1/n)\sum_{\mu=1}^{n} \chi_\mu(g_r^{-1})\,\chi_\mu(g_s) = (1/n)\sum_{\mu=1}^{n} \chi_\mu(g_r^{-1}g_s)$$

$$= \begin{cases} 1, & \text{if } r = s, \\ 0, & \text{if } r \ne s. \end{cases}$$

Consequently, LM is the $n \times n$ identity matrix. This completes the proof. □

THEOREM 7.3. *Let A be an $n \times n$ matrix with components in $\mathcal{F}$. Then, in terms of the context for (7.1)–(7.4), the matrix A is a group-pattern matrix for G and $\mathcal{L}$ if and only if $M^{-1}AM$ is a diagonal matrix. Moreover, if A is a group-pattern matrix given by (7.1), then the diagonal components of $D = M^{-1}AM$ are*

$$(7.11) \qquad \left[D\right]_{k,k} = \sum_{\nu=1}^{n} \chi_k(g_\nu)\,\sigma(g_\nu), \quad \text{for } k = 1, 2, \ldots, n.$$

PROOF. (i). Let A be a group-pattern matrix for G and $\mathscr{L}$ given by (7.1) and set $D = M^{-1}AM$. We use (7.10), (7.1), (7.4), (7.5), (7.7), (7.6), and (7.8) to obtain

$$[D]_{r,s} = [M^{-1}AM]_{r,s} = \sum_{\mu=1}^{n}\sum_{\nu=1}^{n} [M^{-1}]_{r,\mu} [A]_{\mu,\nu} [M]_{\nu,s}$$

$$= \sum_{\mu=1}^{n}\sum_{\nu=1}^{n} (1/n)\, \chi_r(g_\mu^{-1})\, \sigma(g_\mu^{-1}g_\nu)\, \chi_s(g_\nu)$$

$$= \sum_{\mu=1}^{n} (1/n)\chi_r(g_\mu^{-1}) \left[\sum_{\nu=1}^{n} \sigma(g_\mu^{-1}g_\nu)\, \chi_s(g_\mu^{-1}g_\nu)\, \chi_s(g_\mu) \right]$$

$$= \left[\sum_{\mu=1}^{n} \chi_\mu(g_r^{-1})\chi_\mu(g_s)/n \right] \left[\sum_{\nu=1}^{n} \sigma(g_\mu^{-1}g_\nu)\, \chi_s(g_\mu^{-1}g_\nu) \right]$$

$$= \left[\sum_{\mu=1}^{n} \chi_\mu(g_r^{-1}g_s)/n \right] \left[\sum_{\nu=1}^{n} \chi_s(g_\nu)\, \sigma(g_\nu) \right]$$

$$= \begin{cases} \sum_{\nu=1}^{n} \chi_s(g_\nu)\, \sigma(g_\nu), & \text{if } r = s, \\ 0, & \text{if } r \neq s. \end{cases}$$

Thus, D is a diagonal matrix and (7.11) is valid.

(ii). Suppose $M^{-1}AM$ is a diagonal matrix. We set $D = M^{-1}AM$ and use $A = MDM^{-1}$ with D as a diagonal matrix, (7.4), (7.10), (7.6), and (7.5) to deduce

$$[A]_{r,s} = [MDM^{-1}]_{r,s} = \sum_{\mu=1}^{n}\sum_{\nu=1}^{n} [M]_{r,\mu} [D]_{\mu,\nu} [L]_{\nu,s}$$

$$= \sum_{\mu=1}^{n} [M]_{r,\mu} [D]_{\mu,\mu} [L]_{\mu,s} = \sum_{\mu=1}^{n} \chi_r(g_\mu) [D]_{\mu,\mu} (1/n)\chi_\mu(g_s^{-1})$$

$$= \sum_{\mu=1}^{n} (1/n)\chi_\mu(g_s^{-1})\chi_\mu(g_r) [D]_{\mu,\mu} = \sum_{\mu=1}^{n} (1/n)\chi_\mu(g_s^{-1}g_r) [D]_{\mu,\mu}$$

$$= \sum_{\mu=1}^{n} (1/n)\chi_\mu\big((g_r^{-1}g_s)^{-1}\big) [D]_{\mu,\mu} = \sigma(g_r^{-1}g_s),$$

where σ is the function from G to $\mathcal{F}$ defined by

$$\sigma(x) = \sum_{\mu=1}^{n} (1/n)\, \chi_\mu(x^{-1}) [D]_{\mu,\mu}, \quad \text{for each } x \text{ in } G.$$

Thus, A is a group-pattern matrix for G and $\mathscr{L}$. This completes the proof. $\square$

OBSERVATION 7.4. When A is a group-pattern matrix for G and $\mathscr{L}$ in the context of Theorem 7.3, the matrix $D = M^{-1}AM$ and (7.11) yield

$$\det(A) = \det(D) = \prod_{k=1}^{n} \left[\sum_{\nu=1}^{n} \chi_k(g_\nu)\, \sigma(g_\nu) \right].$$

EXAMPLE 7.5. Let p denote a positive integer; let G denote a cyclic group of order $n = 2p$ whose elements are h^{k-1}, for $1 \leq k \leq n$, with $h^n = e$; let $\mathscr{L}$ be the list for G indicated by $\mathscr{L} = (g_1, g_2, \ldots, g_n)$ that has

(7.12)
$$g_k = \begin{cases} h^{k-1}, & \text{for } 1 \leq k \leq p, \\ h^{3p-k}, & \text{for } p+1 \leq k \leq 2p; \end{cases}$$

let σ be a function from G to $\mathcal{F}$; set $a_k = \sigma(g_k)$, for $1 \leq k \leq n$; and let A_n denote the $n \times n$ group-pattern matrix for G and $\mathscr{L}$ having $[A_n]_{r,s} = \sigma(g_r^{-1} g_s)$, for $r, s = 1, 2, \ldots, n$. In particular, for $n = 2, 4, 6$, this yields

$$A_2 = \begin{bmatrix} a_1 & a_2 \\ a_2 & a_1 \end{bmatrix}, \quad A_4 = \begin{bmatrix} a_1 & a_2 & a_3 & a_4 \\ a_3 & a_1 & a_4 & a_2 \\ a_2 & a_4 & a_1 & a_3 \\ a_4 & a_3 & a_2 & a_1 \end{bmatrix}, \quad A_6 = \begin{bmatrix} a_1 & a_2 & a_3 & a_4 & a_5 & a_6 \\ a_4 & a_1 & a_2 & a_5 & a_6 & a_3 \\ a_5 & a_4 & a_1 & a_6 & a_3 & a_2 \\ a_2 & a_3 & a_6 & a_1 & a_4 & a_5 \\ a_3 & a_6 & a_5 & a_2 & a_1 & a_4 \\ a_6 & a_5 & a_4 & a_3 & a_2 & a_1 \end{bmatrix}.$$

PROPOSITION 7.6. *For $p = 1, 2, 3, \ldots$, the matrix A_{2p} of Example 7.5 is symmetric with respect to reflections in the point at its center.*

PROOF. We are to establish that

(7.13) $$[A_n]_{r,s} = [A_n]_{2p+1-r,\, 2p+1-s}, \quad \text{when } 1 \leq r \leq p \text{ and } 1 \leq s \leq 2p.$$

(i) Suppose that $1 \leq r \leq p$ and $1 \leq s \leq p$. Then, we have $g_r = h^{r-1}$, $g_s = h^{s-1}$, $p+1 \leq 2p+1-r \leq 2p$, $p+1 \leq 2p+1-s \leq 2p$, $g_{2p+1-r} = h^{p+r-1}$, $g_{2p+1-s} = h^{p+s-1}$, and

$$[A_n]_{2p+1-r,\, 2p+1-s} = \sigma(g_{2p+1-r}^{-1} g_{2p+1-s}) = \sigma(h^{s-r}) = \sigma(g_r^{-1} g_s) = [A_n]_{r,s}.$$

(ii) Suppose that $1 \leq r \leq p$ and $p+1 \leq s \leq 2p$. Then, we find that $g_r = h^{r-1}$, $g_s = h^{3p-s}$, $p+1 \leq 2p+1-r \leq 2p$, $1 \leq 2p+1-s \leq p$, $g_{2p+1-r} = h^{p+r-1}$, $g_{2p+1-s} = h^{2p-s}$, $h^{2p} = e$, and

$$[A_n]_{2p+1-r,\, 2p+1-s} = \sigma(g_{2p+1-r}^{-1} g_{2p+1-s}) = \sigma(h^{p+1-r-s}) = \sigma(g_r^{-1} g_s) = [A_n]_{r,s}.$$

Thus, (7.13) is valid. This completes the proof.   □

7.2. Diagonalization for the matrix A_{2p} of Example 7.5

We use (7.12) to see that (7.2) and (7.3) require

$$\alpha_k = \begin{cases} k-1, & \text{for } 1 \leq k \leq p, \\ 3p-k, & \text{for } p+1 \leq k \leq 2p, \end{cases}$$

and

(7.14) $$\chi_k(h^\beta) = \begin{cases} \rho^{(k-1)\beta}, & \text{for } 1 \leq k \leq p \text{ and } 0 \leq \beta \leq n-1, \\ \rho^{(3p-k)\beta}, & \text{for } p+1 \leq k \leq 2p \text{ and } 0 \leq \beta \leq n-1. \end{cases}$$

Thus, with respect to L and M defined by (7.10) and (7.4) in terms of (7.14), Theorem 7.3 is applicable to A_n and shows that the matrix

(7.15) $$D = M^{-1} A_n M = L A_n M$$

is an $n \times n$ diagonal matrix such that

$$(7.16) \quad \left[D\right]_{k,k} = \sum_{\nu=1}^{n} \chi_k(g_\nu)\, \sigma(g_\nu) = \sum_{\nu=1}^{p} \chi_k(h^{\nu-1})\, a_\nu + \sum_{\nu=p+1}^{2p} \chi_k(h^{3p-\nu})\, a_\nu$$

$$= \begin{cases} \displaystyle\sum_{\nu=1}^{p} \rho^{(k-1)(\nu-1)}\, a_\nu + \sum_{\nu=p+1}^{2p} \rho^{(k-1)(3p-\nu)}\, a_\nu, & \text{for } 1 \le k \le p, \\[2em] \displaystyle\sum_{\nu=1}^{p} \rho^{(3p-k)(\nu-1)}\, a_\nu + \sum_{\nu=p+1}^{2p} \rho^{(3p-k)(3p-\nu)}\, a_\nu, & \text{for } p+1 \le k \le 2p. \end{cases}$$

EXAMPLE 7.7. When $p=2$, $n=4$, and $\rho = i$ where $i^2 = -1$, (7.16) yields

$$d_1 = \left[D\right]_{1,1} = a_1 + a_2 + a_3 + a_4, \qquad d_2 = \left[D\right]_{2,2} = a_1 + i\,a_2 - i\,a_3 - a_4,$$

$$d_3 = \left[D\right]_{3,3} = a_1 - i\,a_2 + i\,a_3 - a_4, \qquad d_4 = \left[D\right]_{4,4} = a_1 - a_2 - a_3 + a_4.$$

For this situation, we use (7.10) and (7.4) to verify that $M^{-1}A_4 M$ is given by

$$\frac{1}{4} \begin{bmatrix} 1 & 1 & 1 & 1 \\ 1 & -i & i & -1 \\ 1 & i & -i & -1 \\ 1 & -1 & -1 & 1 \end{bmatrix} \begin{bmatrix} a_1 & a_2 & a_3 & a_4 \\ a_3 & a_1 & a_4 & a_2 \\ a_2 & a_4 & a_1 & a_3 \\ a_4 & a_3 & a_2 & a_1 \end{bmatrix} \begin{bmatrix} 1 & 1 & 1 & 1 \\ 1 & i & -i & -1 \\ 1 & -i & i & -1 \\ 1 & -1 & -1 & 1 \end{bmatrix} = \begin{bmatrix} d_1 & 0 & 0 & 0 \\ 0 & d_2 & 0 & 0 \\ 0 & 0 & d_3 & 0 \\ 0 & 0 & 0 & d_4 \end{bmatrix}.$$

CHAPTER 8

Diagonalizations
of Abelian-Group-Pattern Matrices

8.1. Context and principal results

Let G be an abelian group of order n, let $\mathscr{L} = (g_1, g_2, \ldots, g_n)$ be any list of the elements in G, and let $\mathcal{F}$ denote a field that contains a primitive nth root ρ of unity having period n as introduced in Section 6.1. An $n \times n$ matrix A with components in $\mathcal{F}$ is a group-pattern matrix for G and $\mathscr{L}$ if and only if there is a function σ from G to $\mathcal{F}$ such that

$$(8.1) \qquad [A]_{r,s} = \sigma(g_r^{-1} g_s), \quad \text{for } r,\, s = 1,\, 2,\, \ldots,\, n.$$

The basis theorem for finite abelian groups is presented in [**35**, pp. 72–77]. It also appears in standard treatises on group theory. We use it to see that there are elements $h_1, h_2, \ldots, h_d$ in G of respective periods $n_1, n_2, \ldots, n_d$ such that each of the n elements of G has a unique representation

$$(8.2) \qquad h_1^{\beta_1} h_2^{\beta_2} \cdots h_d^{\beta_d}, \quad \text{with } 0 \le \beta_i \le n_i - 1 \text{ for } 1 \le i \le d.$$

Since there are $n_1 n_2 \cdots n_d$ such representations, we have

$$(8.3) \qquad n = n_1 n_2 \cdots n_d.$$

For $k = 1, 2, \ldots, n$, let $\alpha_{1,k}, \alpha_{2,k}, \ldots, \alpha_{d,k}$ denote the unique integers that yield

$$(8.4) \qquad g_k = h_1^{\alpha_{1,k}} h_2^{\alpha_{2,k}} \cdots h_d^{\alpha_{d,k}}, \quad \text{with } 0 \le \alpha_{i,k} \le n_i - 1 \text{ for } 1 \le i \le d.$$

For $i = 1, 2, \ldots, d$, we set $\rho_i = \rho^{n/n_i}$ and observe that the elements $\rho_i, \rho_i^2, \ldots, \rho_i^{n_i}$, with $\rho_i^{n_i} = e$, form a cyclic group of order n_i. For $k = 1, 2, \ldots, n$, we use (8.2) to define a function χ_k from G to $\mathcal{F}^*$ by means of

$$(8.5) \qquad h_1^{\beta_1} h_2^{\beta_2} \cdots h_d^{\beta_d} \mapsto \chi_k\big(h_1^{\beta_1} h_2^{\beta_2} \cdots h_d^{\beta_d}\big) = \rho_1^{\alpha_{1,k}\beta_1} \rho_2^{\alpha_{2,k}\beta_2} \cdots \rho_d^{\alpha_{d,k}\beta_d}.$$

We introduce M as the $n \times n$ matrix over $\mathcal{F}$ having

$$(8.6) \qquad [M]_{r,s} = \chi_r(g_s), \quad \text{for } r,\, s = 1,\, 2,\, \ldots,\, n.$$

THEOREM 8.1. *The functions $\chi_1, \chi_2, \ldots, \chi_n$ from G to $\mathcal{F}^*$ satisfy*

$$(8.7) \qquad \chi_k(xy) = \chi_k(x)\chi_k(y), \quad \text{for } x,\, y \text{ in } G \text{ and } 1 \le k \le n,$$

$$(8.8) \qquad \chi_r(g_s) = \chi_s(g_r), \quad \text{for } r,\, s = 1,\, 2,\, \ldots,\, n,$$

$$(8.9) \qquad \chi_r\big(g_s^{-1}\big) = \big(\chi_r(g_s)\big)^{-1} = \big(\chi_s(g_r)\big)^{-1} = \chi_s\big(g_r^{-1}\big), \quad \text{for } r,\, s = 1,\, 2,\, \ldots,\, n,$$

and, for each x in G,

$$(8.10) \qquad \sum_{k=1}^{n} \chi_k(x) = \begin{cases} n, & \text{if } x = e, \\ 0, & \text{if } x \ne e. \end{cases}$$

PROOF. For any integers $x_1, x_2, \ldots, x_d$ there are integers $q_1, q_2, \ldots, q_d$ and $\beta_1, \beta_2, \ldots, \beta_d$ such that: $x_i = q_i n_i + \beta_i$ and $0 \leq \beta_i \leq n_i - 1$, for $1 \leq i \leq d$. We use this with $h_i^{n_i} = e$ and $\rho_i^{n_i} = 1$, for $1 \leq i \leq d$, to obtain

$$h_1^{x_1} h_2^{x_2} \cdots h_d^{x_d} = h_1^{\beta_1} h_2^{\beta_2} \cdots h_d^{\beta_d}$$

and

$$\rho_1^{\alpha_{1,k} x_1} \rho_2^{\alpha_{2,k} x_2} \cdots \rho_d^{\alpha_{d,k} x_d} = \rho_1^{\alpha_{1,k} \beta_1} \rho_2^{\alpha_{2,k} \beta_2} \cdots \rho_d^{\alpha_{d,k} \beta_d}.$$

Thus, for $1 \leq k \leq n$, we have

$$\chi_k\left(h_1^{x_1} h_2^{x_2} \cdots h_d^{x_d}\right) = \chi_k\left(h_1^{\beta_1} h_2^{\beta_2} \cdots h_d^{\beta_d}\right) = \rho_1^{\alpha_{1,k} \beta_1} \rho_2^{\alpha_{2,k} \beta_2} \cdots \rho_d^{\alpha_{d,k} \beta_d}$$
$$= \rho_1^{\alpha_{1,k} x_1} \rho_2^{\alpha_{2,k} x_2} \cdots \rho_d^{\alpha_{d,k} x_d}$$

and therefore observe that (8.5) remains valid when $\beta_1, \beta_2, \ldots, \beta_d$ are any integers.

For any x and y in G, there are integers $\beta_1, \beta_2, \ldots, \beta_d$ and $\gamma_1, \gamma_2, \ldots, \gamma_d$ such that $x = h_1^{\beta_1} h_2^{\beta_2} \cdots h_d^{\beta_d}$ and $y = h_1^{\gamma_1} h_2^{\gamma_2} \cdots h_d^{\gamma_d}$. Then, for $1 \leq k \leq n$, (8.5) yields

$$\chi_k(xy) = \chi_k\left(h_1^{\beta_1+\gamma_1} h_2^{\beta_2+\gamma_2} \cdots h_d^{\beta_d+\gamma_d}\right) = \rho_1^{\alpha_{1,k}(\beta_1+\gamma_1)} \rho_2^{\alpha_{2,k}(\beta_2+\gamma_2)} \cdots \rho_d^{\alpha_{d,k}(\beta_d+\gamma_d)}$$
$$= \left(\rho_1^{\alpha_{1,k} \beta_1} \rho_2^{\alpha_{2,k} \beta_2} \cdots \rho_d^{\alpha_{d,k} \beta_d}\right)\left(\rho_1^{\alpha_{1,k} \gamma_1} \rho_2^{\alpha_{2,k} \gamma_2} \cdots \rho_d^{\alpha_{d,k} \gamma_d}\right) = \chi_k(x)\,\chi_k(y).$$

Thus, (8.7) is valid.

To verify (8.8), for $r, s = 1, 2, \ldots, n$, we use (8.5) and (8.4) to obtain

$$\chi_r(g_s) = \rho_1^{\alpha_{1,r}\,\alpha_{1,s}} \rho_2^{\alpha_{2,r}\,\alpha_{2,s}} \cdots \rho_d^{\alpha_{d,r}\,\alpha_{d,s}} = \rho_1^{\alpha_{1,s}\,\alpha_{1,r}} \rho_2^{\alpha_{2,s}\,\alpha_{2,r}} \cdots \rho_d^{\alpha_{d,s}\,\alpha_{d,r}} = \chi_s(g_r).$$

Thus, (8.8) is valid.

For $1 \leq r, s \leq n$, we find that (8.7) yields

$$\chi_r(g_s)\,\chi_r\left(g_s^{-1}\right) = \chi_r(e) = \chi_r\left(h_1^0 h_2^0 \cdots h_d^0\right) = \rho_1^0 \rho_2^0 \cdots \rho_d^0 = 1.$$

We use this result and (8.8) to see that (8.9) is valid.

To verify (8.10), let x be given by (8.2). If $x = e$, then $\beta_1 = \beta_2 = \cdots = \beta_d = 0$, (8.5) yields $\chi_k(x) = 1$, for $1 \leq k \leq n$, and we find that the sum in the left member of (8.10) equals n. Suppose that $x \neq e$. Then, there is an integer p such that $1 \leq p \leq d$ and $\beta_p \neq 0$. The representation (8.4) makes it clear that: as k ranges over $\{1, 2, \ldots, n\}$, the d-tuple $(\alpha_{1,k}, \alpha_{2,k}, \ldots, \alpha_{d,k})$ ranges over

$$I = \left\{0, 1, \ldots, (n_1 - 1)\right\} \times \left\{0, 1, \ldots, (n_2 - 1)\right\} \times \cdots \times \left\{0, 1, \ldots, (n_d - 1)\right\}.$$

We use this with (8.5) and the representation (8.2) for x where $\beta_p \neq 0$ to obtain

$$(8.11) \qquad \sum_{k=1}^{n} \chi_k(x) = \sum_{k=1}^{n} \rho_1^{\alpha_{1,k} \beta_1} \rho_2^{\alpha_{2,k} \beta_2} \cdots \rho_d^{\alpha_{d,k} \beta_d}$$
$$= \sum_{(i_1, i_2, \ldots, i_d) \in I} \left(\rho_1^{\beta_1}\right)^{i_1} \left(\rho_2^{\beta_2}\right)^{i_2} \cdots \left(\rho_d^{\beta_d}\right)^{i_d}$$
$$= \sum_{i_1=0}^{n_1-1} \sum_{i_2=0}^{n_2-1} \cdots \sum_{i_d=0}^{n_d-1} \left(\rho_1^{\beta_1}\right)^{i_1} \left(\rho_2^{\beta_2}\right)^{i_2} \cdots \left(\rho_d^{\beta_d}\right)^{i_d}$$
$$= \left[\sum_{i_1=0}^{n_1-1} \left(\rho_1^{\beta_1}\right)^{i_1}\right]\left[\sum_{i_2=0}^{n_2-1} \left(\rho_2^{\beta_2}\right)^{i_2}\right] \cdots \left[\sum_{i_d=0}^{n_d-1} \left(\rho_d^{\beta_d}\right)^{i_d}\right].$$

Since we have $1 \le p \le d$, $\rho_p^{\beta_p} \ne 1$, $\left(\rho_p^{\beta_p}\right)^{n_p} = \left(\rho_p^{n_p}\right)^{\beta_p} = (1)^{\beta_p} = 1$, and

$$\sum_{i_p=0}^{n_p-1} \left(\rho_p^{\beta_p}\right)^{i_p} = 1 + \rho_p^{\beta_p} + \left(\rho_p^{\beta_p}\right)^2 + \cdots + \left(\rho_p^{\beta_p}\right)^{n_p-1} = \frac{\left(\rho_p^{\beta_p}\right)^{n_p} - 1}{\rho_p^{\beta_p} - 1} = \frac{1-1}{\rho_p^{\beta_p} - 1} = 0,$$

the sum in (8.11) equals 0. Thus, (8.10) is valid. This completes the proof. $\square$

To supplement M in (8.6), we introduce L as the $n \times n$ matrix having

$$(8.12) \qquad [L]_{r,s} = (1/n)\,\chi_r(g_s^{-1}), \quad \text{for } r, s = 1, 2, \ldots, n.$$

PROPOSITION 8.2. *The $n \times n$ matrices M and L defined by (8.6) and (8.12) are symmetric, they are nonsingular, and $M^{-1} = L$.*

PROOF. We use (8.8) and (8.9) to see that M and L are symmetric matrices. For $1 \le r, s \le n$, we employ (8.12), (8.6), (8.9), (8.7), and (8.10) to obtain

$$[LM]_{r,s} = \sum_{\mu=1}^n [L]_{r,\mu}[M]_{\mu,s} = \sum_{\mu=1}^n (1/n)\,\chi_r(g_\mu^{-1})\,\chi_\mu(g_s)$$

$$= (1/n)\sum_{\mu=1}^n \chi_\mu(g_r^{-1})\,\chi_\mu(g_s) = (1/n)\sum_{\mu=1}^n \chi_\mu(g_r^{-1}g_s) = \begin{cases} 1, & \text{if } r = s, \\ 0, & \text{if } r \ne s. \end{cases}$$

Consequently, LM is the $n \times n$ identity matrix. This completes the proof. $\square$

THEOREM 8.3. *Let $\mathscr{L} = (g_1, g_2, \ldots g_n)$ be a list of the elements in an abelian group G of order n and let A be an $n \times n$ matrix over $\mathcal{F}$. Then, A is a group-pattern matrix for G and $\mathscr{L}$ if and only if $M^{-1}AM$ is a diagonal matrix.*

PROOF. (i). Suppose that A is a group-pattern matrix for G and $\mathscr{L}$ given by (8.1). We use (8.12), (8.1), (8.6), (8.8), (8.7), (8.9), and (8.10) to obtain

$$(8.13) \quad [M^{-1}AM]_{r,s} = \sum_{\mu=1}^n \sum_{\nu=1}^n [M^{-1}]_{r,\mu}[A]_{\mu,\nu}[M]_{\nu,s}$$

$$= \sum_{\mu=1}^n \sum_{\nu=1}^n (1/n)\,\chi_r(g_\mu^{-1})\,\sigma(g_\mu^{-1}g_\nu)\,\chi_s(g_\nu)$$

$$= \sum_{\mu=1}^n (1/n)\chi_r(g_\mu^{-1})\left[\sum_{\nu=1}^n \sigma(g_\mu^{-1}g_\nu)\,\chi_s(g_\mu^{-1}g_\nu)\,\chi_s(g_\mu)\right]$$

$$= \left[\sum_{\mu=1}^n \chi_\mu(g_r^{-1})\chi_\mu(g_s)/n\right]\left[\sum_{\nu=1}^n \sigma(g_\mu^{-1}g_\nu)\,\chi_s(g_\mu^{-1}g_\nu)\right]$$

$$= \left[\sum_{\mu=1}^n \chi_\mu(g_r^{-1}g_s)/n\right]\left[\sum_{\nu=1}^n \chi_s(g_\nu)\,\sigma(g_\nu)\right]$$

$$= \begin{cases} \sum_{\nu=1}^n \chi_s(g_\nu)\,\sigma(g_\nu), & \text{if } r = s, \\ 0, & \text{if } r \ne s, \end{cases} \qquad \text{for } r, s = 1, 2, \ldots, n.$$

Thus, $M^{-1}AM$ is a diagonal matrix.

(ii). Suppose $M^{-1}AM$ is a diagonal matrix. We set $D = M^{-1}AM$ and use $A = MDM^{-1}$ with D as a diagonal matrix, (8.6), (8.12), (8.8), and (8.7) to deduce

$$\left[A\right]_{r,s} = \left[MDM^{-1}\right]_{r,s} = \sum_{\mu=1}^{n}\sum_{\nu=1}^{n}\left[M\right]_{r,\mu}\left[D\right]_{\mu,\nu}\left[L\right]_{\nu,s}$$

$$= \sum_{\mu=1}^{n}\left[M\right]_{r,\mu}\left[D\right]_{\mu,\mu}\left[L\right]_{\mu,s} = \sum_{\mu=1}^{n}\chi_r(g_\mu)\left[D\right]_{\mu,\mu}(1/n)\chi_\mu\left(g_s^{-1}\right)$$

$$= \sum_{\mu=1}^{n}(1/n)\chi_\mu\left(g_s^{-1}\right)\chi_\mu(g_r)\left[D\right]_{\mu,\mu} = \sum_{\mu=1}^{n}(1/n)\chi_\mu\left(g_s^{-1}g_r\right)\left[D\right]_{\mu,\mu}$$

$$= \sigma\left(g_r^{-1}g_s\right), \quad \text{for } r,\, s = 1,\, 2,\, \ldots,\, n,$$

where σ is the function from G to $\mathcal{F}$ defined by

$$\sigma(x) = \sum_{\mu=1}^{n}(1/n)\chi_\mu\left(x^{-1}\right)\left[D\right]_{\mu,\mu}, \quad \text{for each } x \text{ in } G.$$

Thus, A is a group-pattern matrix for G and $\mathcal{L}$. This completes the proof. □

EXAMPLE 8.4. For the group $G = \mathcal{C}_3 \times \mathcal{C}_3$ of Example 3.12 on page 36 and its list $\mathcal{L} = (g_1, g_2, \ldots, g_9)$, there are elements h_1, h_2 in G such that: $h_1^3 = e$, $h_2^3 = e$, $h_2 h_1 = h_1 h_2$, and each element of G is uniquely expressible as

$$h_1^{\beta_1}h_2^{\beta_2}, \quad \text{for } 0 \le \beta_1 \le 2 \text{ and } 0 \le \beta_2 \le 2.$$

We select ω in $\mathcal{F}$ subject to $\omega^3 = 1$ and $\omega \neq 1$. The elements $g_1, g_2, \ldots, g_9$ of $\mathcal{L}$ and the corresponding functions $\chi_1, \chi_2, \ldots, \chi_9$ for Theorem 8.1 are given by

$$g_1 = h_1^0 h_2^0, \qquad \chi_1\left(h_1^{\beta_1}h_2^{\beta_2}\right) = \omega^{0\beta_1}\omega^{0\beta_2} = 1,$$
$$g_2 = h_1^1 h_2^0, \qquad \chi_2\left(h_1^{\beta_1}h_2^{\beta_2}\right) = \omega^{1\beta_1}\omega^{0\beta_2} = \omega^{\beta_1},$$
$$g_3 = h_1^2 h_2^0, \qquad \chi_3\left(h_1^{\beta_1}h_2^{\beta_2}\right) = \omega^{2\beta_1}\omega^{0\beta_2} = \omega^{2\beta_1},$$
$$g_4 = h_1^0 h_2^1, \qquad \chi_4\left(h_1^{\beta_1}h_2^{\beta_2}\right) = \omega^{0\beta_1}\omega^{1\beta_2} = \omega^{\beta_2},$$
$$g_5 = h_1^1 h_2^1, \qquad \chi_5\left(h_1^{\beta_1}h_2^{\beta_2}\right) = \omega^{1\beta_1}\omega^{1\beta_2} = \omega^{\beta_1+\beta_2},$$
$$g_6 = h_1^2 h_2^1, \qquad \chi_6\left(h_1^{\beta_1}h_2^{\beta_2}\right) = \omega^{2\beta_1}\omega^{1\beta_2} = \omega^{2\beta_1+\beta_2},$$
$$g_7 = h_1^0 h_2^2, \qquad \chi_7\left(h_1^{\beta_1}h_2^{\beta_2}\right) = \omega^{0\beta_1}\omega^{2\beta_2} = \omega^{2\beta_2},$$
$$g_8 = h_1^1 h_2^2, \qquad \chi_8\left(h_1^{\beta_1}h_2^{\beta_2}\right) = \omega^{1\beta_1}\omega^{2\beta_2} = \omega^{\beta_1+2\beta_2},$$
$$g_9 = h_1^2 h_2^2, \qquad \chi_9\left(h_1^{\beta_1}h_2^{\beta_2}\right) = \omega^{2\beta_1}\omega^{2\beta_2} = \omega^{2\beta_1+2\beta_2}.$$

For the matrix M having $\left[M\right]_{r,s} = \chi_r(g_s)$, for $1 \le r,\, s \le 9$, we find that

$$(8.14) \qquad M = \begin{bmatrix} 1 & 1 & 1 & 1 & 1 & 1 & 1 & 1 & 1 \\ 1 & \omega & \omega^2 & 1 & \omega & \omega^2 & 1 & \omega & \omega^2 \\ 1 & \omega^2 & \omega & 1 & \omega^2 & \omega & 1 & \omega^2 & \omega \\ 1 & 1 & 1 & \omega & \omega & \omega & \omega^2 & \omega^2 & \omega^2 \\ 1 & \omega & \omega^2 & \omega & \omega^2 & 1 & \omega^2 & 1 & \omega \\ 1 & \omega^2 & \omega & \omega & 1 & \omega^2 & \omega^2 & \omega & 1 \\ 1 & 1 & 1 & \omega^2 & \omega^2 & \omega^2 & \omega & \omega & \omega \\ 1 & \omega & \omega^2 & \omega^2 & 1 & \omega & \omega & \omega^2 & 1 \\ 1 & \omega^2 & \omega & \omega^2 & \omega & 1 & \omega & 1 & \omega^2 \end{bmatrix}.$$

For the matrix L with $[L]_{r,s} = (1/9)\,\chi_r\big(g_s^{-1}\big)$, for $1 \le r$, $s \le 9$, we obtain

$$(8.15) \qquad L = \frac{1}{9}
\begin{bmatrix}
1 & 1 & 1 & 1 & 1 & 1 & 1 & 1 & 1 \\
1 & \omega^1 & \omega & 1 & \omega^2 & \omega & 1 & \omega^2 & \omega \\
1 & \omega & \omega^2 & 1 & \omega & \omega^2 & 1 & \omega & \omega^2 \\
1 & 1 & 1 & \omega^2 & \omega^2 & \omega^2 & \omega & \omega & \omega \\
1 & \omega^2 & \omega & \omega^2 & \omega & 1 & \omega & 1 & \omega^2 \\
1 & \omega & \omega^2 & \omega^2 & 1 & \omega & \omega & \omega^2 & 1 \\
1 & 1 & 1 & \omega & \omega & \omega & \omega^2 & \omega^2 & \omega^2 \\
1 & \omega^2 & \omega & \omega & 1 & \omega^2 & \omega^2 & \omega & 1 \\
1 & \omega & \omega^2 & \omega & \omega^2 & 1 & \omega^2 & 1 & \omega
\end{bmatrix}.$$

Example 2.15 on page 20 and Example 3.12 on page 36 show that the matrix

$$(8.16) \qquad A =
\begin{bmatrix}
a_1 & a_2 & a_3 & a_4 & a_5 & a_6 & a_7 & a_8 & a_9 \\
a_3 & a_1 & a_2 & a_6 & a_4 & a_5 & a_9 & a_7 & a_8 \\
a_2 & a_3 & a_1 & a_5 & a_6 & a_4 & a_8 & a_9 & a_7 \\
a_7 & a_8 & a_9 & a_1 & a_2 & a_3 & a_4 & a_5 & a_6 \\
a_9 & a_7 & a_8 & a_3 & a_1 & a_2 & a_6 & a_4 & a_5 \\
a_8 & a_9 & a_7 & a_2 & a_3 & a_1 & a_5 & a_6 & a_4 \\
a_4 & a_5 & a_6 & a_7 & a_8 & a_9 & a_1 & a_2 & a_3 \\
a_6 & a_4 & a_5 & a_9 & a_7 & a_8 & a_3 & a_1 & a_2 \\
a_5 & a_6 & a_4 & a_8 & a_9 & a_7 & a_2 & a_3 & a_1
\end{bmatrix}$$

from (2.19) of page 20 is a group-pattern matrix for G and $\mathscr{L}$. We use Theorem 8.3 to see that the $n \times n$ matrix $D = LAM$ is a diagonal matrix and we employ (9.3) of Theorem 9.1 to obtain the diagonal components $d_1, d_2, \ldots, d_n$ of D from

$$[d_1, d_2, d_3, d_4, d_5, d_6, d_7, d_8, d_9,] = [a_1, a_2, a_3, a_4, a_5, a_6, a_7, a_8, a_9,]M$$

as

$$d_1 = [D]_{1,1} = a_1 + a_2 + a_3 + a_4 + a_5 + a_6 + a_7 + a_8 + a_9,$$
$$d_2 = [D]_{2,2} = a_1 + \omega a_2 + \omega^2 a_3 + a_4 + \omega a_5 + \omega^2 a_6 + a_7 + \omega a_8 + \omega^2 a_9,$$
$$d_3 = [D]_{3,3} = a_1 + \omega^2 a_2 + \omega a_3 + a_4 + \omega^2 a_5 + \omega a_6 + a_7 + \omega^2 a_8 + \omega a_9,$$
$$d_4 = [D]_{4,4} = a_1 + a_2 + a_3 + \omega a_4 + \omega a_5 + \omega a_6 + \omega^2 a_7 + \omega^2 a_8 + \omega^2 a_9,$$
$$d_5 = [D]_{5,5} = a_1 + \omega a_2 + \omega^2 a_3 + \omega a_4 + \omega^2 a_5 + a_6 + \omega^2 a_7 + a_8 + \omega a_9,$$
$$d_6 = [D]_{6,6} = a_1 + \omega^2 a_2 + \omega a_3 + \omega a_4 + a_5 + \omega^2 a_6 + \omega^2 a_7 + \omega a_8 + a_9,$$
$$d_7 = [D]_{7,7} = a_1 + a_2 + a_3 + \omega^2 a_4 + \omega^2 a_5 + \omega^2 a_6 + \omega a_7 + \omega a_8 + \omega a_9,$$
$$d_8 = [D]_{8,8} = a_1 + \omega a_2 + \omega^2 a_3 + \omega^2 a_4 + a_5 + \omega a_6 + \omega a_7 + \omega^2 a_8 + a_9,$$
$$d_9 = [D]_{9,9} = a_1 + \omega^2 a_2 + \omega a_3 + \omega^2 a_4 + \omega a_5 + a_6 + \omega a_7 + a_8 + \omega^2 a_9.$$

Of course, the same diagonal matrix is obtained when a version of *Mathematica* such as [48] is used to evaluate LAM. Namely, writing w for ω, we enter and evaluate representations M, L, and A for M in (8.14), L in (8.15), and A in (8.16). Then, we enter and evaluate

```
Expand[L.A.M]  /. {w^4->w, w^3->1, w^2->-1-w}  // Expand
```

to obtain a representation of D in which ω^2 has been replaced by $-1 - w$.

COROLLARY 8.5. *For an abelian group G of order n and a list $\mathscr{L}$ of its elements, suppose that A_1 and A_2 are $n \times n$ matrices over $\mathcal{F}$ that satisfy*

$$(8.17) \qquad\qquad A_1 M = M A_2.$$

Then, A_1 is a group-pattern matrix for G and $\mathscr{L}$ if and only if A_2 is a diagonal matrix. Moreover, A_2 is a group-pattern matrix for G and $\mathscr{L}$ if and only if A_1 is a diagonal matrix.

PROOF. (i). By writing (8.17) in the form $M^{-1}A_1 M = A_2$, we use Theorem 8.3 to see that A_1 is a group-pattern matrix for G and $\mathscr{L}$ if and only if A_2 is a diagonal matrix.

(ii). Since (8.6) and (8.8) show that M is a symmetric matrix, (8.17) yields

$$A_2^T M = A_2^T M^T = (M A_2)^T = (A_1 M)^T = M^T A_1^T = M A_1^T$$

and $M^{-1} A_2^T M = A_1^T$. Thus, Theorem 8.3 shows that A_2^T is a group-pattern matrix for G and $\mathscr{L}$ if and only if A_1^T is a diagonal matrix. Since Theorem 4.1 of page 41 shows that A_2^T is a group-pattern matrix for G and $\mathscr{L}$ if and only if A_2 is a group-pattern matrix for G and $\mathscr{L}$, we conclude that: A_2 is a group-pattern matrix for G and $\mathscr{L}$ if and only if A_1 is a diagonal matrix. This completes the proof. □

OBSERVATION 8.6. The diagonalizations of Theorem 8.3 and its Corollary 8.5 include as special cases the diagonalizations of Chapters 6 and 7. For an explanation of the symmetrical formulation in Corollary 8.5, see Subsection 9.0.3 on page 90.

8.2. The structure of $\mathcal{M}[\mathcal{F}, G, \mathscr{L}]$ when G is abelian

When $\mathfrak{F}$ is a field and $\mathscr{L}$ is a list of the elements in a finite group G, Theorem 4.2 on page 42 establishes that the set $\mathcal{M}[\mathfrak{F}, G, \mathscr{L}]$ of group-pattern matrices for G and $\mathscr{L}$ having components in $\mathfrak{F}$ forms a ring and a vector space over $\mathfrak{F}$ with respect to matrix addition, multiplication, and scalar multiplication. When $\mathfrak{F} = \mathcal{F}$ and G is abelian, more can be asserted. For that purpose, let $\mathfrak{D}_n(\mathcal{F})$ denote the set of $n \times n$ diagonal matrices having components in $\mathcal{F}$. We observe that $\mathfrak{D}_n(\mathcal{F})$ is both a ring and a vector space over $\mathcal{F}$ with respect to the matrix operations of addition, multiplication, and scalar multiplication by elements of $\mathcal{F}$.

THEOREM 8.7. *Suppose that $\mathscr{L}$ is a list of the elements in a finite abelian group G of order n and $\mathcal{F}$ is a field that contains a primitive nth root of unity having period n. Then, the ring and vector-space $\mathcal{M}[\mathcal{F}, G, \mathscr{L}]$ is isomorphic to the ring and vector space $\mathfrak{D}_n(\mathcal{F})$ of $n \times n$ diagonal matrices over $\mathcal{F}$.*

PROOF. After constructing M and $M^{-1} = L$ for Theorem 8.3, we use that result to see that a function Φ from $\mathcal{M}[\mathcal{F}, G, \mathscr{L}]$ onto $\mathfrak{D}_n(\mathcal{F})$ is defined by

$$\Phi(A) = M^{-1} A M, \quad \text{for each } A \text{ in } \mathcal{M}[\mathcal{F}, G, \mathscr{L}].$$

Moreover, Φ is clearly one-to-one. For A_1, A_2 in $\mathcal{M}[\mathcal{F}, G, \mathscr{L}]$ and c in $\mathcal{F}$, we have

$$\Phi(A_1 + A_2) = M^{-1}(A_1 + A_2)M = M^{-1}A_1 M + M^{-1}A_2 M = \Phi(A_1) + \Phi(A_2),$$

$$\Phi(A_1 A_2) = M^{-1}(A_1 A_2)M = (M^{-1}A_1 M)(M^{-1}A_2 M) = \Phi(A_1)\,\Phi(A_2),$$

$$\Phi(c A_1) = M^{-1}(c A_1)M = c\,(M^{-1}A_1 M) = c\,\Phi(A_1).$$

Thus, Φ is both a ring isomorphism of $\mathcal{M}[\mathcal{F}, G, \mathscr{L}]$ onto $\mathfrak{D}_n(\mathcal{F})$ and a vector-space isomorphism of $\mathcal{M}[\mathcal{F}, G, \mathscr{L}]$ onto $\mathfrak{D}_n(\mathcal{F})$. This completes the proof. □

8.3. The matrix $(1/n)M^2$

Most of the properties in Chapter 6 about circulant matrices can be extended to corresponding assertions about group-pattern matrices for abelian groups. Here, we include several. Others can be found in Theorem 17.7 on page 167.

PROPOSITION 8.8. *The $n \times n$ matrix N with components in $\mathcal{F}$ that is defined by*

$$(8.18) \qquad [N]_{r,s} = \begin{cases} 1, & \text{if } g_r g_s = e, \\ 0, & \text{if } g_r g_s \neq e, \end{cases} \qquad \text{for } r, s = 1, 2, \ldots, n,$$

is a permutation matrix such that $N^T = N$, $N^{-1} = N$, and $N = (1/n)M^2$.

PROOF. We use (8.18) to see that each row of N and each column of N has precisely one component equal to 1 and $n-1$ components equal to 0. Thus, N is an $n \times n$ permutation matrix. Since (8.18) yields $[N]_{s,r} = [N]_{r,s}$, for $1 \leq r, s \leq n$, we have $N^T = N$, and $N^{-1} = N^T = N$.

For $1 \leq r, s \leq n$, we find that (8.6), (8.8), (8.7), and (8.10) yield

$$\left[(1/n)M^2\right]_{r,s} = \sum_{k=1}^{n} (1/n) \left[M\right]_{r,k} \left[M\right]_{k,s} = (1/n) \sum_{k=1}^{n} \chi_r(g_k)\, \chi_k(g_s)$$

$$= (1/n) \sum_{k=1}^{n} \chi_k(g_r)\, \chi_k(g_s) = (1/n) \sum_{k=1}^{n} \chi_k(g_r g_s) = \begin{cases} 1, & \text{if } g_r g_s = e, \\ 0, & \text{if } g_r g_s \neq e, \end{cases} = [N]_{r,s}.$$

This shows that $N = (1/n)M^2$ and completes the proof. $\qquad \square$

OBSERVATION 8.9. For any group G of order n, we note that formula (8.18) defines a symmetric $n \times n$ permutation matrix. Additional properties are presented for it in Chapter 17 both for that more general context and for the case where G is abelian. However, for the context of this chapter where N is given by $N = (1/n)M^2$, there are other easily established identities in Exercise 1 on page 87.

PROPOSITION 8.10. *Let A be an $n \times n$ group-pattern matrix over $\mathcal{F}$ for an abelian group G and its list $\mathscr{L} = (g_1, g_2, \ldots, g_n)$. Then, A^T is a group-pattern matrix for G with $\mathscr{L}$ and $NAN = A^T$.*

PROOF. Theorem 4.1 on page 41 shows that A^T is a group-pattern matrix for G and $\mathscr{L}$. There is a function σ from G to $\mathcal{F}$ such that

$$(8.19) \qquad [A]_{r,s} = \sigma\big(g_r^{-1} g_s\big), \quad \text{for } r, s = 1, 2, \ldots, n.$$

Then, for $1 \leq r, s \leq n$, we obtain

$$[NAN]_{r,s} = \sum_{\mu=1}^{n} \sum_{\nu=1}^{n} [N]_{r,\mu} [A]_{\mu,\nu} [N]_{\nu,s} = \sum_{\mu=1}^{n} \sum_{\nu=1}^{n} [N]_{r,\mu} \sigma\big(g_\mu^{-1} g_\nu\big) [N]_{\nu,s}.$$

For $1 \leq \mu, \nu, r, s \leq n$, we have $[N]_{r,\mu} [N]_{\nu,s} \neq 0$ if and only if $g_r g_\mu = e = g_\nu g_s$, or equivalently, if and only if $g_\mu = g_r^{-1}$ and $g_\nu = g_s^{-1}$. With this, along with commutativity of group multiplication and (8.19), we find that

$$[NAN]_{r,s} = \sigma\big(g_r g_s^{-1}\big) = \sigma\big(g_s^{-1} g_r\big) = [A]_{s,r} = [A^T]_{r,s}, \quad \text{for } r, s = 1, 2, \ldots, n.$$

Thus, we have $NAN = A^T$. This completes the proof. $\qquad \square$

8.4. An abelian group of order n has precisely n group characters

Let G be an abelian group of order n.

DEFINITION 8.11. A function χ from G to the multiplicative group $\mathbb{C}^*$ of nonzero complex numbers is a **group character** for G when it satisfies

$$\chi(xy) = \chi(x)\,\chi(y), \quad \text{for each } x, y \text{ in } G.$$

The letter [20, pages 420–421] from Richard Dedekind to Georg Frobenius dated March 25, 1896 shows that Dedekind was already aware by February of 1886 that each finite abelian group G of order n possesses precisely n group characters. In this regard, see the next to last sentence in Subsection 21.0.1 on page 193.

On page 79, we defined n distinct functions $\chi_1, \chi_2, \ldots, \chi_n$ from G to the multiplicative group $\mathcal{F}^*$ of nonzero elements in $\mathcal{F}$. They satisfy

$$\chi_k(xy) = \chi_k(x)\,\chi_k(y), \quad \text{for } x, y \text{ in } G \text{ and } 1 \le k \le n.$$

Thus, when $\mathcal{F}$ is $\mathbb{C}$, the functions $\chi_1, \chi_2, \ldots, \chi_n$ are group characters for G.

PROPOSITION 8.12. *Suppose that χ is a function from G to $\mathcal{F}^*$ that satisfies*

$$(8.20) \qquad \chi(xy) = \chi(x)\,\chi(y), \quad \text{for each } x, y \text{ in } G.$$

Then, χ is one of the functions $\chi_1, \chi_2, \ldots, \chi_n$.

PROOF. We use (8.20) to obtain $\chi(e) = \chi(e\,e) = \chi(e)\,\chi(e)$ and $\chi(e) = 1$. The argument of page 79 shows that each element of G has a unique representation

$$(8.21) \qquad h_1^{\beta_1} h_2^{\beta_2} \cdots h_d^{\beta_d}, \quad \text{with } 0 \le \beta_i \le n_i - 1 \text{ for } 1 \le i \le d.$$

We set $\theta_i = \chi(h_i)$, for $1 \le i \le d$, and apply (8.20) with $(h_i)^{n_i} = e$ to deduce

$$(\theta_i)^{n_i} = \big(\chi(h_i)\big)^{n_i} = \chi\big(h_i^{n_i}\big) = \chi(e) = 1, \quad \text{for } i = 1, 2, \ldots, d.$$

Since $X^{n_i} - 1 = 0$ has n_i distinct roots $(\rho_i)^0, (\rho_i)^1, (\rho_i)^2, \ldots, (\rho_i)^{n_i-1}$ in $\mathcal{F}^*$, there are integers $\gamma_1, \gamma_2, \ldots, \gamma_d$ such that

$$\theta_i = (\rho_i)^{\gamma_i}, \quad \text{with } 0 \le \gamma_i \le n_i - 1 \text{ for } 1 \le i \le d.$$

Formulas (8.2) and (8.4) show that there is an integer k such that $1 \le k \le n$ and

$$\alpha_{1,k} = \gamma_1, \quad \alpha_{2,k} = \gamma_2, \quad \ldots, \quad \alpha_{d,k} = \gamma_d$$

For the function χ_k among $\chi_1, \chi_2, \ldots, \chi_n$ and for each $h_1^{\beta_1} h_2^{\beta_2} \cdots h_d^{\beta_d}$ of (8.21), we employ (8.5) and (8.20) to deduce

$$\chi_k\Big(h_1^{\beta_1} h_2^{\beta_2} \cdots h_d^{\beta_d}\Big) = \big(\rho_1^{\alpha_{1,k}}\big)^{\beta_1} \big(\rho_2^{\alpha_{2,k}}\big)^{\beta_2} \cdots \big(\rho_d^{\alpha_{d,k}}\big)^{\beta_d} = \big(\rho_1^{\gamma_1}\big)^{\beta_1} \big(\rho_2^{\gamma_2}\big)^{\beta_2} \cdots \big(\rho_d^{\gamma_d}\big)^{\beta_d}$$

$$= (\theta_1)^{\beta_1} (\theta_2)^{\beta_2} \cdots (\theta_d)^{\beta_d} = \big(\chi(h_1)\big)^{\beta_1} \big(\chi(h_2)\big)^{\beta_2} \cdots \big(\chi(h_d)\big)^{\beta_d}$$

$$= \chi\Big(h_1^{\beta_1} h_2^{\beta_2} \cdots h_d^{\beta_d}\Big).$$

This establishes that $\chi = \chi_k$ and completes the proof. $\qquad\square$

By using Proposition 8.12, we conclude that an abelian group G of order n has precisely n group characters. They are the functions $\chi_1, \chi_2, \ldots, \chi_n$ from G to $\mathbb{C}^*$ defined by (8.5) on page 79 when $\mathcal{F}$ is $\mathbb{C}$.

8.5. Factors for the determinant
of an abelian-group-pattern matrix

Let $\mathscr{L} = (g_1, g_2, \ldots, g_n)$ be a list of the elements in an abelian group G of order n. Let $\mathfrak{R}$ denote the the ring $\mathbb{C}[X_1, X_2, \ldots, X_n]$ of polynomials in the n variables $X_1, X_2, \ldots, X_n$ over $\mathbb{C}$. For an application of Theorem 8.3, let $\mathcal{F}$ be any field that contains $\mathfrak{R}$. Then, the components of M are elements of $\mathbb{C}$ and $\chi_1, \chi_2, \ldots, \chi_n$ are the n group characters for G. Let A be the $n \times n$ matrix having

$$(8.22) \qquad [A]_{r,s} = \sigma(g_r^{-1} g_s), \quad \text{for } r, s = 1, 2, \ldots, n,$$

where σ is the function from G to $\mathfrak{R}$ having $\sigma(g_k) = X_k$, for $1 \le k \le n$. For this, Theorem 8.3 and (8.13) yield the diagonalization $D = M^{-1}AM$ and

$$\det(A) = \det(MDM^{-1}) = \det(D) = \prod_{k=1}^{n} [D]_{k,k} = \prod_{k=1}^{n} \left[\sum_{\nu=1}^{n} \chi_k(g_\nu) X_\nu \right].$$

Thus, when $\psi_1, \psi_2, \ldots, \psi_n$ are the n group characters for G in some order, the determinant of A is equal to the product of the factors

$$\sum_{\nu=1}^{n} \psi_k(g_\nu) X_\nu, \quad \text{for } k = 1, 2, \ldots, n.$$

This corresponds to the right-hand member of the second formula on page 193.

In ;particular, if $g_1 = e$, then the first row of A is $(X_1, X_2, \ldots, X_n)$ and A is the standard group-pattern matrix for G and $\mathscr{L}$.

8.6. Brief observations

The diagonalization in Theorem 8.3 for group-pattern matrices requires the finite group to be abelian. It is the principal result presented in this chapter.

A generalization of Theorem 8.3 to the situation where G is any finite group is presented as Theorem 13.3 on page 125 for the situation where $\mathcal{F}$ is replaced by a field $\mathbb{F}$ that is either $\mathbb{C}$ or a proper field extension of $\mathbb{C}$.

By restricting the context for (8.22) so that $g_1 = e$, we see that each standard group-pattern matrix for an abelian group is diagonalizable. But, standard group-pattern matrices of nonabelian groups are not diagonalizable. That is shown by Theorem 2.18 on page 23 with $g_1 = e$.

EXERCISES

1. Use $N = (1/n)M^2$ and other properties of N, M, and L to verify that

$$N = nL^2, \quad M = nNL = nLN, \quad L = (1/n)NM = (1/n)MN,$$

$NMN = M$, $NLN = L$, $MNL = N$, and $LNM = N$.

2. Note that the matrix T_5 in (5.24) is equal to the matrix M in (8.14).

Specify suitable lists for the elements in the groups C_2, $C_2 \times C_2$, $C_2 \times C_2 \times C_2$, and $C_2 \times C_2 \times C_2 \times C_2$ such that M in (8.6) yields T_1 in (5.12), T_2 in (5.14), T_3 in (5.17), and T_4 in (5.19).

CHAPTER 9

Supplement to Chapter 8

9.0.1. A group-pattern matrix for G and $\mathscr{L}$ is conveniently specified by its first row when the first element of $\mathscr{L}$ is e. Namely, let A denote a group-pattern matrix for a finite group G of order n and a list $\mathscr{L} = (g_1, g_2, \ldots, g_n)$ of the n elements in G such that $g_1 = e$. Then, we have

$$(9.1) \qquad [A]_{r,s} = \sigma(g_r^{-1} g_s), \quad \text{for } r,\, s = 1,\, 2,\, \ldots,\, n,$$

where σ is some function from G to a set S. By letting $(a_1, a_2, \ldots, a_n)$ denote the first row of A and setting $r = 1$ in (9.1), we see that σ for (9.1) can be identified with the function from G to $S = \{a_1, a_2, \ldots, a_n\}$ having

$$(9.2) \qquad \sigma(g_k) = a_k, \quad \text{for } k = 1,\, 2,\, \ldots,\, n.$$

Of course, an $n \times n$ diagonal matrix D is uniquely specified by the $1 \times n$ matrix $[d_1, d_2, \ldots, d_n]$ where $d_k = [D]_{k,k}$, for $1 \le k \le n$.

Next, we show how these observations enable Theorem 8.3 to be restated.

9.0.2. Supplement for Theorem 8.3 on page 81.

THEOREM 9.1. *For an abelian group G of order n, let $\mathscr{L} = (g_1, g_2, \ldots, g_n)$ be a list for G such that $g_1 = e$. If an $n \times n$ matrix A over $\mathcal{F}$ is a group-pattern matrix for G and $\mathscr{L}$, then $M^{-1}AM$ is the diagonal matrix D given by*

$$(9.3) \qquad [d_1, d_2, \ldots, d_n] = [a_1, a_2, \ldots, a_n] M,$$

where $(a_1, a_2, \ldots, a_n)$ is the first row of A. Moreover, if an $n \times n$ matrix D over $\mathcal{F}$ is a diagonal matrix, then MDM^{-1} is the group-pattern matrix A for G and $\mathscr{L}$ whose first row is provided by

$$(9.4) \qquad [a_1, a_2, \ldots, a_n] = [d_1, d_2, \ldots, d_n] M^{-1}.$$

PROOF. (i) Let A be a group-pattern matrix for G and $\mathscr{L}$. Then, Theorem 8.3 shows that $D = M^{-1}AM$ is a diagonal matrix. Thus, (8.13), (8.6), and (9.2) yield

$$(9.5) \qquad d_k = [D]_{k,k} = \sum_{\nu=1}^{n} \chi_k(g_\nu)\, \sigma(g_\nu) = \sum_{\nu=1}^{n} [M]_{k,\nu}\, a_\nu, \quad \text{for } k = 1,\, 2,\, \ldots,\, n.$$

We express (9.5) as an equality of two $n \times 1$ matrices to obtain

$$(9.6) \qquad [d_1, d_2, \ldots, d_n]^T = M[a_1, a_2, \ldots, a_n]^T.$$

Since the definition (8.6) of M yields $M^T = M$, we transpose (9.6) to obtain (9.3).

(ii) Suppose D is a diagonal matrix. Theorem 8.3 shows that the matrix A defined by $A = MDM^{-1}$ is a group-pattern matrix for G and $\mathscr{L}$. Since we have $D = M^{-1}AM$, Part (i) of this proof shows that (9.3) is valid. By multiplying (9.3) on the right by M^{-1}, we obtain (9.4) and complete the proof. $\qquad \square$

9.0.3. Explanation for the symmetrical formulation of Corollary 8.5.

In terms of (8.2)–(8.5), we introduce $f(r, s) = \sum_{k=1}^{d} \dfrac{n\,\alpha_{k,r}\,\alpha_{k,s}}{n_k}$, for $1 \le r, s \le n$.

Let $F(\rho)$ be the $n \times n$ matrix having $\big[F(\rho)\big]_{r,s} = \rho^{f(r,s)}$, for $1 \le r, s, \le n$.

PROPOSITION 9.2. *For M, L of* Chapter 8, $M = F(\rho)$ *and* $L = (1/n)\,F(1/\rho)$.

PROOF. We use (8.6), (8.5), and $g_s = h_1^{\alpha_{1,s}} h_2^{\alpha_{2,s}} \cdots h_d^{\alpha_{d,s}}$ to obtain

$$
\begin{aligned}
\big[M\big]_{r,s} &= \chi_r(g_s) = \rho_1^{\alpha_{1,r}\alpha_{1,s}} \rho_2^{\alpha_{2,r}\alpha_{2,s}} \cdots \rho_d^{\alpha_{d,r}\alpha_{d,s}} \\
&= \left(\rho^{n/n_1}\right)^{\alpha_{1,r}\alpha_{1,s}} \left(\rho^{n/n_2}\right)^{\alpha_{2,r}\alpha_{2,s}} \cdots \left(\rho^{n/n_d}\right)^{\alpha_{d,r}\alpha_{d,s}} \\
&= \rho^{f(r,s)} = \big[F(\rho)\big]_{r,s}, \quad \text{for } 1 \le r, s \le n.
\end{aligned}
$$

This yields $M = F(\rho)$

We employ (8.12), (8.5), and $g_s^{-1} = h_1^{-\alpha_{1,s}} h_2^{-\alpha_{2,s}} \cdots h_d^{-\alpha_{d,s}}$ with the observation at the top of page 80 to deduce that

$$
\begin{aligned}
\big[nL\big]_{r,s} &= \chi_r(g_s^{-1}) = \rho_1^{-\alpha_{1,r}\alpha_{1,s}} \rho_2^{-\alpha_{2,r}\alpha_{2,s}} \cdots \rho_d^{-\alpha_{d,r}\alpha_{d,s}} \\
&= \left(\rho^{n/n_1}\right)^{-\alpha_{1,r}\alpha_{1,s}} \left(\rho^{n/n_2}\right)^{-\alpha_{2,r}\alpha_{2,s}} \cdots \left(\rho^{n/n_d}\right)^{-\alpha_{d,r}\alpha_{d,s}} \\
&= \rho^{-f(r,s)} = \big[F(1/\rho)\big]_{r,s}, \quad \text{for } 1 \le r, s \le n.
\end{aligned}
$$

Thus, we have $nL = F(1/\rho)$ and $L = (1/n)\,F(1/\rho)$. This completes the proof. $\square$

The key observation is that: *the primitive nth root ρ of unity in $\mathcal{F}$ having period n specifies $1/\rho$ as a primitive nth root of unity in $\mathcal{F}$ having period n.* Thus, the results of Chapter 8 remain valid when ρ is replaced throughout with $\widehat{\rho} = 1/\rho$.

EXPLANATION. For an $n \times n$ matrix A over $\mathcal{F}$ and $M^{-1} = L$, we observe that

(9.7) $M^{-1}AM = (1/n)\,F(1/\rho)\,A\,F(\rho)$ and $(1/n)\,F(\rho)\,A\,F(1/\rho) = MAM^{-1}$.

For the context of Theorem 8.3, A is a group-pattern matrix for G and $\mathcal{L}$ if and only if $M^{-1}AM$ is a diagonal matrix. Since $(1/n)\,F(1/\rho)\,A\,F(\rho)$ is a diagonal matrix if and only if $(1/n)\,F(\rho)\,A\,F(1/\rho)$ is a diagonal matrix, we use (9.7) to see that: A is a group-patten matrix for G and $\mathcal{L}$ if and only if MAM^{-1} is a diagonal matrix.

CHAPTER 10

Historical Perspective, Part 2

10.1. Gruppendeterminanten of Dedekind and Frobenius

Richard Dedekind described in two letters to Georg Frobenius dated March 25 of 1896 and April 6 of 1896 results of his from 1886. For each finite group of order n, he had introduced a homogeneous polynomial H of degree n in n variables that he called the Gruppendeterminante of G. For the letters, see pages 193–195.

We use the transpose of the matrix for H on page 193, to rewrite H as

$$(10.1) \qquad H = \begin{vmatrix} X_{11'} & X_{12'} & \cdots & X_{1n'} \\ X_{21'} & X_{22'} & \cdots & X_{2n'} \\ \vdots & \vdots & \vdots & \vdots \\ X_{n1'} & X_{n2'} & \cdots & X_{nn'} \end{vmatrix}.$$

Dedekind denoted in some order the elements of G by $1, 2, \ldots, n$ as symbols (not as integers) and he denoted their respective inverses by $1', 2', \ldots, n'$. The elements $1, 2, \ldots, n$ of G were then employed as subscripts to create corresponding variables $X_1, X_2, \ldots, X_n$ over the field $\mathbb{C}$ of complex numbers. For a product $rs' = t$ in G, the symbol $X_{rs'}$ was defined to be X_t. Then, the expansion of H in (10.1) yields H as a homogeneous polynomial of degree n in the variables $X_1, X_2, \ldots, X_n$ over $\mathbb{C}$.

The notation that Georg Frobenius employed to define Gruppendeterminanten was presented in [**24**, page 1343] and may be viewed on pages 197. He used letters $A, B, C, \ldots$ to represent the elements in a group G of order n and introduced corresponding variables $X_A, X_B, X_C, \ldots$. For elements M, N in G he defined X_{MN} by $X_{MN} = X_L$, where L is the product $L = MN$ in G. Then, with respect to $X_{P,Q} = X_{PQ^{-1}}$, he used the formula

$$(10.2) \qquad \Theta = \left| X_{P,Q} \right| = \left| X_{PQ^{-1}} \right|$$

to define the *Gruppendeterminante* for G as the homogeneous polynomial Θ of degree n in the variables $X_A, X_B, X_C, \ldots$ over $\mathbb{C}$.

Of course, for (10.2) to be applied, a definite list such as $(h_1, h_2, \ldots, h_n)$ is needed for the n elements of G. Then, we have

$$(10.3) \qquad \Theta = \det(B), \quad \text{where} \quad B = \begin{bmatrix} X_{h_1 h_1^{-1}} & X_{h_1 h_2^{-1}} & \cdots & X_{h_1 h_n^{-1}} \\ X_{h_2 h_1^{-1}} & X_{h_2 h_2^{-1}} & \cdots & X_{h_2 h_n^{-1}} \\ \vdots & \vdots & \vdots & \vdots \\ X_{h_n h_1^{-1}} & X_{h_n h_2^{-1}} & \cdots & X_{h_n h_n^{-1}} \end{bmatrix}.$$

The notation (10.2) implies that Θ is well-defined by (10.2) independently of the particular list $(h_1, h_2, \ldots, h_n)$ used to specify the Gruppenmatrix B in (10.3). A verification of that is provided by the next result.

THEOREM 10.1. *A unique polynomial Θ is well defined by means of (10.2) and any list $(h_1, h_2, \ldots, h_n)$ of the element in G for (10.3).*

PROOF. In addition to the list $\mathscr{L}_1 = (h_1, h_2, \ldots, h_n)$ that specifies B in (10.3), let $\mathscr{L}_2 = (g_1, g_2, \ldots, g_n)$ be any list of the elements in G. Then, Θ is well defined if and only if Θ in (10.3) is also given by

$$(10.4) \qquad \Theta = \det(A), \quad \text{where} \quad A = \begin{bmatrix} X_{g_1 g_1^{-1}} & X_{g_1 g_2^{-1}} & \cdots & X_{g_1 g_n^{-1}} \\ X_{g_2 g_1^{-1}} & X_{g_2 g_2^{-1}} & \cdots & X_{g_2 g_n^{-1}} \\ \vdots & \vdots & \vdots & \vdots \\ X_{g_n g_1^{-1}} & X_{g_n g_2^{-1}} & \cdots & X_{g_n g_n^{-1}} \end{bmatrix}.$$

We use $\mathscr{L}_1$ and $\mathscr{L}_2$ to define a permutation π of $\{1, 2, \ldots, n\}$ such that

$$(10.5) \qquad g_k = h_{\pi(k)}, \quad \text{for } k = 1, 2, \ldots, n.$$

Let P be the $n \times n$ permutation matrix that is defined in terms of (10.5) by

$$(10.6) \qquad [P]_{r,s} = \delta\big(\pi(r), s\big) = \begin{cases} 1, & \text{if } \pi(r) = s, \\ 0, & \text{if } \pi(r) \neq s, \end{cases} \qquad \text{for } r, s = 1, 2, \ldots, n.$$

For $r, s = 1, 2, \ldots, n$, we use (10.6), (10.3), (10.5), and (10.4) to obtain

$$\begin{aligned} \left[PBP^T\right]_{r,s} &= \sum_{\mu=1}^{n} \sum_{\nu=1}^{n} [P]_{r,\mu} [B]_{\mu,\nu} [P^T]_{\nu,s} = \sum_{\mu=1}^{n} \sum_{\nu=1}^{n} [P]_{r,\mu} [B]_{\mu,\nu} [P]_{s,\nu} \\ &= \sum_{\mu=1}^{n} \sum_{\nu=1}^{n} \delta\big(\pi(r), \mu\big) [B]_{\mu,\nu} \, \delta\big(\pi(s), \nu\big) = [B]_{\pi(r), \pi(s)} \\ &= X_{h_{\pi(r)} h_{\pi(s)}^{-1}} = X_{g_r g_s^{-1}} = [A]_{r,s}. \end{aligned}$$

Hence, we have $A = PBP^T$ and $\det(A) = \det(P)\det(B)\det(P^{-1}) = \det(B)$. This completes the proof. $\qquad\qquad\square$

EXAMPLE 10.2. With multiplication, the set $\{1, i, -1, -i\}$ of four complex numbers forms a cyclic group G of order 4. Its elements specify corresponding variables denoted by X_1, X_i, X_{-1}, X_{-i}. The lists

$$(h_1, h_2, h_3, h_4) = (1, i, -1, -i) \quad \text{and} \quad (g_1, g_2, g_3, g_4) = (1, i, -i, -1)$$

specify B for (10.3) and A for (10.4) as

$$B = \begin{bmatrix} X_1 & X_{-i} & X_{-1} & X_i \\ X_i & X_1 & X_{-i} & X_{-1} \\ X_{-1} & X_i & X_1 & X_{-i} \\ X_{-i} & X_{-1} & X_i & X_1 \end{bmatrix} \quad \text{and} \quad A = \begin{bmatrix} X_1 & X_{-i} & X_i & X_{-1} \\ X_i & X_1 & X_{-1} & X_{-i} \\ X_{-i} & X_{-1} & X_1 & X_i \\ X_{-1} & X_i & X_{-i} & X_1 \end{bmatrix}.$$

The corresponding determinants of these 4×4 matrices are given by

$$\begin{aligned} \det(B) = \, &X_1^4 + X_{-1}^4 - X_i^4 - X_{-i}^4 - 2X_1^2 X_{-1}^2 + 2X_i^2 X_{-i}^2 \\ &- 4X_1^2 X_i X_{-1} - 4X_{-1}^2 X_i X_{-i} + 4X_i^2 X_1 X_{-1} + 4X_{-i}^2 X_1 X_{-1} \end{aligned}$$

and $\det(A) = \det(B)$. We note that A is obtained from B by interchanging the last two rows of B and the last two columns of B. This example illustrates the notation of Georg Frobenius and the argument for Theorem 10.1.

10.2. Dedekind's factorization of the Gruppendeterminante for $\mathfrak{S}_3$

To learn about the discoveries involving (10.1) made by Richard Dedekind, Subsections 21.0.1 and 21.0.2 on pages 193–195 may be read before continuing. It is necessary to point out that Dedekind and Frobenius computed the product of two permutation symbols by reading them from left to right.

The notation Richard Dedekind introduced for (10.1) was employed in his letter [20, pages 423–425], dated April 6 of 1896, to deduce the Gruppendeterminante H for the nonabelian symmetric group $\mathfrak{S}_3$ consisting of the six permutations of three objects. As indicated on page 194, he obtained

$$(10.7) \quad H = \begin{vmatrix} x_1 & x_3 & x_2 & x_4 & x_5 & x_6 \\ x_2 & x_1 & x_3 & x_5 & x_6 & x_4 \\ x_3 & x_2 & x_1 & x_6 & x_4 & x_5 \\ x_4 & x_5 & x_6 & x_1 & x_3 & x_2 \\ x_5 & x_6 & x_4 & x_2 & x_1 & x_3 \\ x_6 & x_4 & x_5 & x_3 & x_2 & x_1 \end{vmatrix} = (u+v)(u-v)(u_1 u_2 - v_1 v_2)^2,$$

where

$$(10.8) \quad \begin{cases} u = x_1 + x_2 + x_3, & v = x_4 + x_5 + x_6, \\ u_1 = x_1 + \omega x_2 + \omega^2 x_3, & v_1 = x_4 + \omega x_5 + \omega^2 x_6, \\ u_2 = x_1 + \omega^2 x_2 + \omega x_3, & v_2 = x_4 + \omega^2 x_5 + \omega x_6 \end{cases}$$

with $\omega^2 + \omega + 1 = 0$ and $\omega^3 = 1$. Dedekind indicated that a verification of (10.7) can be made by multiplying H with $\det Q$ where

$$(10.9) \quad Q = \begin{bmatrix} 1 & 1 & 1 & 1 & 1 & 1 \\ 1 & \omega & \omega^2 & 1 & \omega & \omega^2 \\ 1 & \omega^2 & \omega & 1 & \omega^2 & \omega \\ 1 & 1 & 1 & -1 & -1 & -1 \\ 1 & \omega & \omega^2 & -1 & -\omega & -\omega^2 \\ 1 & \omega^2 & \omega & -1 & -\omega^2 & -\omega \end{bmatrix} \quad \text{and} \quad \det(Q) = 216 \neq 0.$$

To explain one way that can be done, we define the matrix

$$(10.10) \quad R = \begin{bmatrix} u+v & 0 & 0 & 0 & 0 & 0 \\ 0 & u_1 & v_2 & 0 & 0 & 0 \\ 0 & v_1 & u_2 & 0 & 0 & 0 \\ 0 & 0 & 0 & u-v & 0 & 0 \\ 0 & 0 & 0 & 0 & u_1 & -v_2 \\ 0 & 0 & 0 & 0 & -v_1 & u_2 \end{bmatrix}$$

in terms of (10.8) and verify that

$$(10.11) \qquad H \det(Q) = \det(QR) = \det(Q) \det(R)$$

is a valid identity. It yields $H = \det(R)$. Thus, (10.7) is valid.

PROPOSITION 10.3. *The quadratic polynomial factor*

$$(10.12) \qquad u_1 u_2 - v_1 v_2 = + x_1^2 + x_2^2 + x_3^2 - x_1 x_2 - x_1 x_3 - x_2 x_3$$
$$- x_4^2 - x_5^2 - x_6^2 + x_4 x_5 + x_4 x_6 + x_5 x_6$$

in (10.7) *is irreducible over* $\mathbb{C}$.

PROOF. For an indirect argument, suppose (10.12) is not irreducible over $\mathbb{C}$. Then, there are complex numbers α_1, α_2, α_3, $\ldots$, α_6, β_1, β_2, β_3, $\ldots$, β_6 such that

$$(10.13) \qquad u_1 u_2 - v_1 v_2 = \left(\alpha_1 x_1 + \alpha_2 x_2 + \alpha_3 x_3 + \cdots + \alpha_6 x_6 \right)$$
$$\times \left(\beta_1 x_1 + \beta_2 x_2 + \beta_3 x_3 + \cdots + \beta_6 x_6 \right).$$

The coefficients of x_1^2 in (10.12) and (10.13) show that $\alpha_1 \beta_1 = 1$. By respectively multiplying the two factors in (10.13) with $1/\alpha_1$ and α_1, we henceforth assume that $\alpha_1 = \beta_1 = 1$. Next, we equate the coefficients of x_2^2 and $x_1 x_2$ in (10.12) and (10.13) to deduce that $\alpha_2 \beta_2 = 1$ and $\alpha_2 + \beta_2 = -1$. This requires $\alpha_2^2 + \alpha_2 + 1 = 0$ and $\alpha_2 \beta_2 = 1$. Thus, we henceforth assume the notation for (10.13) has been selected so that $\alpha_1 = \beta_1 = 1$, $\alpha_2 = \omega$, and $\beta_2 = \omega^2$. Now, we equate the coefficients of x_4^2, $x_1 x_4$, and $x_2 x_4$ to obtain the contradiction that the conditions

$$\alpha_4 \beta_4 = -1,$$
$$\alpha_4 + \beta_4 = 0,$$
$$\omega^2 \alpha_4 + \omega \beta_4 = 0$$

are satisfied by complex numbers α_4 and β_4. Thus, the polynomial in (10.12) is irreducible over $\mathbb{C}$. This completes the proof. $\qquad\square$

10.3. A reinterpretation for the preceding results of Dedekind

Rather than focus only on determinants, it is natural to introduce the matrix

$$(10.14) \qquad X = \begin{bmatrix} x_1 & x_3 & x_2 & x_4 & x_5 & x_6 \\ x_2 & x_1 & x_3 & x_5 & x_6 & x_4 \\ x_3 & x_2 & x_1 & x_6 & x_4 & x_5 \\ x_4 & x_5 & x_6 & x_1 & x_3 & x_2 \\ x_5 & x_6 & x_4 & x_2 & x_1 & x_3 \\ x_6 & x_4 & x_5 & x_3 & x_2 & x_1 \end{bmatrix}$$

so that, in place of (10.11), we have $XQ = QR$ and therefore

$$(10.15) \qquad Q^{-1} X Q = R,$$

with respect to Q in (10.9) and R in (10.10). We employ the notation

$$(10.16) \qquad \begin{cases} h_1 = \begin{pmatrix} 1\,2\,3 \\ 1\,2\,3 \end{pmatrix}, \quad h_2 = \begin{pmatrix} 1\,2\,3 \\ 2\,3\,1 \end{pmatrix}, \quad h_3 = \begin{pmatrix} 1\,2\,3 \\ 3\,1\,2 \end{pmatrix}, \\[2mm] h_4 = \begin{pmatrix} 1\,2\,3 \\ 1\,3\,2 \end{pmatrix}, \quad h_5 = \begin{pmatrix} 1\,2\,3 \\ 3\,2\,1 \end{pmatrix}, \quad h_6 = \begin{pmatrix} 1\,2\,3 \\ 2\,1\,3 \end{pmatrix} \end{cases}$$

for the elements of the group $\mathfrak{S}_3$ of Section 1.5. Their multiplication table

$$(10.17)$$

$\cdot$	h_1	h_3	h_2	h_4	h_5	h_6
h_1	h_1	h_3	h_2	h_4	h_5	h_6
h_2	h_2	h_1	h_3	h_5	h_6	h_4
h_3	h_3	h_2	h_1	h_6	h_4	h_5
h_4	h_4	h_5	h_6	h_1	h_3	h_2
h_5	h_5	h_6	h_4	h_2	h_1	h_3
h_6	h_6	h_4	h_5	h_3	h_2	h_1

is based upon evaluating products of permutation symbols by reading them from left to right so that (10.17) is consistent with Dedekind's multiplication table for the first example of Subsection 21.0.2. For $k = 1, 2, \ldots, 6$, let P_k denote the 6×6 permutation matrix that is obtained from X in (10.14) by replacing each x_k in X with 1 and replacing each x_j having $1 \leq j \leq 6$ and $j \neq k$ with 0. This yields

$$P_1 = \begin{bmatrix} 1 & 0 & 0 & 0 & 0 & 0 \\ 0 & 1 & 0 & 0 & 0 & 0 \\ 0 & 0 & 1 & 0 & 0 & 0 \\ 0 & 0 & 0 & 1 & 0 & 0 \\ 0 & 0 & 0 & 0 & 1 & 0 \\ 0 & 0 & 0 & 0 & 0 & 1 \end{bmatrix}, \quad P_2 = \begin{bmatrix} 0 & 0 & 1 & 0 & 0 & 0 \\ 1 & 0 & 0 & 0 & 0 & 0 \\ 0 & 1 & 0 & 0 & 0 & 0 \\ 0 & 0 & 0 & 0 & 0 & 1 \\ 0 & 0 & 0 & 1 & 0 & 0 \\ 0 & 0 & 0 & 0 & 1 & 0 \end{bmatrix}, \quad P_3 = \begin{bmatrix} 0 & 1 & 0 & 0 & 0 & 0 \\ 0 & 0 & 1 & 0 & 0 & 0 \\ 1 & 0 & 0 & 0 & 0 & 0 \\ 0 & 0 & 0 & 0 & 1 & 0 \\ 0 & 0 & 0 & 0 & 0 & 1 \\ 0 & 0 & 0 & 1 & 0 & 0 \end{bmatrix},$$

$$P_4 = \begin{bmatrix} 0 & 0 & 0 & 1 & 0 & 0 \\ 0 & 0 & 0 & 0 & 0 & 1 \\ 0 & 0 & 0 & 0 & 1 & 0 \\ 1 & 0 & 0 & 0 & 0 & 0 \\ 0 & 0 & 1 & 0 & 0 & 0 \\ 0 & 1 & 0 & 0 & 0 & 0 \end{bmatrix}, \quad P_5 = \begin{bmatrix} 0 & 0 & 0 & 0 & 1 & 0 \\ 0 & 0 & 0 & 1 & 0 & 0 \\ 0 & 0 & 0 & 0 & 0 & 1 \\ 0 & 1 & 0 & 0 & 0 & 0 \\ 1 & 0 & 0 & 0 & 0 & 0 \\ 0 & 0 & 1 & 0 & 0 & 0 \end{bmatrix}, \quad P_6 = \begin{bmatrix} 0 & 0 & 0 & 0 & 0 & 1 \\ 0 & 0 & 0 & 0 & 1 & 0 \\ 0 & 0 & 0 & 1 & 0 & 0 \\ 0 & 0 & 1 & 0 & 0 & 0 \\ 0 & 1 & 0 & 0 & 0 & 0 \\ 1 & 0 & 0 & 0 & 0 & 0 \end{bmatrix}.$$

When $1 \leq i, j, k \leq 6$, the relation $h_i h_j = h_k$ is satisfied if and only if $P_i P_j = P_k$; e.g., modify Theorem 2.16 to this context. For Q in (10.9), we set

(10.18) $$R_k = Q^{-1} P_k Q, \quad \text{for } k = 1, 2, \ldots, 6.$$

With blocks $Z_1, Z_2, \ldots, Z_{12}$ of zeros, the block-diagonal matrices R_k are given by

(10.19) $$R_k = \left[\begin{array}{c|c|c|c} U_k & Z_1 & Z_2 & Z_3 \\ \hline Z_4 & W_k & Z_5 & Z_6 \\ \hline Z_7 & Z_8 & V_k & Z_9 \\ \hline Z_{10} & Z_{11} & Z_{12} & T_k \end{array} \right], \quad \text{for } k = 1, 2, \ldots, 6,$$

where we have

(10.20) $$U_k = [1], \quad \text{for } k = 1, 2, \ldots, 6,$$

(10.21) $$V_k = \begin{cases} [1], & \text{for } k = 1, 2, 3, \\ [-1], & \text{for } k = 4, 5, 6, \end{cases}$$

(10.22) $$\begin{cases} W_1 = \begin{bmatrix} 1 & 0 \\ 0 & 1 \end{bmatrix}, \quad W_2 = \begin{bmatrix} \omega^2 & 0 \\ 0 & \omega \end{bmatrix}, \quad W_3 = \begin{bmatrix} \omega & 0 \\ 0 & \omega^2 \end{bmatrix}, \\ W_4 = \begin{bmatrix} 0 & 1 \\ 1 & 0 \end{bmatrix}, \quad W_5 = \begin{bmatrix} 0 & \omega^2 \\ \omega & 0 \end{bmatrix}, \quad W_6 = \begin{bmatrix} 0 & \omega \\ \omega^2 & 0 \end{bmatrix}, \end{cases}$$

(10.23) $$T_k = \begin{cases} W_k, & \text{for } k = 1, 2, 3, \\ -W_k, & \text{for } k = 4, 5, 6. \end{cases}$$

We use (10.18) and properties of $P_1, P_2, \ldots, P_6$ to deduce that, for $1 \leq i, j, k \leq 6$, the relation $h_i h_j = h_k$ is satisfied if and only if $R_i R_j = R_k$. We apply this with (10.19)–(10.23) to verify that: for $1 \leq i, j, k \leq 6$, if $h_i h_j = h_k$, then $U_i U_j = U_k$, $V_i V_j = V_k$, $W_i W_j = W_k$, and $T_i T_j = T_k$. Thus, each of the assignments $h_k \mapsto P_k$, $h_k \mapsto R_k$, $h_k \mapsto U_k$, $h_k \mapsto V_k$, $h_k \mapsto W_k$, $h_k \mapsto T_k$, for $1 \leq k \leq n$, specifies a *matrix representation* for $\mathfrak{S}_3$; e.g., see Definition 10.4 and Example 10.6.

10.4. Matrix identity for (10.7) developed by Frobenius

When Georg Frobenius presented in [**25**, pages 1007–1008] the factorization (10.7) of Section 10.2 as a result that Richard Dedekind had given him in a letter dated April 6, 1896 (see page 198), he did so by emphasizing a remarkable relation involving the matrix

$$(10.24) \qquad X = \begin{bmatrix} x_1 & x_3 & x_2 & x_4 & x_5 & x_6 \\ x_2 & x_1 & x_3 & x_5 & x_6 & x_4 \\ x_3 & x_2 & x_1 & x_6 & x_4 & x_5 \\ x_4 & x_5 & x_6 & x_1 & x_3 & x_2 \\ x_5 & x_6 & x_4 & x_2 & x_1 & x_3 \\ x_6 & x_4 & x_5 & x_3 & x_2 & x_1 \end{bmatrix}$$

and the matrix

$$(10.25) \qquad L = \begin{bmatrix} 1 & -1 & 1 & 0 & 0 & 1 \\ 1 & -1 & \omega^2 & 0 & 0 & \omega \\ 1 & -1 & \omega & 0 & 0 & \omega^2 \\ 1 & 1 & 0 & 1 & 1 & 0 \\ 1 & 1 & 0 & \omega & \omega^2 & 0 \\ 1 & 1 & 0 & \omega^2 & \omega & 0 \end{bmatrix}, \quad \text{with } \det(L) = 54.$$

Namely, L and X yield

$$(10.26) \qquad L^{-1}XL = \begin{bmatrix} u+v & 0 & 0 & 0 & 0 & 0 \\ 0 & u-v & 0 & 0 & 0 & 0 \\ 0 & 0 & u_1 & v_1 & 0 & 0 \\ 0 & 0 & v_2 & u_2 & 0 & 0 \\ 0 & 0 & 0 & 0 & u_1 & v_1 \\ 0 & 0 & 0 & 0 & v_2 & u_2 \end{bmatrix},$$

where u, v, u_1, u_2, v_1, v_2, are defined by (10.8) on page 93. Of course, this gives $\det(X) = \det(L^{-1}XL) = (u+v)(u-v)(u_1 u_2 - v_1 v_2)^2$ as the factorization (10.7).

We observe that the thirty-six components of L are the thirty-six components of the matrices U_k, V_k, W_k, for $1 \leq k \leq 6$, in (10.20), (10.21), and (10.22).

It is natural to wonder whether the construction by Frobenius of the matrix L to obtain the remarkable identity (10.26) can be generalized and made algorithmic. The answer is yes and Theorem 13.3 of page 125 provides the details.

10.5. Matrix representations of finite groups and some properties

The proof of Theorem 13.3 is based upon results about matrix representations for finite groups that we present in this section.

Historical details in [**17**] explain how the research of Frobenius in [**23, 24, 25**] or [**26**] of 1896–1897 led to extensive investigations abut group representations; e.g., see [**18**] and [**31**, Chapter 18]. However, to present the main ideas clearly, it suffices to consider only matrix representations that have complex components.

DEFINITION 10.4. A function f from a group G to a group G_1 is said to be a **homomorphism** of G into G_1 when $f(xy) = f(x)\,f(y)$, for each x, y in G. A **matrix representation** for G is a homomorphism of G into a group G_1 where the elements of G_1 are nonsingular $d \times d$ matrices over $\mathbb{C}$ and the multiplication for G_1 is that of matrices. The positive integer d is the **degree** of the representation.

EXAMPLE 10.5. Let G be the group $\mathfrak{S}_3$ that consists of the six permutations

(10.27)
$$\begin{cases} h_1 = \begin{pmatrix} 1\,2\,3 \\ 1\,2\,3 \end{pmatrix}, & h_2 = \begin{pmatrix} 1\,2\,3 \\ 2\,3\,1 \end{pmatrix}, & h_3 = \begin{pmatrix} 1\,2\,3 \\ 3\,1\,2 \end{pmatrix}, \\[2mm] h_4 = \begin{pmatrix} 1\,2\,3 \\ 1\,3\,2 \end{pmatrix}, & h_5 = \begin{pmatrix} 1\,2\,3 \\ 3\,2\,1 \end{pmatrix}, & h_6 = \begin{pmatrix} 1\,2\,3 \\ 2\,1\,3 \end{pmatrix}. \end{cases}$$

They have the multiplication table (10.17) when products are computed from left to right. For $1 \le k \le 6$, let P_k denote the 3×3 permutation matrix obtained from I_3 by performing the row replacements for I_3 specified by h_k. Thus, we obtain

(10.28)
$$\begin{cases} P_1 = \begin{bmatrix} 1 & 0 & 0 \\ 0 & 1 & 0 \\ 0 & 0 & 1 \end{bmatrix}, & P_2 = \begin{bmatrix} 0 & 1 & 0 \\ 0 & 0 & 1 \\ 1 & 0 & 0 \end{bmatrix}, & P_3 = \begin{bmatrix} 0 & 0 & 1 \\ 1 & 0 & 0 \\ 0 & 1 & 0 \end{bmatrix}, \\[4mm] P_4 = \begin{bmatrix} 1 & 0 & 0 \\ 0 & 0 & 1 \\ 0 & 1 & 0 \end{bmatrix}, & P_5 = \begin{bmatrix} 0 & 0 & 1 \\ 0 & 1 & 0 \\ 1 & 0 & 0 \end{bmatrix}, & P_6 = \begin{bmatrix} 0 & 1 & 0 \\ 1 & 0 & 0 \\ 0 & 0 & 1 \end{bmatrix}. \end{cases}$$

A direct computation shows that that the six matrices of (10.28) form a group G_1 of order 6 and that the function F from G onto G_1 defined by

(10.29)
$$F(h_k) = P_k, \quad \text{for } 1 \le k \le 6,$$

satisfies

$$F(h_i h_j) = F(h_i)F(h_j), \quad \text{for } 1 \le i, j \le 6.$$

Consequently, F is a matrix representation for $\mathfrak{S}_3$ and its degree is 3.

EXAMPLE 10.6. Let G_U, G_V, and G_W denote the groups of order 1, 2, and 6 respectively formed of the matrices in (10.20), (10.21), and (10.22). Then, examples of matrix representations for $\mathfrak{S}_3$ are given by: the function Φ_1 from $\mathfrak{S}_3$ to G_U having $\Phi_1(h_k) = U_k$, for $1 \le k \le 6$; the function Φ_2 from $\mathfrak{S}_3$ to G_V having $\Phi_2(h_k) = V_k$, for $1 \le k \le 6$; and the function Φ_3 from $\mathfrak{S}_3$ to G_W having $\Phi_3(h_k) = W_k$, for $1 \le k \le 6$. The degree of Φ_1 is 1, the degree of Φ_2 is 1, and the degree of Φ_3 is 2. Of course, for $1 \le i, j, k \le 6$ and $h_i h_j = h_k$, the consequences $U_i U_j = U_k$, $V_i V_j = V_k$, and $W_i W_j = W_k$ are the homomorphism properties $\Phi_1(h_i h_j) = \Phi_1(h_i)\Phi_1(h_j)$, $\Phi_2(h_i h_j) = \Phi_2(h_i)\Phi_2(h_j)$, and $\Phi_3(h_i h_j) = \Phi_3(h_i)\Phi_3(h_j)$ for Φ_1, Φ_2, and Φ_3.

DEFINITION 10.7. Two matrix representations F_1 and F_2 of degrees n_1 and n_2 for a finite group G are said to be *equivalent* when $n_1 = n_2$ and, with $n_0 = n_1 = n_2$, there is a nonsingular $n_0 \times n_0$ matrix S over $\mathbb{C}$ such that

(10.30)
$$S^{-1}F_1(g)S = F_2(g), \quad \text{for each } g \text{ in } G.$$

Otherwise, F_1 and F_2 are said to be *inequivalent*.

EXAMPLE 10.8. Clearly, no two of the three matrix representations Φ_1, Φ_2, Φ_3 for $\mathfrak{S}_3$ are equivalent. However, there is the additional matrix representation Ψ for $\mathfrak{S}_3$ that is defined in terms of (10.23) by $\Psi(h_k) = T_k$, for $1 \le k \le 6$. We employ (10.23) and (10.22) to verify that

$$\begin{bmatrix} 1 & 0 \\ 0 & -1 \end{bmatrix}^{-1} \Psi(h_k) \begin{bmatrix} 1 & 0 \\ 0 & -1 \end{bmatrix} = \begin{bmatrix} 1 & 0 \\ 0 & -1 \end{bmatrix} T_k \begin{bmatrix} 1 & 0 \\ 0 & -1 \end{bmatrix} = W_k = \Phi_3(h_k), \quad \text{for } 1 \le k \le 6.$$

A comparison of this with (10.30) shows that Ψ and Φ_3 are equivalent.

DEFINITION 10.9. Let G be a group of order n whose elements are g_1, g_2, ..., g_n and let F denote a matrix representation for G of degree n_0. Then, F is said to be *reducible* if and only if there are positive integers p, q with $p + q = n_0$ and there is a nonsingular $n_0 \times n_0$ matrix S_0 over $\mathbb{C}$ such that

$$(10.31) \qquad S_0^{-1} F(g_k) S_0 = \left[\begin{array}{c|c} \widehat{A}_k & Z_1 \\ \hline \widehat{C}_k & \widehat{B}_k \end{array} \right], \quad \text{for } k = 1, 2, \ldots, n,$$

where each $\widehat{A}_k$ is a $p \times p$ matrix, each $\widehat{B}_k$ is a $q \times q$ matrix, each $\widehat{C}_k$ is a $q \times p$ matrix, and Z_1 is the $p \times q$ zero matrix. Otherwise, F is said to be *irreducible*.

EXAMPLE 10.10. We let ω denote a root in $\mathbb{C}$ of $X^2 + X + 1 = 0$ and, with $\omega^3 = 1$, we introduce a nonsingular matrix M and its inverse through

$$M = \begin{bmatrix} 1 & 1 & 1 \\ 1 & \omega & \omega^2 \\ 1 & \omega^2 & \omega \end{bmatrix} \quad \text{and} \quad M^{-1} = \frac{1}{3} \begin{bmatrix} 1 & 1 & 1 \\ 1 & \omega^2 & \omega \\ 1 & \omega & \omega^2 \end{bmatrix}.$$

The matrix representation F for $\mathfrak{S}_3$ in Example 10.5 is defined by (10.27)–(10.29). In terms of M and the matrices P_1, P_2, ... P_6 of (10.28), we set

$$Q_k = M^{-1} P_k M, \quad \text{for } k = 1, 2, \ldots, 6,$$

and repeat the observation made in [**7**] that Q_1, Q_2, ..., Q_6 are given by

$$Q_1 = \left[\begin{array}{c|cc} 1 & 0 & 0 \\ \hline 0 & 1 & 0 \\ 0 & 0 & 1 \end{array} \right], \quad Q_2 = \left[\begin{array}{c|cc} 1 & 0 & 0 \\ \hline 0 & \omega & 0 \\ 0 & 0 & \omega^2 \end{array} \right], \quad Q_3 = \left[\begin{array}{c|cc} 1 & 0 & 0 \\ \hline 0 & \omega^2 & 0 \\ 0 & 0 & \omega \end{array} \right],$$

$$Q_4 = \left[\begin{array}{c|cc} 1 & 0 & 0 \\ \hline 0 & 0 & 1 \\ 0 & 1 & 0 \end{array} \right], \quad Q_5 = \left[\begin{array}{c|cc} 1 & 0 & 0 \\ \hline 0 & 0 & \omega \\ 0 & \omega^2 & 0 \end{array} \right], \quad Q_6 = \left[\begin{array}{c|cc} 1 & 0 & 0 \\ \hline 0 & 0 & \omega^2 \\ 0 & \omega & 0 \end{array} \right].$$

This example shows that the representation F for $\mathfrak{S}_3$ in Example 10.5 is reducible. It also shows that a matrix representation f for $\mathfrak{S}_3$ of degree 2 is defined by

$$(10.32) \quad \begin{cases} f(h_1) = \begin{bmatrix} 1 & 0 \\ 0 & 1 \end{bmatrix}, \quad f(h_2) = \begin{bmatrix} \omega & 0 \\ 0 & \omega^2 \end{bmatrix}, \quad f(h_3) = \begin{bmatrix} \omega^2 & 0 \\ 0 & \omega \end{bmatrix}, \\[2ex] f(h_4) = \begin{bmatrix} 0 & 1 \\ 1 & 0 \end{bmatrix}, \quad f(h_5) = \begin{bmatrix} 0 & \omega \\ \omega^2 & 0 \end{bmatrix}, \quad f(h_6) = \begin{bmatrix} 0 & \omega^2 \\ \omega & 0 \end{bmatrix}. \end{cases}$$

By replacing ω with ω^2 throughout this example, we see that the modification for (10.32) yields the matrix representation $h_k \mapsto W_k$, for $1 \leq k \leq 6$, of (10.22).

THEOREM 10.11. *Suppose that F is a reducible matrix representation having degree n_0 for a group G of order n whose elements are g_1, g_2, ..., g_n. Then, in term of positive integers p, q with $p + q = n_0$ that are suitable for the formulation (10.31), there is a nonsingular $n_0 \times n_0$ matrix S over $\mathbb{C}$ such that*

$$(10.33) \qquad S^{-1} F(g_k) S = \left[\begin{array}{c|c} A_k & Z_1 \\ \hline Z_2 & B_k \end{array} \right], \quad \text{for } k = 1, 2, \ldots, n,$$

where each A_k is a $p \times p$ matrix, each B_k is a $q \times q$ matrix, while Z_1 and Z_2 are the zero matrices of respective sizes $p \times q$ and $q \times p$.

PROOF. For a verification, see [**37**, pages 14–15]. □

EXAMPLE 10.12. The matrix representations Φ_1 and Φ_2 for $\mathfrak{S}_3$ in Example 10.6 have degree 1. Thus, Definition 10.9 shows that both Φ_1 and Φ_2 are irreducible.

To verify that Φ_3 for $\mathfrak{S}_3$ in Example 10.6 is also irreducible, suppose that Φ_3 is reducible. Then, with $n_0 = 2$, we use Definition 10.9 and (10.33) of Theorem 10.11 to see that there is a nonsingular 2×2 matrix S over $\mathbb{C}$ such that each of

$$S^{-1}\Phi_3(w_k)S, \quad \text{for } k = 1, 2, \ldots, 6,$$

is a 2×2 diagonal matrix. But, this yields the contradiction that multiplication for the matrices $\Phi_3(h_1)$, $\Phi_3(h_2)$, $\ldots$, $\Phi_3(h_6)$ is commutative whereas multiplication for W_1, W_2, $\ldots$, W_6 in (10.22) is not commutative. Thus, Φ_3 is irreducible.

DEFINITION 10.13. A set $\{F_1, F_2, \ldots, F_p\}$ of matrix representations for a finite group G is *complete* when, for any irreducible matrix representation F of G, F is equivalent to some one of $F_1, F_2, \ldots, F_p$.

THEOREM 10.14. *If G is a finite group, then there is a complete set of pairwise-inequivalent irreducible matrix representations for G.*

PROOF. For a verification, see [**37**, page 25–26]. □

DEFINITION 10.15. Elements x, y of a finite group G are said to be *conjugate* when there is an element u in G such that $u^{-1}xu = y$. This defines a relation on G that is easily checked to be an equivalence relation. The equivalence classes of this equivalence relation are called *conjugacy classes*.

THEOREM 10.16. *Suppose that $\{F_1, F_2, \ldots, F_p\}$ is a compete set of pairwise-inequivalent irreducible matrix representations for a finite group G of order n that has q as the number of its conjugacy classes. Then, $p = q$ and the respective degrees $n_1, n_2, \ldots, n_q$ of $F_1, F_2, \ldots, F_q$ satisfy*

(10.34) $$n_1^2 + n_2^2 + \cdots + n_q^2 = n.$$

PROOF. For a verification, see [**37**, page 25–26]. □

THEOREM 10.17. *Suppose $\{F_1, F_2, \ldots, F_p\}$ is a set of pairwise-inequivalent irreducible matrix representations for a finite group G of order n such that the respective degrees $n_1, n_2, \ldots, n_p$ of $F_1, F_2, \ldots, F_p$ satisfy*

(10.35) $$n_1^2 + n_2^2 + \cdots + n_p^2 = n.$$

Then, $\{F_1, F_2, \ldots, F_p\}$ is a complete set of matrix representations for G.

PROOF. For a verification, see [**37**, page 25]. □

EXAMPLE 10.18. The set $\mathscr{C} = \{\Phi_1, \Phi_2, \Phi_3\}$ consists of pairwise-inequivalent irreducible matrix representations for the group $\mathfrak{S}_3$ of order $n = 6$ in Example 10.6. We note that the degree of Φ_1 is $n_1 = 1$, the degree of Φ_2 is $n_2 = 1$, and the degree of Φ_3 is $n_3 = 2$. This gives $n_1^2 + n_2^2 + n_3^2 = 1^2 + 1^2 + 2^2 = 6 = n$. With (10.35) satisfied, we specialize Theorem 10.17 to conclude that $\mathscr{C}$ is complete. Thus, each irreducible matrix representation for $\mathfrak{S}_3$ is equivalent to one of Φ_1, Φ_2, and Φ_3.

This also shows that $\mathfrak{S}_3$ has $q = 3$ conjugacy classes. They are given in terms of (10.27) by $\{h_1\}$, $\{h_2, h_3\}$, and $\{h_4, h_5, h_6\}$. In particular, we have $h_4^{-1}h_2h_4 = h_3$, $h_2^{-1}h_4h_2 = h_5$, and $h_3^{-1}h_4h_3 = h_6$.

THEOREM 10.19. *Let F be an irreducible matrix representation of degree n_0 for a group G of order n. Then, n_0 divides n and, for any integers i, j, r, s subject to $1 \leq i, j, r, s \leq n_0$, the components for F satisfy*

$$(10.36) \qquad \sum_{g \in G} \left[F(g^{-1}) \right]_{i,j} \left[F(g) \right]_{r,s} = \begin{cases} n/n_0, & \text{when } i = s \text{ and } j = r, \\ 0, & \text{otherwise.} \end{cases}$$

PROOF. For a verification, see [**37**, pages 16–17].   □

EXAMPLE 10.20. For Example 10.6 where $n = 6$ and Φ_1 has $n_1 = 1$, we obtain

$$\sum_{k=1}^{6} \left[\Phi_1(h_k^{-1}) \right]_{1,1} \left[\Phi_1(h_k) \right]_{1,1} = 1 + 1 + 1 + 1 + 1 + 1 = 6.$$

For Φ_2 in Example 10.6 with $n = 6$ and $n_2 = 1$, we find that

$$\sum_{k=1}^{6} \left[\Phi_2(h_k^{-1}) \right]_{1,1} \left[\Phi_2(h_k) \right]_{1,1} = 1 + 1 + 1 + 1 + 1 + 1 = 6.$$

For Φ_3 in in Example 10.6 with $n = 6$ and $n_3 = 2$, we observe that

$$\sum_{k=1}^{6} \left[\Phi_3(h_k^{-1}) \right]_{1,1} \left[\Phi_3(h_k) \right]_{1,1} = 1 + \omega^3 + \omega^3 + 0 + 0 + 0 = 3,$$

and

$$\sum_{k=1}^{6} \left[\Phi_3(h_k^{-1}) \right]_{1,1} \left[\Phi_3(h_k) \right]_{1,2} = 0 + 0 + 0 + 0 + 0 + 0 = 0,$$

as well as fourteen more expressions of a similar kind. Four of the sixteen expressions have a sum of 3 and the other twelve have a sum of 0.

Thus, the matrix representations Φ_1, Φ_2, and Φ_3 for $\mathfrak{S}_3$ illustrate (10.36).

THEOREM 10.21. *Let F_1 of degree n_1 and F_2 of degree n_2 denote inequivalent irreducible representations for a group G of order n. Then, the components satisfy*

$$(10.37) \qquad \sum_{g \in G} \left[F_1(g^{-1}) \right]_{i,j} \left[F_2(g) \right]_{r,s} = 0, \quad \text{for } 1 \leq i, j \leq n_1 \text{ and } 1 \leq r, s \leq n_2.$$

PROOF. For a verification, see [**37**, page 15].   □

EXAMPLE 10.22. In Example 10.6, we see that Φ_1 and Φ_2 satisfy

$$\sum_{k=1}^{6} \left[\Phi_1(h_k^{-1}) \right]_{1,1} \left[\Phi_2(h_k) \right]_{1,1} = 1 + 1 + 1 - 1 - 1 - 1 = 0.$$

The matrix representations Φ_1 and Φ_3 in Example 10.6 yield

$$\sum_{k=1}^{6} \left[\Phi_1(h_k^{-1}) \right]_{1,1} \left[\Phi_3(h_k) \right]_{1,1} = 1 + \omega^2 + \omega + 0 + 0 + 0 = 0,$$

and three more analogous expressions having a sum of 0.

For the matrix representations Φ_2 and Φ_3 in Example 10.6, we obtain

$$\sum_{k=1}^{6} \left[\Phi_2(h_k^{-1}) \right]_{1,1} \left[\Phi_3(h_k) \right]_{1,1} = 1 + \omega^2 + \omega - 0 - 0 - 0 = 0,$$

and three more analogous expressions having a sum of 0.

Thus, Φ_1, Φ_2, and Φ_3 of Example 10.6 serve to illustrate (10.37).

THEOREM 10.23. *Suppose* $\{F_1, F_2, \ldots, F_p\}$ *is a set of pairwise-inequivalent irreducible matrix representations for a finite group G of order n. Let the respective degrees of $F_1, F_2, \ldots, F_p$ be $n_1, n_2, \ldots, n_p$. Then, for $1 \leq \kappa, \lambda \leq p$, $1 \leq i, j \leq n_\kappa$, and $1 \leq r, s \leq n_\lambda$, the components satisfy*

$$(10.38) \quad \sum_{g \in G} \left[F_\kappa(g^{-1}) \right]_{i,j} \left[F_\lambda(g) \right]_{r,s} = \begin{cases} (n/n_\kappa), & \text{if } \kappa = \lambda, \ i = s, \text{ and } j = r, \\ 0, & \text{otherwise.} \end{cases}$$

PROOF. This is a consequence of (10.36) and (10.37). $\square$

CHAPTER 11

Viewpoints about Notation

Before demonstrating in Chapter 13 how a complete set of pairwise-inequivalent irreducible matrix representations for a finite group G of order n can be used to explicitly construct an $n \times n$ matrix that has properties analogous to those of L in (10.25), we provide a suitable context.

11.1. The notation for group-matrices

Theorem 10.1 on page 92 shows that the definition of the polynomial Θ in (10.2) on page 91 is independent of the list selected for the elements of G. Thus, if the list $(g_1, g_2, \ldots, g_n)$ used to specify Θ in (10.4) on page 92 had instead been $(g_1^{-1}, g_2^{-1}, \ldots, g_n^{-1})$, then the same Θ would be given by

$$\Theta = \det(C), \quad \text{where} \quad C = \begin{bmatrix} X_{g_1^{-1}g_1} & X_{g_1^{-1}g_2} & \cdots & X_{g_1^{-1}g_n} \\ X_{g_2^{-1}g_1} & X_{g_2^{-1}g_2} & \cdots & X_{g_2^{-1}g_n} \\ \vdots & \vdots & \vdots & \vdots \\ X_{g_n^{-1}g_1} & X_{g_n^{-1}g_2} & \cdots & X_{g_n^{-1}g_n} \end{bmatrix}.$$

Thus, when Georg Frobenius used notation like $|X_{PQ^{-1}}|$ to indicate his concept of a Gruppendeterminante, he could just as well have written $|X_{P^{-1}Q}|$. Definition 2.2 on page 12 emphasizes notation where it is better for a list $\mathscr{L} = (g_1, g_2, \ldots, g_n)$ to be identified with column indices of a multiplication table rather than row indices.

PROPOSITION 11.1. *The number of group-matrices for G is $(n-1)!$.*

PROOF. Proposition 2.7 on page 13 shows that each group-matrix for G is specified by a list whose first element is e. Let $g_1, g_2, \ldots, g_n$ denote the n elements in G such that $g_1 = e$. We introduce the set $S = \{X_{g_1}, X_{g_2}, \ldots, X_{g_n}\}$ of variables; and we let σ denote the one-to-one function from G to S having $\sigma(g_k) = X_{g_k}$, for $k = 1, 2, \ldots, n$. Thus, the group-matrices for G are given by

$$(11.1) \qquad \left[\sigma\left(g_{\pi(r)}^{-1} g_{\pi(s)}\right)\right]_{\substack{1 \le r \le n \\ 1 \le s \le n}}$$

as π ranges over the permutations of $\{1, 2, \ldots, n\}$ having $\pi(1) = 1$. The first row of (11.1) is $(X_{g_{\pi(1)}}, X_{g_{\pi(2)}}, X_{g_{\pi(3)}}, \ldots, X_{g_{\pi(n)}})$. Since the group-matrices for G are in one-to-one correspondence with the permutations π of $\{1, 2, \ldots, n\}$ having $\pi(1) = 1$, there are $(n-1)!$ of them. This completes the proof. □

OBSERVATION 11.2. Each group-matrix given by (11.1) specifies a standard group-pattern matrix that is obtained by replacing $X_{g_{\pi(k)}}$ with X_k, for $1 \le k \le n$. The proof of Theorem 3.7 shows that each standard group-pattern matrix for G is obtained in this manner from m group-matrices, where $m = |Aut(G)|$.

11.2. A numbering for triples of integers

11.2.1. A one-to-one correspondence. To provide a general technique that will be specialized in Sections 11.3 and 11.4 to yield the components of the Frobenius matrix L in (10.25), we introduce the following construction.

THEOREM 11.3. *Let n_1, n_2, ..., n_q denote positive integers and set*

$$(11.2) \qquad n = n_1^2 + n_2^2 + \cdots + n_q^2.$$

Then, the function $\boldsymbol{\Gamma}$ defined on the set

$$\mathcal{T} = \Big\{ (\lambda, \mu, \nu) : \quad 1 \le \lambda \le q \ and \ 1 \le \mu, \nu \le n_\lambda \Big\}$$

by

$$(11.3) \qquad \boldsymbol{\Gamma}(\lambda, \mu, \nu) = \left(\sum_{k=1}^{\lambda-1} n_k^2 \right) + (\mu - 1) n_\lambda + \nu, \quad for \ (\lambda, \mu, \nu) \ in \ \mathcal{T},$$

provides a one-to-one correspondence between $\mathcal{T}$ and the set $\mathcal{S} = \big\{ 1, 2, \ldots, n \big\}$.

PROOF. For (λ, μ, ν) in $\mathcal{T}$, we observe that

$$1 \le \nu \le \left(\sum_{k=1}^{\lambda-1} n_k^2 \right) + (\mu - 1) n_\lambda + \nu \le \left(\sum_{k=1}^{\lambda-1} n_k^2 \right) + (n_\lambda - 1) n_\lambda + n_\lambda = \sum_{k=1}^{\lambda} n_k^2 \le n.$$

Thus, $\boldsymbol{\Gamma}$ is a function from $\mathcal{T}$ to $\mathcal{S}$.

To prove that $\boldsymbol{\Gamma}$ is a function from $\mathcal{T}$ onto S, let β denote an integer in $\mathcal{S}$. Then, there is a unique integer λ_0 such that $1 \le \lambda_0 \le q$ and

$$(11.4) \qquad \sum_{k=1}^{\lambda_0-1} n_k^2 < \beta \le \sum_{k=1}^{\lambda_0} n_k^2.$$

We set

$$(11.5) \qquad D = \beta - \sum_{k=1}^{\lambda_0-1} n_k^2.$$

In view of (11.4), the integer D satisfies $1 \le D \le n_{\lambda_0}^2$. If D is not divisible by n_{λ_0}, let Q and R denote the unique integers such that

$$D = Q \, n_{\lambda_0} + R, \quad with \ 1 \le R < n_{\lambda_0} \ and \ 0 \le Q \le n_{\lambda_0} - 1.$$

If D is divisible by n_{λ_0}, let Q and R denote the unique integers such that

$$D = Q \, n_{\lambda_0} + R, \quad with \ 1 \le R = n_{\lambda_0} \ and \ 0 \le Q \le n_{\lambda_0} - 1.$$

We set $\mu_0 = Q + 1$ and $\nu_0 = R$. Therefore, by combining this with (11.5), we obtain

$$(11.6) \quad \beta = \sum_{k=1}^{\lambda_0-1} n_k^2 + (\mu_0 - 1) \, n_{\lambda_0} + \nu_0, \quad where \ 1 \le \lambda_0 \le q \ and \ 1 \le \mu_0, \nu_0 \le n_\lambda.$$

This shows that $\big(\lambda_0, \mu_0, \nu_0\big)$ is an element of $\mathcal{T}$ and yields $\boldsymbol{\Gamma}(\lambda_0, \mu_0, \nu_0) = \beta$. Thus, the function $\boldsymbol{\Gamma}$ is onto.

To show that $\boldsymbol{\Gamma}$ is one-to-one, let $\big(\lambda_1, \mu_1, \nu_1\big)$ and $\big(\lambda_2, \mu_2, \nu_2\big)$ denote any two elements of $\mathcal{T}$ such that $\big(\lambda_1, \mu_1, \nu_1\big) \ne \big(\lambda_2, \mu_2, \nu_2\big)$. There are three cases.

(i) Suppose that $\lambda_1 \neq \lambda_2$. We may assume $\lambda_1 < \lambda_2$. Then, we have

$$\boldsymbol{\Gamma}(\lambda_1, \mu_1, \nu_1) = \left(\sum_{k=1}^{\lambda_1 - 1} n_k^2 \right) + (\mu_1 - 1) n_{\lambda_1} + \nu_1$$

$$\leq \left(\sum_{k=1}^{\lambda_1 - 1} n_k^2 \right) + n_{\lambda_1}^2 = \sum_{k=1}^{\lambda_1} n_k^2 \leq \sum_{k=1}^{\lambda_2 - 1} n_k^2 < \boldsymbol{\Gamma}(\lambda_2, \mu_2, \nu_2).$$

(ii) Suppose that $\lambda_1 = \lambda_2$ and $\mu_1 \neq \mu_2$. We may assume $\mu_1 < \mu_2$. Since we have $1 \leq \nu_1, \nu_2 \leq n_{\lambda_2}$, this yields

$$\boldsymbol{\Gamma}(\lambda_2, \mu_2, \nu_2) - \boldsymbol{\Gamma}(\lambda_1, \mu_1, \nu_1) = \left((\mu_2 - 1) n_{\lambda_2} + \nu_2 \right) - \left((\mu_1 - 1) n_{\lambda_2} + \nu_1 \right)$$

$$= (\mu_2 - \mu_1) n_{\lambda_2} + (\nu_2 - \nu_1)$$

$$\geq n_{\lambda_2} + (1 - n_{\lambda_2}) = 1 > 0.$$

(iii) Suppose that $\lambda_1 = \lambda_2$, $\mu_1 = \mu_2$, and $\nu_1 \neq \nu_2$. We may assume $\nu_1 < \nu_2$. Then,, we find that $\boldsymbol{\Gamma}(\lambda_2, \mu_2, \nu_2) - \boldsymbol{\Gamma}(\lambda_1, \mu_1, \nu_1) = \nu_2 - \nu_1 > 0$.

Thus, $\boldsymbol{\Gamma}$ is one-to-one. This completes the proof. $\square$

11.2.2. The functions $\lambda(s)$, $\mu(s)$, and $\nu(s)$. The one-to-one function $\boldsymbol{\Gamma}$ from $\mathcal{T}$ onto $\mathcal{S}$ defined by (11.3) has an inverse function from $\mathcal{S}$ onto $\mathcal{T}$ that specifies three functions $\lambda(s)$, $\mu(s)$, $\nu(s)$ on $\mathcal{S}$ such that

(11.7) $1 \leq \lambda(s) \leq q$,   and $1 \leq \mu(s), \nu(s) \leq n_{\lambda(s)}$,   for each s in S,

and

(11.8) $\left(\sum_{k=1}^{\lambda(s)-1} n_k^2 \right) + \left(\mu(s) - 1 \right) n_{\lambda(s)} + \nu(s) = s$,   for each s in $\mathcal{S}$.

To visualize our first application for the triple $\left(\lambda(s), \mu(s), \nu(s) \right)$ as s ranges over the integers from 1 through n, consider matrices

$$A_1 \text{ of size } n_1 \times n_1, \; A_2 \text{ of size } n_2 \times n_2, \; \ldots, \; A_q \text{ of size } n_q \times n_q$$

and observe that the triple $\left(\lambda(s), \mu(s), \nu(s) \right)$ specifies the component $\left[A_{\lambda(s)} \right]_{\mu(s), \nu(s)}$ in the $\mu(s)$th row and $\nu(s)$th column of the matrix $A_{\lambda(s)}$. For instance, when $s = 1$, the triple $\left(\lambda(1), \mu(1), \nu(1) \right)$ equals $(1, 1, 1)$ and specifies the leftmost component in the first row of the matrix A_1. As s varies through the consecutive integers from 1 through n, we see that the corresponding components range from left to right across the consecutive rows of A_1 from first to last, then from left to right across the consecutive rows of A_2 from first to last, $\ldots$, then from left to right across the consecutive rows of A_q from first to last.

OBSERVATION 11.4. The 36 components of the matrix L in (10.25) are the $6 + 6 + 24$ components of the matrices in (10.20), (10.21), and (10.22) on page 95. They are the 36 components for the matrix representations Φ_1, Φ_2, and Φ_3 in Example 10.18 on page 99 for the group $\mathfrak{S}_3$ of order $n = 6$. Since the degrees of Φ_1, Φ_2, and Φ_3 are respectively $n_1 = 1$, $n_2 = 1$, and $n_3 = 2$ with $n_1^2 + n_2^2 + n_3^2 = 6 = n$, there are corresponding functions $\lambda(s)$, $\mu(s)$, $\nu(s)$ defined on $\mathcal{S} = \{1, 2, \ldots, 6\}$ The immediate goal is to show how this $\lambda(s)$, $\mu(s)$, $\nu(s)$ enables the components for Φ_1, Φ_2, and Φ_3 to be arranged in the form of a useful matrix like L in (10.25).

11.3. Selection of components for construction of a matrix like L

We recall that Richard Dedekind and Georg Frobenius computed the product of two juxtaposed permutation symbols by reading them from left to right to form their product. Thus, for the elements

$$h_1 = \begin{pmatrix} 1\,2\,3 \\ 1\,2\,3 \end{pmatrix}, \qquad h_2 = \begin{pmatrix} 1\,2\,3 \\ 2\,3\,1 \end{pmatrix}, \qquad h_3 = \begin{pmatrix} 1\,2\,3 \\ 3\,1\,2 \end{pmatrix},$$

$$h_4 = \begin{pmatrix} 1\,2\,3 \\ 1\,3\,2 \end{pmatrix}, \qquad h_5 = \begin{pmatrix} 1\,2\,3 \\ 3\,2\,1 \end{pmatrix}, \qquad h_6 = \begin{pmatrix} 1\,2\,3 \\ 2\,1\,3 \end{pmatrix}$$

of $\mathfrak{S}_3$, a complete set of pairwise-inequivalent irreducible matrix representations for $\mathfrak{S}_3$ is given by $\{\Phi_1, \Phi_2, \Phi_3\}$ from Example 10.6 of page 97 where

$$
(11.9) \quad
\begin{cases}
\Phi_1(h_k) = [1], \quad \text{for } k = 1, 2, 3, 4, 5, 6, \\[4pt]
\Phi_2(h_k) = [1], \quad \text{for } k = 1, 2, 3, \text{ and } \quad \Phi_2(h_k) = [-1], \quad \text{for } k = 4, 5, 6, \\[4pt]
\Phi_3(h_1) = \begin{bmatrix} 1 & 0 \\ 0 & 1 \end{bmatrix}, \quad \Phi_3(h_2) = \begin{bmatrix} \omega^2 & 0 \\ 0 & \omega \end{bmatrix}, \quad \Phi_3(h_3) = \begin{bmatrix} \omega & 0 \\ 0 & \omega^2 \end{bmatrix}, \\[12pt]
\Phi_3(h_4) = \begin{bmatrix} 0 & 1 \\ 1 & 0 \end{bmatrix}, \quad \Phi_3(h_5) = \begin{bmatrix} 0 & \omega^2 \\ \omega & 0 \end{bmatrix}, \quad \Phi_3(h_6) = \begin{bmatrix} 0 & \omega \\ \omega^2 & 0 \end{bmatrix}.
\end{cases}
$$

The entries in the first column of the matrix

$$
(11.10) \quad L_1 = \begin{bmatrix}
1 & 1 & 1 & 0 & 0 & 1 \\
1 & 1 & \omega^2 & 0 & 0 & \omega \\
1 & 1 & \omega & 0 & 0 & \omega^2 \\
1 & -1 & 0 & 1 & 1 & 0 \\
1 & -1 & 0 & \omega & \omega^2 & 0 \\
1 & -1 & 0 & \omega^2 & \omega & 0
\end{bmatrix}, \quad \text{with } \det(M) = -54,
$$

are the six components $[\Phi_1(h_k)]_{1,1}$, for $1 \le k \le 6$. The second column of L_1 consists of the six components $[\Phi_2(h_k)]_{1,1}$, for $1 \le k \le 6$. The third column of L_1 consists of the six components $[\Phi_3(h_k)]_{1,1}$, for $1 \le k \le 6$. The fourth column of L_1 consists of the six components $[\Phi_3(h_k)]_{2,1}$, for $1 \le k \le 6$. The fifth column of L_1 consists of the six components $[\Phi_3(h_k)]_{1,2}$, for $1 \le k \le 6$. The sixth column of L_1 consists of the six components $[\Phi_3(h_k)]_{2,2}$, for $1 \le k \le 6$.

A comparison of L_1 in (11.10) with L in (10.25) on page 96 shows that: L *can be obtained from* L_1 *by multiplying the second column of* L_1 *by* -1. We use computer algebra to check that X in (10.24) on page 96 and L_1 in (11.10) yield

$$
(11.11) \quad L_1^{-1} X L_1 = \begin{bmatrix}
u+v & 0 & 0 & 0 & 0 & 0 \\
0 & u-v & 0 & 0 & 0 & 0 \\
0 & 0 & u_1 & v_1 & 0 & 0 \\
0 & 0 & v_2 & u_2 & 0 & 0 \\
0 & 0 & 0 & 0 & u_1 & v_1 \\
0 & 0 & 0 & 0 & v_2 & u_2
\end{bmatrix} = L^{-1} X L,
$$

in agreement with (10.26) where u, v, u_1, u_2, v_1, v_2 are defined by (10.8) on page 93.

With $n_1 = 1$, $n_2 = 1$, $n_3 = 2$ and $n_1^2 + n_2^2 + n_3^2 = 6 = n$, we apply the context for Theorem 11.3 to solve (11.3) for $(\lambda(s), \mu(s), \nu(s))$ in terms of s when $1 \le s \le 6$. That gives the information of the table

	$s=1$	$s=2$	$s=3$	$s=4$	$s=5$	$s=6$
$\big(\lambda(s),\,\mu(s),\,\nu(s)\big)$	$(1,1,1)$	$(2,1,1)$	$(3,1,1)$	$(3,1,2)$	$(3,2,1)$	$(3,2,2)$
$\big(\lambda(s),\,\nu(s),\,\mu(s)\big)$	$(1,1,1)$	$(2,1,1)$	$(3,1,1)$	$(3,2,1)$	$(3,1,2)$	$(3,2,2)$

and enables us to see that the the element in the rth row and sth column of L_1 is

$$(11.12) \qquad [L_1]_{r,s} = [\Phi_{\lambda(s)}(g_r)]_{\nu(s),\mu(s)}, \quad \text{for } r, s = 1, 2, \ldots, 6.$$

11.4. Contextual improvement

We prefer to specify each group-pattern in terms of its column labels. Thus, since the matrix X of (10.14) on page 94 is based on the multiplication table

(11.13)

$\cdot$	h_1^{-1}	h_2^{-1}	h_3^{-1}	h_4^{-1}	h_5^{-1}	h_6^{-1}
h_1	h_1	h_3	h_2	h_4	h_5	h_6
h_2	h_2	h_1	h_3	h_5	h_6	h_4
h_3	h_3	h_2	h_1	h_6	h_4	h_5
h_4	h_4	h_5	h_6	h_1	h_3	h_2
h_5	h_5	h_6	h_4	h_2	h_1	h_3
h_6	h_6	h_4	h_5	h_3	h_2	h_1

from (10.17), we set $\mathscr{L} = \big(h_1^{-1},\, h_2^{-1},\, h_3^{-1},\, h_4^{-1},\, h_5^{-1},\, h_6^{-1}\big)$ as well as $g_k = h_k^{-1}$ and $h_k = g_k^{-1}$, for $1 \le k \le 6$, in order to regard X as the group-pattern matrix for $\mathfrak{S}_3$ and $\mathscr{L} = \big(g_1,\, g_2,\, g_3,\, g_4,\, g_5,\, g_6\big)$. Thus, (11.13) yields

(11.14)

$\cdot$	g_1	g_2	g_3	g_4	g_5	g_6
g_1^{-1}	g_1	g_2	g_3	g_4	g_5	g_6
g_2^{-1}	g_3	g_1	g_2	g_5	g_6	g_4
g_3^{-1}	g_2	g_3	g_1	g_6	g_4	g_5
g_4^{-1}	g_4	g_5	g_6	g_1	g_2	g_3
g_5^{-1}	g_5	g_6	g_4	g_3	g_1	g_2
g_6^{-1}	g_6	g_4	g_5	g_2	g_3	g_1

where an identification of (11.14) with (1.5) on page 4 can be made when $g_1 = e$, $g_2 = \alpha$, $g_3 = \alpha^2$, $g_4 = \beta_1$, $g_5 = \beta_2$, and $g_6 = \beta_3$. We introduce

$$(11.15) \qquad A = \left[\begin{array}{ccc|ccc} x_1 & x_2 & x_3 & x_4 & x_5 & x_6 \\ x_3 & x_1 & x_2 & x_5 & x_6 & x_4 \\ x_2 & x_3 & x_1 & x_6 & x_4 & x_5 \\ \hline x_4 & x_5 & x_6 & x_1 & x_2 & x_3 \\ x_5 & x_6 & x_4 & x_3 & x_1 & x_2 \\ x_6 & x_4 & x_5 & x_2 & x_3 & x_1 \end{array}\right].$$

as a typical group-pattern matrix for $\mathfrak{S}_3$ and $\mathscr{L} = \big(g_1, g_2, \ldots, g_6\big)$.

To see how this modifies (11.12), we first employ (11.9) to introduce

(11.16) $F_\lambda(g_k) = \Phi_\lambda(g_k) = \Phi_\lambda(h_k^{-1})$,   for $\lambda = 1, 2, 3$ and $1 \le k \le 6$.

Then, with respect to $\mathscr{L} = (g_1, g_2, \ldots, g_6)$, we use (11.16) and (11.9) to see that a complete set of pairwise-inequivalent irreducible matrix representations for $\mathfrak{S}_3$ is given by

(11.17) $\begin{cases} F_1(g_k) = [1], & \text{for } k = 1, 2, 3, 4, 5, 6, \\[4pt] F_2(g_k) = [1], & \text{for } k = 1, 2, 3 \quad \text{and} \quad F_2(g_k) = [-1], \quad \text{for } k = 4, 5, 6, \\[6pt] F_3(g_1) = \begin{bmatrix} 1 & 0 \\ 0 & 1 \end{bmatrix}, \quad F_3(g_2) = \begin{bmatrix} \omega & 0 \\ 0 & \omega^2 \end{bmatrix}, \quad F_3(g_3) = \begin{bmatrix} \omega^2 & 0 \\ 0 & \omega \end{bmatrix}, \\[12pt] F_3(g_4) = \begin{bmatrix} 0 & 1 \\ 1 & 0 \end{bmatrix}, \quad F_3(g_5) = \begin{bmatrix} 0 & \omega^2 \\ \omega & 0 \end{bmatrix}, \quad F_3(g_6) = \begin{bmatrix} 0 & \omega \\ \omega^2 & 0 \end{bmatrix}. \end{cases}$

In particular, the degree of F_1 is $n_1 = 1$, the degree of F_2 is $n_2 = 1$, and the degree of F_3 is $n_3 = 2$. Using the natural correspondence

(11.18)

	$s = 1$	$s = 2$	$s = 3$	$s = 4$	$s = 5$	$s = 6$
$(\lambda(s),\, \mu(s),\, \nu(s))$	$(1,1,1)$	$(2,1,1)$	$(3,1,1)$	$(3,1,2)$	$(3,2,1)$	$(3,2,2)$

provided by Theorem 11.3, we see that the 6×6 matrix M having

(11.19) $[M]_{r,s} = [F_{\lambda(s)}(g_r)]_{\mu(s),\nu(s)}$,   for $r, s = 1, 2, \ldots, 6$.

is given by

(11.20) $M = \begin{bmatrix} 1 & 1 & 1 & 0 & 0 & 1 \\ 1 & 1 & \omega & 0 & 0 & \omega^2 \\ 1 & 1 & \omega^2 & 0 & 0 & \omega \\ 1 & -1 & 0 & 1 & 1 & 0 \\ 1 & -1 & 0 & \omega^2 & \omega & 0 \\ 1 & -1 & 0 & \omega & \omega^2 & 0 \end{bmatrix}$

with $\det(M) = -54$. Here, we find that

(11.21) $M^{-1}AM = \begin{bmatrix} u+v & 0 & 0 & 0 & 0 & 0 \\ 0 & u-v & 0 & 0 & 0 & 0 \\ 0 & 0 & u_1 & v_2 & 0 & 0 \\ 0 & 0 & v_1 & u_2 & 0 & 0 \\ 0 & 0 & 0 & 0 & u_1 & v_2 \\ 0 & 0 & 0 & 0 & v_1 & u_2 \end{bmatrix}$,

where u, v, u_1, v_1, u_2, v_2 continue to be defined by (10.8) on page 93.

In view of $\det(A) = (\det M)^{-1} \det(A) \det(M) = \det(M^{-1}AM)$, (11.21) yields

(11.22) $\det(A) = (u+v)(u-v)(u_1u_2 - v_1v_2)^2$.

Since the matrices A in (11.15) and X in (10.24) are transposes of one another, they satisfy $\det(X) = \det(A)$. Thus, the factorization in (11.22) is the same as that for $\det(X)$ following (10.26) and it is the same as that for H in (10.7).

11.5. Machine computations for $\lambda(s)$, $\mu(s)$, and $\nu(s)$

11.5.1. Technique. Using the context of Theorem 11.3 on page 104, we know that: for any integer s satisfying $1 \le s \le n$, there are unique integers $\lambda(s)$, $\mu(s)$, and $\nu(s)$ such that

$$(11.23) \qquad \left(\sum_{k=1}^{\lambda(s)-1} n_k^2 \right) + \big(\mu(s) - 1\big)\, n_{\lambda(s)} + \nu(s) = s,$$

$$\text{with } 1 \le \lambda(s) \le q \text{ and } 1 \le \mu(s),\, \nu(s) \le n_{\lambda(s)}.$$

Consequently, $\lambda(s)$ is the unique integer that satisfies

$$\sum_{k=1}^{\lambda(s)-1} n_k^2 < s \le \sum_{k=1}^{\lambda(s)} n_k^2.$$

In terms of this $\lambda(s)$, we introduce

$$\mathcal{D} = s - \sum_{k=1}^{\lambda(s)-1} n_k^2 = \big(\mu(s) - 1\big)\, n_{\lambda(s)} + \nu(s), \quad \text{where } 1 \le \mathcal{D} \le n_{\lambda(s)}^2.$$

Unique integers $\mathcal{Q}$ and $\mathcal{R}$ exist such that $\mathcal{D} = \mathcal{Q} n_{\lambda(s)} + \mathcal{R}$ and $0 \le \mathcal{R} < n_{\lambda(s)}$. There are two cases. (i) If $\mathcal{R} = 0$, we set $\mu = \mathcal{Q}$ and have $\mathcal{D} = (\mu - 1) n_{\lambda(s)} + n_{\lambda(s)}$. (ii) If $\mathcal{R} \ne 0$, we set $\mu = \mathcal{Q} + 1$ and have $\mathcal{D} = (\mu - 1) n_{\lambda(s)} + \mathcal{R}$, where $1 \le \mathcal{R} < n_{\lambda(s)}$. After $\lambda(s)$ and $\mu(s)$ are found, we then solve (11.23) for $\nu(s)$.

11.5.2. Computer implementation. With a version of *Mathematica* such as [**48**], we first enter an assignment of the given values for q, n_1, n_2, ..., n_q to the corresponding symbols q, n[1], n[2], ..., n[q]. After their evaluation, we enter

```
sum = Sum[n[k]^2, {k, 1, q}]

lambda[s_ /; 1 <= s && s <= sum] := Catch[Do[If[
    Sum[n[k]^2, {k, 1, u - 1}] < s <= Sum[n[k]^2, {k,1,u}],
    Throw[u]], {u, 1, q}]]

mu[s_ /; 1 <= s && s <= sum] :=
    Module[{u, v}, u = s - Sum[n[k]^2, {k,1,lambda[s]-1}];
    v = Mod[u, n[lambda[s]]]; Catch[If[v == 0,
    Throw[Quotient[u, n[lambda[s]]]],
    Throw[Quotient[u, n[lambda[s]]] + 1]]]]

nu[s_ /; 1 <= s && s <= sum] :=
    (s - Sum[n[k]^2, {k, 1, lambda[s] - 1}]
        - (mu[s] - 1)*n[lambda[s]] )
```

and also evaluate these last four commands.

EXAMPLE 11.5. To illustrate how machine computations can be advantageously employed in more complicated situations, we apply them here to the simple task of deducing M in (11.20) from (11.19).

Using a version of *Mathematica* such as [**48**], we enter the input instructions

```
q = 3;  n[1] = 1;    n[2] = 1;    n[3] = 2;

sum = Sum[ n[k]^2, {k,1,3}];

lambda[s_ /; 1 <= s && s <= sum] := Catch[Do[If[
    Sum[n[k]^2, {k, 1, u - 1}] < s <= Sum[n[k]^2, {k,1,u}],
    Throw[u]], {u, 1, q}]]

mu[s_ /; 1 <= s && s <= sum] :=
    Module[{u, v}, u = s - Sum[n[k]^2, {k,1,lambda[s]-1}];
    v = Mod[u, n[lambda[s]]]; Catch[If[v == 0,
    Throw[Quotient[u, n[lambda[s]]]],
    Throw[Quotient[u, n[lambda[s]]] + 1]]]]

nu[s_ /; 1 <= s && s <= sum] :=
    (s - Sum[n[k]^2, {k, 1, lambda[s] - 1}]
      - (mu[s] - 1)*n[lambda[s]] )

F[1,k_] := {{1}}    /; 1 <= k <= 6

F[2,k_] := {{1}}    /; 1 <= k <= 3

F[2,k_] := {{-1}}  /; 4 <= k <= 6

F[3,1] = {{1,0}, {0,1}};     F[3,2] = {{w,0}, {0,w^2}};

F[3,3] = {{w^2,0}, {0,w}};   F[3,4] = {{0,1}, {1,0}};

F[3,5] = {{0,w^2}, {w,0}};   F[3,6] = {{0,w}, {w^2,0}};

m[r_,s_] := ( F[lambda[s],r][[mu[s],nu[s]]]
            /;  1 <= r <= 6  &&  1 <= s <= 6 )

M = Table[ m[r,s], {r,1,6}, {s,1,6}]

M  // MatrixForm
```

and evaluate them. The expression for M yields (11.20) when w is replaced by ω.

An Instructive Frobenius Block-Diagonalization

12.1. The group $\mathfrak{T}$ of tetrahedral rotational symmetries

For a regular tetrahedron

(12.1)

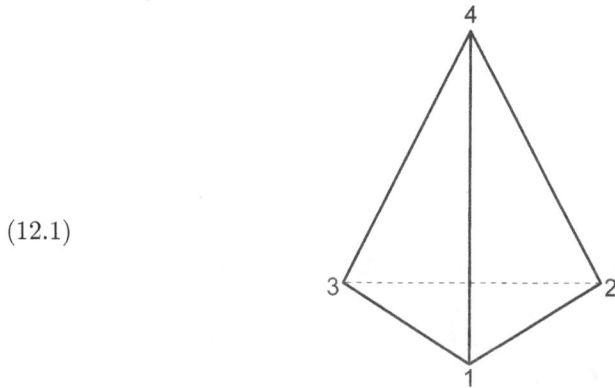

whose vertices are labeled 1, 2, 3 and 4, two of its twelve rotational symmetries are represented by the permutation symbols

(12.2)
$$\alpha = \begin{pmatrix} 1\,2\,3\,4 \\ 2\,1\,4\,3 \end{pmatrix} \quad \text{and} \quad \beta = \begin{pmatrix} 1\,2\,3\,4 \\ 3\,1\,2\,4 \end{pmatrix}.$$

We note that α specifies a rotation of the tetrahedron through 180 degrees about the line through the midpoints of the edges 12 and 34. Moreover, β represents a rotation through 120 degrees about the line through vertex 4 and the center of triangle 123.

The twelve rotational symmetries for (12.1) are given by

$$g_1 = e = \begin{pmatrix} 1\,2\,3\,4 \\ 1\,2\,3\,4 \end{pmatrix}, \qquad g_2 = \beta = \begin{pmatrix} 1\,2\,3\,4 \\ 3\,1\,2\,4 \end{pmatrix}, \qquad g_3 = \beta^2 = \begin{pmatrix} 1\,2\,3\,4 \\ 2\,3\,1\,4 \end{pmatrix},$$

$$g_4 = \alpha = \begin{pmatrix} 1\,2\,3\,4 \\ 2\,1\,4\,3 \end{pmatrix}, \qquad g_5 = \alpha\beta = \begin{pmatrix} 1\,2\,3\,4 \\ 1\,3\,4\,2 \end{pmatrix}, \qquad g_6 = \alpha\beta^2 = \begin{pmatrix} 1\,2\,3\,4 \\ 3\,2\,4\,1 \end{pmatrix},$$

$$g_7 = \beta^2\alpha\beta = \begin{pmatrix} 1\,2\,3\,4 \\ 3\,4\,1\,2 \end{pmatrix}, \qquad g_8 = \alpha\beta\alpha = \begin{pmatrix} 1\,2\,3\,4 \\ 2\,4\,3\,1 \end{pmatrix}, \qquad g_9 = \beta^2\alpha = \begin{pmatrix} 1\,2\,3\,4 \\ 1\,4\,2\,3 \end{pmatrix},$$

$$g_{10} = \beta\alpha\beta^2 = \begin{pmatrix} 1\,2\,3\,4 \\ 4\,3\,2\,1 \end{pmatrix}, \qquad g_{11} = \beta\alpha = \begin{pmatrix} 1\,2\,3\,4 \\ 4\,2\,1\,3 \end{pmatrix}, \qquad g_{12} = \beta\alpha\beta = \begin{pmatrix} 1\,2\,3\,4 \\ 4\,1\,3\,2 \end{pmatrix}$$

when two juxtaposed symbols are read from left to right for their composite effect.

The preceding twelve symmetries have the multiplication table

(12.3)

$\cdot$	g_1	g_2	g_3	g_4	g_5	g_6	g_7	g_8	g_9	g_{10}	g_{11}	g_{12}
$g_1 = g_1^{-1}$	g_1	g_2	g_3	g_4	g_5	g_6	g_7	g_8	g_9	g_{10}	g_{11}	g_{12}
$g_3 = g_2^{-1}$	g_3	g_1	g_2	g_9	g_7	g_8	g_{12}	g_{10}	g_{11}	g_6	g_4	g_5
$g_2 = g_3^{-1}$	g_2	g_3	g_1	g_{11}	g_{12}	g_{10}	g_5	g_6	g_4	g_8	g_9	g_7
$g_4 = g_4^{-1}$	g_4	g_5	g_6	g_1	g_2	g_3	g_{10}	g_{11}	g_{12}	g_7	g_8	g_9
$g_9 = g_5^{-1}$	g_9	g_7	g_8	g_3	g_1	g_2	g_6	g_4	g_5	g_{12}	g_{10}	g_{11}
$g_{11} = g_6^{-1}$	g_{11}	g_{12}	g_{10}	g_2	g_3	g_1	g_8	g_9	g_7	g_5	g_6	g_4
$g_7 = g_7^{-1}$	g_7	g_8	g_9	g_{10}	g_{11}	g_{12}	g_1	g_2	g_3	g_4	g_5	g_6
$g_{12} = g_8^{-1}$	g_{12}	g_{10}	g_{11}	g_6	g_4	g_5	g_3	g_1	g_2	g_9	g_7	g_8
$g_5 = g_9^{-1}$	g_5	g_6	g_4	g_8	g_9	g_7	g_2	g_3	g_1	g_{11}	g_{12}	g_{10}
$g_{10} = g_{10}^{-1}$	g_{10}	g_{11}	g_{12}	g_7	g_8	g_9	g_4	g_5	g_6	g_1	g_2	g_3
$g_6 = g_{11}^{-1}$	g_6	g_4	g_5	g_{12}	g_{10}	g_{11}	g_9	g_7	g_8	g_3	g_1	g_2
$g_8 = g_{12}^{-1}$	g_8	g_9	g_7	g_5	g_6	g_4	g_{11}	g_{12}	g_{10}	g_2	g_3	g_1

that specifies a group $\mathfrak{T}$ of order 12.

12.2. A matrix representation for $\mathfrak{T}$ of degree 3

The matrices $\mathcal{A} = \begin{bmatrix} 1 & 0 & 0 \\ 0 & -1 & 0 \\ 0 & 0 & -1 \end{bmatrix}$ and $\mathcal{B} = \begin{bmatrix} 0 & 0 & 1 \\ 1 & 0 & 0 \\ 0 & 1 & 0 \end{bmatrix}$ from [**37**, page 31]

satisfy $\mathcal{A}^2 = I$, $\mathcal{B}^3 = I$, and $(\mathcal{A}\mathcal{B})^3 = I$ in correspondence with $\alpha^2 = e$, $\beta^3 = e$, and $(\alpha\beta)^3 = e$ for α and β in (12.2). We set

$$h_1 = I, \qquad h_2 = \mathcal{B}, \qquad h_3 = \mathcal{B}^2,$$
$$h_4 = \mathcal{A}, \qquad h_5 = \mathcal{A}\mathcal{B}, \qquad h_6 = \mathcal{A}\mathcal{B}^2,$$
$$h_7 = \mathcal{B}^2\mathcal{A}\mathcal{B}, \qquad h_8 = \mathcal{A}\mathcal{B}\mathcal{A}, \qquad h_9 = \mathcal{B}^2\mathcal{A},$$
$$h_{10} = \mathcal{B}\mathcal{A}\mathcal{B}^2, \qquad h_{11} = \mathcal{B}\mathcal{A}, \qquad h_{12} = \mathcal{B}\mathcal{A}\mathcal{B}$$

and use computer algebra to see that these twelve matrices are given by

(12.4)
$$
\begin{cases}
h_1 = \begin{bmatrix} 1 & 0 & 0 \\ 0 & 1 & 0 \\ 0 & 0 & 1 \end{bmatrix}, \quad
h_2 = \begin{bmatrix} 0 & 0 & 1 \\ 1 & 0 & 0 \\ 0 & 1 & 0 \end{bmatrix}, \quad
h_3 = \begin{bmatrix} 0 & 1 & 0 \\ 0 & 0 & 1 \\ 1 & 0 & 0 \end{bmatrix}, \\[12pt]
h_4 = \begin{bmatrix} 1 & 0 & 0 \\ 0 & -1 & 0 \\ 0 & 0 & -1 \end{bmatrix}, \quad
h_5 = \begin{bmatrix} 0 & 0 & 1 \\ -1 & 0 & 0 \\ 0 & -1 & 0 \end{bmatrix}, \quad
h_6 = \begin{bmatrix} 0 & 1 & 0 \\ 0 & 0 & -1 \\ -1 & 0 & 0 \end{bmatrix}, \\[12pt]
h_7 = \begin{bmatrix} -1 & 0 & 0 \\ 0 & -1 & 0 \\ 0 & 0 & 1 \end{bmatrix}, \quad
h_8 = \begin{bmatrix} 0 & 0 & -1 \\ -1 & 0 & 0 \\ 0 & 1 & 0 \end{bmatrix}, \quad
h_9 = \begin{bmatrix} 0 & -1 & 0 \\ 0 & 0 & -1 \\ 1 & 0 & 0 \end{bmatrix}, \\[12pt]
h_{10} = \begin{bmatrix} -1 & 0 & 0 \\ 0 & 1 & 0 \\ 0 & 0 & -1 \end{bmatrix}, \quad
h_{11} = \begin{bmatrix} 0 & 0 & -1 \\ 1 & 0 & 0 \\ 0 & -1 & 0 \end{bmatrix}, \quad
h_{12} = \begin{bmatrix} 0 & -1 & 0 \\ 0 & 0 & 1 \\ -1 & 0 & 0 \end{bmatrix}.
\end{cases}
$$

They have the multiplication table

$\cdot$	h_1	h_2	h_3	h_4	h_5	h_6	h_7	h_8	h_9	h_{10}	h_{11}	h_{12}
$h_1 = h_1^{-1}$	h_1	h_2	h_3	h_4	h_5	h_6	h_7	h_8	h_9	h_{10}	h_{11}	h_{12}
$h_3 = h_2^{-1}$	h_3	h_1	h_2	h_9	h_7	h_8	h_{12}	h_{10}	h_{11}	h_6	h_4	h_5
$h_2 = h_3^{-1}$	h_2	h_3	h_1	h_{11}	h_{12}	h_{10}	h_5	h_6	h_4	h_8	h_9	h_7
$h_4 = h_4^{-1}$	h_4	h_5	h_6	h_1	h_2	h_3	h_{10}	h_{11}	h_{12}	h_7	h_8	h_9
$h_9 = h_5^{-1}$	h_9	h_7	h_8	h_3	h_1	h_2	h_6	h_4	h_5	h_{12}	h_{10}	h_{11}
$h_{11} = h_6^{-1}$	h_{11}	h_{12}	h_{10}	h_2	h_3	h_1	h_8	h_9	h_7	h_5	h_6	h_4
$h_7 = h_7^{-1}$	h_7	h_8	h_9	h_{10}	h_{11}	h_{12}	h_1	h_2	h_3	h_4	h_5	h_6
$h_{12} = h_8^{-1}$	h_{12}	h_{10}	h_{11}	h_6	h_4	h_5	h_3	h_1	h_2	h_9	h_7	h_8
$h_5 = h_9^{-1}$	h_5	h_6	h_4	h_8	h_9	h_7	h_2	h_3	h_1	h_{11}	h_{12}	h_{10}
$h_{10} = h_{10}^{-1}$	h_{10}	h_{11}	h_{12}	h_7	h_8	h_9	h_4	h_5	h_6	h_1	h_2	h_3
$h_6 = h_{11}^{-1}$	h_6	h_4	h_5	h_{12}	h_{10}	h_{11}	h_9	h_7	h_8	h_3	h_1	h_2
$h_8 = h_{12}^{-1}$	h_8	h_9	h_7	h_5	h_6	h_4	h_{11}	h_{12}	h_{10}	h_2	h_3	h_1

(12.5)

that can be regarded as obtained from (12.3) by replacing g with h throughout. Thus, a one-to-one matrix representation for $\mathfrak{T}$ is given by $g_k \mapsto h_k$, for $1 \leq k \leq 12$.

12.3. The matrix representations of degree 1 for $\mathfrak{T}$

Suppose that $g_k \mapsto \big[f(g_k)\big]$, for $1 \leq k \leq 12$, is a matrix representation for $\mathfrak{T}$ of degree 1. Then, f is a group-homomorphism of $\mathfrak{T}$ into $\mathbb{C}^*$ and the definitions on page 111 for $g_1, g_2, \ldots, g_{12}$ in terms of α and β yield

$$f(g_1) = 1, \qquad f(g_2) = f(\beta), \qquad f(g_3) = \big(f(\beta)\big)^2,$$

$$f(g_4) = f(\alpha), \qquad f(g_5) = f(\alpha)f(\beta), \qquad f(g_6) = f(\alpha)\big(f(\beta)\big)^2,$$

$$f(g_7) = \big(f(\beta)\big)^2 f(\alpha)f(\beta), \quad f(g_8) = f(\alpha)f(\beta)f(\alpha), \quad f(g_9) = \big(f(\beta)\big)^2 f(\alpha),$$

$$f(g_{10}) = f(\beta)f(\alpha)\big(f(\beta)\big)^2, \quad f(g_{11}) = f(\beta)f(\alpha), \qquad f(g_{12}) = f(\beta)f(\alpha)f(\beta).$$

In view of $\alpha^2 = e$ and $\beta^3 = e$, we have $\big(f(\alpha)\big)^2 = 1$ and $\big(f(\beta)\big)^3 = 1$. Thus, there are six possibilities for f depending on whether $u = f(\alpha)$ is 1 or -1 and whether $v = f(\beta)$ is 1 or ω or ω^2. The table

$f_{u,v}(g_k)$	g_1	g_2	g_3	g_4	g_5	g_6	g_7	g_8	g_9	g_{10}	g_{11}	g_{12}
$f_{1,1}$	1	1	1	1	1	1	1	1	1	1	1	1
$f_{1,\omega}$	1	ω	ω^2	1	ω	ω^2	1	ω	ω^2	1	ω	ω^2
f_{1,ω^2}	1	ω^2	ω	1	ω^2	ω	1	ω^2	ω	1	ω^2	ω
$f_{-1,1}$	1	1	1	-1	-1	-1	-1	1	-1	-1	-1	-1
$f_{-1,\omega}$	1	ω	ω^2	-1	$-\omega$	$-\omega^2$	-1	ω	$-\omega^2$	-1	$-\omega$	$-\omega^2$
f_{-1,ω^2}	1	ω^2	ω	-1	$-\omega^2$	$-\omega$	-1	ω^2	$-\omega$	-1	$-\omega^2$	$-\omega$

(12.6)

provides the details. We observe that $g_t \mapsto \big[f_{u,v}(g_t)\big]$, for $1 \leq t \leq 12$, is a matrix representation for $\mathfrak{T}$ of degree 1 if the condition

$$f_{u,v}(g_i)\, f_{u,v}(g_j) - f_{u,v}(g_k) = 0$$

is satisfied whenever i, j, k are integers subject to $1 \leq i, j, k \leq 12$ and $g_i g_j = g_k$.

PROPOSITION 12.1. *The group $\mathfrak{T}$ has precisely three matrix representations of degree 1. They are given with respect to* (12.6) *by*

$$(12.7) \qquad g_t \mapsto \big[f_{1,1}(g_t)\big], \quad g_t \mapsto \big[f_{1,\omega}(g_t)\big], \quad g_t \mapsto \big[f_{1,\omega^2}(g_t)\big], \quad for \ 1 \le t \le 12.$$

PROOF. For $1 \le r \le 6$ and $1 \le s \le 12$, let $F_{r,s}$ be the complex number in the rth row and sth column of (12.6). Also, we introduce the numbering

$$(12.8) \qquad (1) \ f_{1,1}, \quad (2) \ f_{1,\omega}, \quad (3) \ f_{1,\omega^2}, \quad (4) \ f_{-1,1}, \quad (5) \ f_{-1,\omega}, \quad (6) \ f_{-1,\omega^2}.$$

for the six functions of that table. Then, for $1 \le r \le 6$, the rth function in (12.8) specifies a matrix representation for $\mathfrak{T}$ of degree 1 if the condition

$$(12.9) \qquad\qquad\qquad F_{r,i}F_{r,j} - F_{r,k} = 0$$

is satisfied whenever i, j, k are integers such that $1 \le i, j, k \le 12$ and $h_i h_j - h_k$ is the 3×3 zero matrix.

To perform this check with a version of *Mathematica* such as [48], enter the matrices of (12.4) with names h[1], h[2], ..., h[12] and, while writing w for ω, enter each $F_{r,s}$ for (12.6) with the corresponding name F[r,s]. Then, enter the command

```
repQ[r_] := ( w = Cos[2*Pi/3] + I*Sin[2*Pi/3];
  Do[ If[h[i].h[j]-h[k]=={{0,0,0},{0,0,0},{0,0,0}},
     Print[ "F[", r, ",", i, "].F[", r, ",", j,
          "] - F[", r, ",", k, "] = ",
          Expand[ F[r,i]*F[r,j]-F[r,k] ]], {}],
  {i,1,12}, {j,1,12}, {k,1,12}] )
```

that incorporates the logic for (12.9). After evaluating each of the preceding input commands, we individually evaluate each of repQ[1], repQ[2], repQ[3] and see that each of the individual outputs is 0. Thus, the functions $f_{1,1}$, $f_{1,\omega}$, f_{1,ω^2} of (12.6) specify three distinct matrix representations for $\mathfrak{T}$ of degree 1. We find that each of the separate evaluations of repQ[4], repQ[5], repQ[6] has some nonzero outputs. Thus, there are precisely three matrix representations of degree 1 for $\mathfrak{T}$ and they are given by (12.7). This completes the proof. $\qquad\square$

12.3.1. A suitable set of matrix representations for $\mathfrak{T}$. In terms of the component $F_{r,s}$ in the rth row and sth column of (12.6) for $1 \le r \le 3$ and $1 \le s \le 12$ as well as the twelve 3×3 matrices $h_1, h_2, \ldots, h_{12}$ of (12.4), we observe that four matrix representations F_1, F_2, F_3, F_4 for the group $\mathfrak{T}$ are given by

$$(12.10) \qquad \begin{cases} F_1(g_k) = \big[F_{1,k}\big], & \text{for } 1 \le k \le 12, \\ F_2(g_k) = \big[F_{2,k}\big], & \text{for } 1 \le k \le 12, \\ F_3(g_k) = \big[F_{3,k}\big], & \text{for } 1 \le k \le 12, \\ F_4(g_k) = h_k, & \text{for } 1 \le k \le 12. \end{cases}$$

PROPOSITION 12.2. *The set $\{F_1, F_2, F_3, F_4\}$ is a complete set of pairwise-inequivalent irreducible matrix representations for $\mathfrak{T}$.*

PROOF. There are precisely three pairwise-inequivalent matrix representations of degree 1 for $\mathfrak{T}$. They are given by F_1, F_2, and F_3 in (12.10). Theorems 10.14 and 10.16 show that there is a complete set $\{F_1, F_2, F_3, \ldots\}$ of pairwise-inequivalent irreducible matrix representations for $\mathfrak{T}$ having $1^2 + 1^2 + 1^2 + X = 12$, where $X = 12 - 3 = 9$ is a sum of squares of integers ≥ 2. We can not have $X = 2^2 + Y$, where $Y = 5$ is a sum of squares of integers ≥ 2. Consequently, there are no irreducible matrix representations of $\mathfrak{T}$ having degree 2.

To verify that F_4 is irreducible, suppose F_4 is reducible. Then, Theorem 10.11 on page 98 shows that there is a nonsingular 3×3 matrix S_1 with components in $\mathbb{C}$ and there are elements a_k, α_k, β_k, γ_k, δ_k in $\mathbb{C}$, for $1 \leq k \leq 12$, such that

$$S_1^{-1} F_4(g_k)\, S_1 = S_1^{-1} h_k\, S_1 = \left[\begin{array}{c|cc} a_k & 0 & 0 \\ \hline 0 & \alpha_k & \beta_k \\ 0 & \gamma_k & \delta_k \end{array}\right], \quad \text{for } k = 1, 2, \ldots, 12.$$

Since the matrix representation $g_k \mapsto \begin{bmatrix} \alpha_k & \beta_k \\ \gamma_k & \delta_k \end{bmatrix}$, for $1 \leq k \leq 12$, has degree 2, it is reducible. Therefore, there is a nonsingular 3×3 matrix S_2 with components in $\mathbb{C}$ for which $S_2^{-1} h_k S_2$ is a diagonal matrix when $1 \leq k \leq 12$. However, this gives the contradiction that matrix multiplication for the matrices $h_1, h_2, \ldots, h_{12}$ in (12.4) is commutative. For example, we have $h_2 h_4 = h_{11} \neq h_5 = h_4 h_2$. Thus, F_4 is an irreducible matrix representation for $\mathfrak{T}$.

The pairwise-inequivalent irreducible matrix representations F_1, F_2, F_3, F_4 have respective degrees n_1, n_2, n_3, n_4 that satisfy

$$n_1^2 + n_2^2 + n_3^2 + n_4^2 = 1^2 + 1^2 + 1^2 + 3^2 = 12 = n.$$

Therefore, Theorem 10.17 on page 99 shows that they form a complete set of matrix representations for $\mathfrak{T}$. This completes the proof. $\qquad\square$

12.4. Block-diagonalization of a group-pattern matrix for $\mathfrak{T}$

With respect to the polynomial ring $\mathbb{C}[x_1, x_2, \ldots, x_{12}]$ in twelve independent variables $x_1, x_2, \ldots, x_{12}$ over $\mathbb{C}$, let σ be the function form $\mathfrak{T}$ to $\mathbb{C}[x_1, x_2, \ldots, x_{12}]$ defined by $\sigma(g_k) = x_k$, for $1 \leq k \leq 12$. Let A be the corresponding group-pattern matrix for $\mathfrak{T}$ and $\mathscr{L} = (g_1, g_2, \ldots, g_{12})$ such that $[A]_{r,s} = \sigma(g_r^{-1} g_s)$, for $r, s = 1, 2, \ldots, 12$. It is given by

$$(12.11)\quad A = \left[\begin{array}{ccc|ccc|ccc|ccc}
x_1 & x_2 & x_3 & x_4 & x_5 & x_6 & x_7 & x_8 & x_9 & x_{10} & x_{11} & x_{12} \\
x_3 & x_1 & x_2 & x_9 & x_7 & x_8 & x_{12} & x_{10} & x_{11} & x_6 & x_4 & x_5 \\
x_2 & x_3 & x_1 & x_{11} & x_{12} & x_{10} & x_5 & x_6 & x_4 & x_8 & x_9 & x_7 \\
\hline
x_4 & x_5 & x_6 & x_1 & x_2 & x_3 & x_{10} & x_{11} & x_{12} & x_7 & x_8 & x_9 \\
x_9 & x_7 & x_8 & x_3 & x_1 & x_2 & x_6 & x_4 & x_5 & x_{12} & x_{10} & x_{11} \\
x_{11} & x_{12} & x_{10} & x_2 & x_3 & x_1 & x_8 & x_9 & x_7 & x_5 & x_6 & x_4 \\
\hline
x_7 & x_8 & x_9 & x_{10} & x_{11} & x_{12} & x_1 & x_2 & x_3 & x_4 & x_5 & x_6 \\
x_{12} & x_{10} & x_{11} & x_6 & x_4 & x_5 & x_3 & x_1 & x_2 & x_9 & x_7 & x_8 \\
x_5 & x_6 & x_4 & x_8 & x_9 & x_7 & x_2 & x_3 & x_1 & x_{11} & x_{12} & x_{10} \\
\hline
x_{10} & x_{11} & x_{12} & x_7 & x_8 & x_9 & x_4 & x_5 & x_6 & x_1 & x_2 & x_3 \\
x_6 & x_4 & x_5 & x_{12} & x_{10} & x_{11} & x_9 & x_7 & x_8 & x_3 & x_1 & x_2 \\
x_8 & x_9 & x_7 & x_5 & x_6 & x_4 & x_{11} & x_{12} & x_{10} & x_2 & x_3 & x_1
\end{array}\right].$$

The matrix representations F_1, F_2, F_3, and F_4 of (12.10) have the respective degrees 1, 1, 1, and 3. By specializing Theorem 11.3 to $n_1 = 1$, $n_2 = 1$, $n_3 = 1$, and $n_4 = 3$ with $q = 4$, we find that the table

(12.12)

s	1	2	3	4	5	6	7	8	9	10	11	12
$\lambda(s)$	1	2	3	4	4	4	4	4	4	4	4	4
$\mu(s)$	1	1	1	1	1	1	2	2	2	3	3	3
$\nu(s)$	1	1	1	1	2	3	1	2	3	1	2	3

specifies the corresponding functions $\lambda(s)$, $\mu(s)$, $\nu(s)$, for $1 \le s \le 12$.

The symbolism $\left[F_{\lambda(s)}(g_r)\right]_{\mu(s),\nu(s)}$ was specialized in (11.19) to the context of Section 11.4. Now we apply it to our present context where $\lambda(s)$, $\mu(s)$, $\nu(s)$ and F_k are given by (12.12) and (12.10). This enables us to define a 12×12 matrix M by

$$(12.13) \qquad \left[M\right]_{r,s} = \left[F_{\lambda(s)}(g_r)\right]_{\mu(s),\nu(s)}, \quad \text{for } r, s = 1, 2, \ldots, 12.$$

Machine computations presented in Section 12.5 show that M is given by

$$(12.14) \quad M = \begin{bmatrix}
1 & 1 & 1 & 1 & 0 & 0 & 0 & 1 & 0 & 0 & 0 & 1 \\
1 & \omega & \omega^2 & 0 & 0 & 1 & 1 & 0 & 0 & 0 & 1 & 0 \\
1 & \omega^2 & \omega & 0 & 1 & 0 & 0 & 1 & 1 & 0 & 0 \\
1 & 1 & 1 & 1 & 0 & 0 & 0 & -1 & 0 & 0 & 0 & -1 \\
1 & \omega & \omega^2 & 0 & 0 & 1 & -1 & 0 & 0 & 0 & -1 & 0 \\
1 & \omega^2 & \omega & 0 & 1 & 0 & 0 & 0 & -1 & -1 & 0 & 0 \\
1 & 1 & 1 & -1 & 0 & 0 & 0 & -1 & 0 & 0 & 0 & 1 \\
1 & \omega & \omega^2 & 0 & 0 & -1 & -1 & 0 & 0 & 0 & 1 & 0 \\
1 & \omega^2 & \omega & 0 & -1 & 0 & 0 & 0 & -1 & 1 & 0 & 0 \\
1 & 1 & 1 & -1 & 0 & 0 & 0 & 1 & 0 & 0 & 0 & -1 \\
1 & \omega & \omega^2 & 0 & 0 & -1 & 1 & 0 & 0 & 0 & -1 & 0 \\
1 & \omega^2 & \omega & 0 & -1 & 0 & 0 & 0 & 1 & -1 & 0 & 0
\end{bmatrix}$$

and M is nonsingular. They also establish the main result of this section. Namely, *M transforms the group-pattern matrix A of* (12.11) *into the block-diagonal form*

$$(12.15) \qquad\qquad\qquad M^{-1}AM = B$$

where

$$(12.16) \quad B = \begin{bmatrix}
d_1 & 0 & 0 & 0 & 0 & 0 & 0 & 0 & 0 & 0 & 0 & 0 \\
0 & d_2 & 0 & 0 & 0 & 0 & 0 & 0 & 0 & 0 & 0 & 0 \\
0 & 0 & d_3 & 0 & 0 & 0 & 0 & 0 & 0 & 0 & 0 & 0 \\
0 & 0 & 0 & d_4 & d_5 & d_6 & 0 & 0 & 0 & 0 & 0 & 0 \\
0 & 0 & 0 & d_7 & d_8 & d_9 & 0 & 0 & 0 & 0 & 0 & 0 \\
0 & 0 & 0 & d_{10} & d_{11} & d_{12} & 0 & 0 & 0 & 0 & 0 & 0 \\
0 & 0 & 0 & 0 & 0 & 0 & d_4 & d_5 & d_6 & 0 & 0 & 0 \\
0 & 0 & 0 & 0 & 0 & 0 & d_7 & d_8 & d_9 & 0 & 0 & 0 \\
0 & 0 & 0 & 0 & 0 & 0 & d_{10} & d_{11} & d_{12} & 0 & 0 & 0 \\
0 & 0 & 0 & 0 & 0 & 0 & 0 & 0 & 0 & d_4 & d_5 & d_6 \\
0 & 0 & 0 & 0 & 0 & 0 & 0 & 0 & 0 & d_7 & d_8 & d_9 \\
0 & 0 & 0 & 0 & 0 & 0 & 0 & 0 & 0 & d_{10} & d_{11} & d_{12}
\end{bmatrix}$$

and the components $d_1, d_2, \ldots, d_{12}$ of B are given by

$$d_1 = \sum_{k=1}^{12} x_k,$$

$$d_2 = \left[\sum_{k=0}^{3} x_{1+3k}\right] + \omega \left[\sum_{k=0}^{3} x_{2+3k}\right] + \omega^2 \left[\sum_{k=0}^{3} x_{3+3k}\right],$$

$$d_3 = \left[\sum_{k=0}^{3} x_{1+3k}\right] + \omega^2 \left[\sum_{k=0}^{3} x_{2+3k}\right] + \omega \left[\sum_{k=0}^{3} x_{3+3k}\right],$$

$$d_4 = x_1 + x_4 - x_7 - x_{10}, \quad d_5 = x_3 + x_6 - x_9 - x_{12}, \quad d_6 = x_2 + x_5 - x_8 - x_{11},$$

$$d_7 = x_2 - x_5 - x_8 + x_{11}, \quad d_8 = x_1 - x_4 - x_7 + x_{10}, \quad d_9 = x_3 - x_6 - x_9 + x_{12},$$

$$d_{10} = x_3 - x_6 + x_9 - x_{12}, \quad d_{11} = x_2 - x_5 + x_8 - x_{11}, \quad d_{12} = x_1 - x_4 + x_7 - x_{10}.$$

PROPOSITION 12.3. *The matrix M of (12.13) is nonsingular and has as its inverse the 12×12 matrix L that is defined by*

$$(12.17) \qquad [L]_{r,s} = (n_{\lambda(r)}/12)[F_{\lambda(r)}(g_s^{-1})]_{\nu(r),\mu(r)}, \quad \text{for } r, s = 1, 2, \ldots, 12.$$

PROOF. For $1 \leq r, s \leq 12$, we apply (12.17), (12.13), and Theorem 10.23 to deduce

$$[LM]_{r,s} = \sum_{k=1}^{12} [L]_{r,k}[M]_{k,s} = \sum_{k=1}^{12} (n_{\lambda(r)}/12)[F_{\lambda(r)}(g_k^{-1})]_{\nu(r),\mu(r)} [F_{\lambda(s)}(g_k)]_{\mu(s),\nu(s)}$$

$$= \begin{cases} 1, & \text{if } \lambda(r) = \lambda(s), \ \mu(r) = \mu(s), \text{ and } \nu(r) = \nu(s), \\ 0, & \text{otherwise,} \end{cases} = \begin{cases} 1, & \text{if } r = s, \\ 0, & \text{if } r \neq s. \end{cases}$$

Thus, M is nonsingular and its inverse is L. This completes the proof. $\qquad \square$

Machine computations of Section 12.5 show that (12.10) and (12.17) yield

$$(12.18) \qquad L = \frac{1}{12} \begin{bmatrix} 1 & 1 & 1 & 1 & 1 & 1 & 1 & 1 & 1 & 1 & 1 & 1 \\ 1 & \omega^2 & \omega & 1 & \omega^2 & \omega & 1 & \omega^2 & \omega & 1 & \omega^2 & \omega \\ 1 & \omega & \omega^2 & 1 & \omega & \omega^2 & 1 & \omega & \omega^2 & 1 & \omega & \omega^2 \\ 3 & 0 & 0 & 3 & 0 & 0 & -3 & 0 & 0 & -3 & 0 & 0 \\ 0 & 0 & 3 & 0 & 0 & 3 & 0 & 0 & -3 & 0 & 0 & -3 \\ 0 & 3 & 0 & 0 & 3 & 0 & 0 & -3 & 0 & 0 & -3 & 0 \\ 0 & 3 & 0 & 0 & -3 & 0 & 0 & -3 & 0 & 0 & 3 & 0 \\ 3 & 0 & 0 & -3 & 0 & 0 & -3 & 0 & 0 & 3 & 0 & 0 \\ 0 & 0 & 3 & 0 & 0 & -3 & 0 & 0 & -3 & 0 & 0 & 3 \\ 0 & 0 & 3 & 0 & 0 & -3 & 0 & 0 & 3 & 0 & 0 & -3 \\ 0 & 3 & 0 & 0 & -3 & 0 & 0 & 3 & 0 & 0 & -3 & 0 \\ 3 & 0 & 0 & -3 & 0 & 0 & 3 & 0 & 0 & -3 & 0 & 0 \end{bmatrix}$$

as well as

$$\det(M) = 12,288\sqrt{3}\,i \quad \text{and} \quad \det(L) = \frac{-i}{12,288\sqrt{3}}, \quad \text{where } i^2 = -1.$$

OBSERVATION 12.4. The formula $F_{\lambda(r)}(g_s^{-1}) = (F_{\lambda(r)}(g_s))^{-1}$ is available for applications of (12.17). However, for Section 12.5, we find it more convenient to employ $F_{\lambda(r)}(g_s^{-1}) = F_{\lambda(r)}(g_{\pi(s)})$, where π is the permutation of $\{1, 2, \ldots, 12\}$ that yields $g_{\pi(s)} = g_s^{-1}$, for $1 \leq s \leq 12$.

12.5. Computer-algebra verifications for the preceding section

We use a version of *Mathematica* such as [48] and enter the input statements

```
n[1] = 1;  n[2] = 1;  n[3] = 1;  n[4] = 3;  q = 4;
sum = Sum[ n[k]^2, {k,1,q}];

lambda[s_ /; 1 <= s && s <= sum] := Catch[ Do[ If[
    Sum[n[k]^2, {k, 1, u - 1}] < s <= Sum[n[k]^2, {k, 1, u}],
    Throw[u] ], {u, 1, q}]  ]

mu[s_ /; 1 <= s && s <= sum] :=
    Module[{u, v}, u = s - Sum[n[k]^2, {k, 1, lambda[s] - 1}];
    v = Mod[u, n[lambda[s]]]; Catch[ If[ v == 0,
    Throw[ Quotient[u, n[lambda[s]]]],
    Throw[ Quotient[u, n[lambda[s]]] + 1 ] ] ] ]

nu[s_ /; 1 <= s && s <= sum] :=
    ( s - Sum[n[k]^2, {k, 1, lambda[s]-1}]
        - (mu[s]-1)* n[lambda[s]] )
```

in a *Mathematica* notebook. After successively evaluating each of them, we enter

```
Do[ Print[ "(lambda[", k, "], mu[", k, "], nu[", k, "]) = (",
lambda[k], ",", mu[k], ",", nu[k], ")"], {k, 1, sum}]
```

and find that its evaluation serves as a check by yielding (12.12). We input the matrices of (12.4) by either using *notebook palettes* or by entering the statements

```
h[1] = {{1,0,0}, {0,1,0}, {0,0,1}};

h[2] = {{0,0,1}, {1,0,0}, {0,1,0}};

h[3] = {{0,1,0}, {0,0,1}, {1,0,0}};

h[4] = {{1,0,0}, {0,-1,0}, {0,0,-1}};

h[5] = {{0,0,1}, {-1,0,0}, {0,-1,0}};

h[6] = {{0,1,0}, {0,0,-1}, {-1,0,0}};

h[7] = {{-1,0,0}, {0,-1,0}, {0,0,1}};

h[8] = {{0,0,-1}, {-1,0,0}, {0,1,0}};

h[9] = {{0,-1,0}, {0,0,-1}, {1,0,0}};

h[10] = {{-1,0,0}, {0,1,0}, {0,0,-1}};

h[11] = {{0,0,-1}, {1,0,0}, {0,-1,0}};

h[12] = {{0,-1,0}, {0,0,1}, {-1,0,0}};
```

and evaluating them. For the complete set of pairwise-inequivalent irreducible matrix representations, we enter the input

```
F[1,k_] := {{1}}    /;   1 <= k <= 12

F[2,k_] := {{1}}    /;   (k == 1 || k == 4 || k == 7 || k == 10)

F[2,k_] := {{w}}    /;   (k == 2 || k == 5 || k == 8 || k == 11)

F[2,k_] := {{w^2}} /;   (k == 3 || k == 6 || k == 9 || k == 12)

F[3,k_] := {{1}}    /;   (k == 1 || k == 4 || k == 7 || k == 10)

F[3,k_] := {{w^2}} /;   (k == 2 || k == 5 || k == 8 || k == 11)

F[3,k_] := {{w}}    /;   (k == 3 || k == 6 || k == 9 || k == 12)

F[4,k_] := h[k]     /;   1 <= k <= 12
```

and evaluate it. We use (12.13) to see that the input and evaluation of

```
m[r_,s_] := F[lambda[s], r][[mu[s], nu[s]]]
            /;  ( 1 <= r <= 12  &&  1 <= s <= 12 )

M =  Table[m[r,s], {r,1,12}, {s,1,12}];   M // MatrixForm


w = Cos[2Pi/3] + I*Sin[2Pi/3]

Det[M]  //  FullSimplify
```

yields a representation M of M and shows that M is nonsingular. To represent the group-pattern matrix A for $\mathfrak{T}$ and $\mathscr{L}$ that is given in (12.11), we enter and evaluate

```
A = {{x1,  x2,  x3,  x4,  x5,  x6,  x7,  x8,  x9,  x10, x11, x12},
     {x3,  x1,  x2,  x9,  x7,  x8,  x12, x10, x11, x6,  x4,  x5 },
     {x2,  x3,  x1,  x11, x12, x10, x5,  x6,  x4,  x8,  x9,  x7 },
     {x4,  x5,  x6,  x1,  x2,  x3,  x10, x11, x12, x7,  x8,  x9 },
     {x9,  x7,  x8,  x3,  x1,  x2,  x6,  x4,  x5,  x12, x10, x11},
     {x11, x12, x10, x2,  x3,  x1,  x8,  x9,  x7,  x5,  x6,  x4 },
     {x7,  x8,  x9,  x10, x11, x12, x1,  x2,  x3,  x4,  x5,  x6 },
     {x12, x10, x11, x6,  x4,  x5,  x3,  x1,  x2,  x9,  x7,  x8 },
     {x5,  x6,  x4,  x8,  x9,  x7,  x2,  x3,  x1,  x11, x12, x10},
     {x10, x11, x12, x7,  x8,  x9,  x4,  x5,  x6,  x1,  x2,  x3 },
     {x6,  x4,  x5,  x12, x10, x11, x9,  x7,  x8,  x3,  x1,  x2 },
     {x8,  x9,  x7,  x5,  x6,  x4,  x11, x12, x10, x2,  x3,  x1 }};
```

or, more easily, use a *notebook palette*. To represent the block-diagonal matrix B in (12.16), we input and evaluate

```
x[1] = x1; x[2]  = x2;   x[3] = x3;   x[4] = x4;
x[5] = x5; x[6]  = x6;   x[7] = x7;   x[8] = x8;
x[9] = x9; x[10] = x10; x[11] = x11; x[12] = x12;

d1 = Sum[x[k], {k,1,12}]

d2 = ( Sum[ x[1+3k], {k,0,3}] + w*Sum[ x[2+3k], {k,0,3}]
          + w^2*Sum[ x[3+3k], {k,0,3}] )

d3 = ( Sum[ x[1+3k],{k,0,3}] + w^2*Sum[ x[2+3k],{k,0,3}]
          + w*Sum[ x[3+3k], {k,0,3}] )

d4  = x1+x4-x7-x10;  d5  = x3+x6-x9-x12;  d6  = x2+x5-x8-x11;
d7  = x2-x5-x8+x11;  d8  = x1-x4-x7+x10;  d9  = x3-x6-x9+x12;
d10 = x3-x6+x9-x12;  d11 = x2-x5+x8-x11;  d12 = x1-x4+x7-x10;

B = {{d1, 0,  0,  0,   0,   0,   0,   0,   0,   0,   0,   0},
     {0, d2,  0,  0,   0,   0,   0,   0,   0,   0,   0,   0},
     {0,  0, d3,  0,   0,   0,   0,   0,   0,   0,   0,   0},
     {0,  0,  0, d4,  d5,  d6,   0,   0,   0,   0,   0,   0},
     {0,  0,  0, d7,  d8,  d9,   0,   0,   0,   0,   0,   0},
     {0,  0,  0, d10, d11, d12,  0,   0,   0,   0,   0,   0},
     {0,  0,  0,  0,   0,   0,  d4,  d5,  d6,   0,   0,   0},
     {0,  0,  0,  0,   0,   0,  d7,  d8,  d9,   0,   0,   0},
     {0,  0,  0,  0,   0,   0,  d10, d11, d12, 0,   0,   0},
     {0,  0,  0,  0,   0,   0,   0,   0,   0,  d4,  d5,  d6},
     {0,  0,  0,  0,   0,   0,   0,   0,   0,  d7,  d8,  d9},
     {0,  0,  0,  0,   0,   0,   0,   0,   0,  d10, d11, d12}};
```

where, at the last step, a *notebook palette* can be used to input B. .

Next, as a climax, the input and evaluation of

```
FullSimplify[ Inverse[M].A.M - B ]
```

yields the 12×12 zero matrix as output and therefore establishes the key result that $M^{-1}AM = B$ in (12.15).

To obtain L in (12.18), we enter and evaluate the statements

```
pi[1]  = 1; pi[2] = 3; pi[3] = 2; pi[4] = 4; pi[5] = 9;
pi[6]  = 11; pi[7] = 7; pi[8] = 12; pi[9] = 5;
pi[10] = 10; pi[11] = 6; pi[12] = 8;
ell[r_,s_] := ( n[lambda[r]]*
                F[lambda[r], pi[s]][[nu[r], mu[r]]]
                /; (1 <= r <= 12 && 1 <= s <= 12) )

L0 = Table[ ell[r,s], {r,1,12}, {s,1,12}]; L0 // MatrixForm
```

and find that (12.18) is given by L = (1/12)L0.

12.6. Further details dependent on Chapter 14

With respect to M in (12.14) and Theorem 14.1 on page 136 , the formula

$$[d_1, d_2, d_3, d_4, d_5, d_6, d_7, d_8, d_9, d_{10}, d_{11}, d_{12}]$$
$$= [x_1, x_2, x_3, x_4, x_5, x_6, x_7, x_8, x_9, x_{10}, x_{11}, x_{12}]M$$

makes it easy to deduce the components $d_1, d_2, \ldots, d_{12}$ of B in (12.16).

For the matrix A defined by (12.11), we use $M^{-1}AM = B$ and (12.16) to obtain $\det(A) = (\det M)^{-1}\det(A)\det(M) = \det(M^{-1}AM) = \det(B)$ and

$$(12.19) \qquad \det(A) = d_1 d_2 d_3 \mathscr{B}_4^3, \quad \text{with} \quad \mathscr{B}_4 = \begin{vmatrix} d_4 & d_5 & d_6 \\ d_7 & d_8 & d_9 \\ d_{10} & d_{11} & d_{12} \end{vmatrix},$$

where $d_1, d_2, d_3, d_4, \ldots, d_{12}$ are given on page 117. Theorem 14.2 on page 137 shows that $\mathscr{B}_4$ is an irreducible polynomial in the variables $x_1, x_2, \ldots, x_{12}$ over $\mathbb{C}$.

By using [48], we find that the expansion of $\mathscr{B}_4$ has 124 terms and the expansion of $\det(A)$ based on (12.19) has $417,577$ terms.

For general results about this particular type of block-diagonalizations that we term a Frobenius block-diagonalization, see Theorem 13.3 on page 125.

CHAPTER 13

Frobenius Block-Diagonalizations
for Group-Pattern Matrices

In accordance with Theorem 10.14, let $\{F_1, F_2, \ldots, F_q\}$ denote a complete set of pairwise-inequivalent irreducible matrix representations with components in $\mathbb{C}$ for a finite group G of order n having $\mathscr{L} = (g_1, g_2, \ldots, g_n)$ as a list of its elements. We also let $\mathbb{F}$ be $\mathbb{C}$ or any proper field extension of $\mathbb{C}$ and we introduce $\mathfrak{A}$ as the set of group-pattern matrices over $\mathbb{F}$ for G and $\mathscr{L}$. Thus, a given $n \times n$ matrix A is an element of $\mathfrak{A}$ if and only if there is a function σ from G to $\mathbb{F}$ such that

$$(13.1) \qquad \left[A\right]_{r,s} = \sigma\left(g_r^{-1} g_s\right), \quad \text{for } r, s = 1, 2, \ldots, n.$$

Our goal is to define an $n \times n$ nonsingular matrix M over $\mathbb{C}$ analogous to M in (12.13) for (12.14)–(12.16) and to introduce a set $\mathfrak{B}$ of special $n \times n$ block-diagonal matrices analogous to (12.16) such that: for any $n \times n$ matrix A over $\mathbb{F}$, A is an element of $\mathfrak{A}$ if and only if $M^{-1}AM$ is an element of $\mathfrak{B}$.

13.1. The set $\mathfrak{B}$ of Frobenius block-diagonal matrices for G

To define $\mathfrak{B}$ uniquely, we assume that $F_1, F_2, \ldots, F_q$ have been named so their respective degrees $n_1, n_2, \ldots, n_q$ satisfy $1 = n_1 \leq n_2 \leq \cdots \leq n_q$. We have $n_1 = 1$; namely, the matrix representation $x \mapsto [1]$, for each x in G, has degree 1.

Let B denote an $n \times n$ matrix over $\mathbb{F}$. To specify block submatrices of B, we rewrite the relation $n_1^2 + \cdots + n_\ell^2 + \cdots + n_q^2 = n$ given by Theorem 10.16 as

$$(\overbrace{n_1}^{n_1=1}) + \cdots + (\overbrace{n_\ell + n_\ell + \cdots + n_\ell}^{n_\ell \text{ terms}}) + \cdots + (\overbrace{n_q + n_q + \cdots + n_q}^{n_q \text{ terms}}) = n$$

to visualize the blocks as submatrices of B located along its principal diagonal. Thus, along that diagonal, we consider the square submatrices

$$(13.2) \qquad \overbrace{\mathscr{B}_{1,1}}^{1 \times 1}, \ldots, \overbrace{\mathscr{B}_{\ell,1}, \mathscr{B}_{\ell,2}, \ldots, \mathscr{B}_{\ell,n_\ell}}^{n_\ell \text{ blocks of size } n_\ell \times n_\ell}, \ldots, \overbrace{\mathscr{B}_{q,1}, \mathscr{B}_{q,2}, \ldots, \mathscr{B}_{q,n_q}}^{n_q \text{ blocks of size } n_q \times n_q}$$

where, for $1 \leq \ell \leq q$ and $1 \leq m \leq n_\ell$, $\mathscr{B}_{\ell,m}$ is defined in terms of

$$\kappa = \sum_{k=1}^{\ell-1} n_k^2 + (m-1)n_\ell$$

as the $n_\ell \times n_\ell$ submatrix having

$$\left[\mathscr{B}_{\ell,m}\right]_{r,s} = \left[B\right]_{\kappa+r,\,\kappa+s}, \quad \text{for } r, s = 1, 2, \ldots, n_\ell.$$

We specialize Theorem 11.3 of page 104 to our present context and see that the functions $\lambda(s)$ and $\mu(s)$ that it defines with respect to (11.7) and (11.8) have the

123

property that: when $[B]_{r,s}$, with $1 \leq r,\ s \leq n$, is any component of B, then $[B]_{r,s}$ is positioned as a component in some one of the submatrices (13.2) of B if and only if $\lambda(r) = \lambda(s)$ and $\mu(r) = \mu(s)$. Thus, $[B]_{r,s}$ is positioned in B outside the submatrices (13.2) of B if and only if either $\lambda(r) \neq \lambda(s)$ or $\mu(r) \neq \mu(s)$.

Using the function δ from $\mathbb{C} \times \mathbb{C}$ to $\mathbb{C}$ having

$$(13.3) \qquad \delta(x,\ y) = \begin{cases} 1, & \text{if } x = y, \\ 0, & \text{if } x \neq y, \end{cases}$$

we see that: for $1 \leq r,\ s \leq n$, the component $[B]_{r,s}$ of B in its rth row and sth column is outside the submatrices of (13.2) if and only if

$$\delta\big(\lambda(r),\ \lambda(s)\big) \cdot \delta\big(\mu(r),\ \mu(s)\big) = 0.$$

Since we also have $\delta\big(\lambda(r),\ \lambda(s)\big) \cdot \delta\big(\mu(r),\ \mu(s)\big) = 1$ if and only if $[B]_{r,s}$ is positioned in some submatrix of (13.2) for B, we arrive at the following conclusion.

In order for the components of B outside the submatrices of (13.2) to be zero, it is necessary and sufficient that

$$(13.4) \qquad [B]_{r,s} = \delta\big(\lambda(r), \lambda(s)\big) \cdot \delta\big(\mu(r),\ \mu(s)\big) \cdot [B]_{r,s}, \quad \text{for } r,\ s = 1,\ 2,\ \ldots,\ n.$$

Let $\mathfrak{B}$ denote the set of $n \times n$ matrices B over $\mathbb{F}$ such that B satisfies (13.4) and its submatrices (13.2) satisfy

$$(13.5) \qquad \mathscr{B}_{\ell,1} = \mathscr{B}_{\ell,2} = \cdots = \mathscr{B}_{\ell,n_\ell}, \quad \text{for } \ell = 1,\ 2,\ \ldots,\ q.$$

NOTATION 13.1. For B in $\mathfrak{B}$ and its submatrices given by (13.2) and (13.5), we introduce $\mathscr{B}_\ell = \mathscr{B}_{\ell,1}$, when $1 \leq \ell \leq q$.

For the set $\mathfrak{A}$ of group-pattern matrices over $\mathbb{F}$ for G and $\mathscr{L}$, see (13.1).

13.2. Employment of a matrix M to relate $\mathfrak{A}$ and $\mathfrak{B}$

We continue the specialization of Theorem 11.3 on page 104 to the situation where $n_1, n_2, \ldots, n_q$ are the respective degrees for the given matrix representations $F_1, F_2, \ldots, F_q$ for G. That yields three integer-valued functions $\lambda(s)$, $\mu(s)$, $\nu(s)$ defined for $1 \leq s \leq n$. We use all three of them with $F_1, F_2, \ldots, F_q$ and the list $\mathscr{L} = \big(g_1, g_2, \ldots, g_n\big)$ for G to define an $n \times n$ matrix M by

$$(13.6) \qquad [M]_{r,s} = \Big[F_{\lambda(s)}(g_r)\Big]_{\mu(s),\nu(s)}, \quad \text{for } r,\ s = 1,\ 2,\ \ldots,\ n.$$

THEOREM 13.2. *The matrix M is nonsingular and its inverse is the $n \times n$ matrix L defined by*

$$(13.7) \qquad [L]_{r,s} = \frac{n_{\lambda(r)}}{n}\Big[F_{\lambda(r)}(g_s^{-1})\Big]_{\nu(r),\mu(r)}, \quad \text{for } r,\ s = 1,\ 2,\ \ldots,\ n.$$

PROOF. For $1 \leq r,\ s \leq n$, we use (13.7), (13.6), and (10.38) to obtain

$$[LM]_{r,s} = \sum_{k=1}^{n} [L]_{r,k}\,[M]_{k,s} = \sum_{k=1}^{n} \frac{n_{\lambda(r)}}{n}\Big[F_{\lambda(r)}(g_k^{-1})\Big]_{\nu(r),\mu(r)}\Big[F_{\lambda(s)}(g_k)\Big]_{\mu(s),\nu(s)}$$

$$= \begin{cases} 1, & \text{if } \lambda(r) = \lambda(s),\ \mu(r) = \mu(s),\ \text{and } \nu(r) = \nu(s), \\ 0, & \text{otherwise,} \end{cases} = \begin{cases} 1, & \text{if } r = s, \\ 0, & \text{if } r \neq s. \end{cases}$$

Thus, M is nonsingular and L is its inverse. This completes the proof. $\square$

THEOREM 13.3. *An $n \times n$ matrix A having components in $\mathbb{F}$ is an element of $\mathfrak{A}$ if and only if $M^{-1}AM$ is an element of $\mathfrak{B}$. Moreover, if A in $\mathfrak{A}$ is defined by (13.1) and the blocks of the matrix $B = M^{-1}AM$ in $\mathfrak{B}$ are designated as*

$$(13.8) \qquad \overbrace{\mathscr{B}_1}^{1 \times 1}, \ldots, \overbrace{\mathscr{B}_\ell, \mathscr{B}_\ell, \ldots, \mathscr{B}_\ell}^{n_\ell \text{ blocks of size } n_\ell \times n_\ell}, \ldots, \overbrace{\mathscr{B}_q, \mathscr{B}_q, \ldots, \mathscr{B}_q}^{n_q \text{ blocks of size } n_q \times n_q},$$

then, with respect to $\{F_1, F_2, \ldots, F_q\}$ and (13.1), these blocks are given by

$$(13.9) \qquad \mathscr{B}_\ell = \sum_{\nu=1}^{n} \sigma(g_\nu)\, F_\ell(g_\nu), \quad \text{for } \ell = 1, 2, \ldots, q.$$

PROOF. (i) Suppose that A is an element of $\mathfrak{A}$ and set $B = M^{-1}AM$. Then, there is a function σ from G to $\mathbb{F}$ such that

$$(13.10) \qquad \left[A\right]_{r,s} = \sigma\big(g_r^{-1}g_s\big), \quad \text{for } r, s = 1, 2, \ldots, n.$$

We observe that

$$(13.11) \quad \left[B\right]_{r,s} = \left[M^{-1}(AM)\right]_{r,s} = \sum_{i=1}^{n} \left[L\right]_{r,i}\left[AM\right]_{i,s}, \quad \text{for } r, s = 1, 2, \ldots, n.$$

For $1 \le i, s \le n$, we use (13.10) and (13.6) to verify that

$$(13.12) \qquad \left[AM\right]_{i,s} = \sum_{j=1}^{n} \left[A\right]_{i,j}\left[M\right]_{j,s} = \sum_{j=1}^{n} \sigma\big(g_i^{-1}g_j\big)\left[F_{\lambda(s)}(g_j)\right]_{\mu(s),\,\nu(s)}.$$

The elements of G are $g_i g_1, g_i g_2, \ldots, g_i g_n$. Thus, we can replace g_j at each of its occurrences in (13.12) with $g_i g_j$ to deduce, for $1 \le i, s \le n$, that

$$\left[AM\right]_{i,s} = \sum_{j=1}^{n} \sigma(g_j)\left[F_{\lambda(s)}(g_i g_j)\right]_{\mu(s),\,\nu(s)} = \sum_{j=1}^{n} \sigma(g_j)\left[F_{\lambda(s)}(g_i)\, F_{\lambda(s)}(g_j)\right]_{\mu(s),\,\nu(s)}$$

$$= \sum_{j=1}^{n} \sigma(g_j) \sum_{t=1}^{n_{\lambda(s)}} \left[F_{\lambda(s)}(g_i)\right]_{\mu(s),t}\left[F_{\lambda(s)}(g_j)\right]_{t,\,\nu(s)}.$$

We use (13.11), (13.7), the preceding formula, (10.38), and (13.3) to obtain

$$\left[B\right]_{r,s} = \sum_{i=1}^{n} \frac{n_{\lambda(r)}}{n}\left[F_{\lambda(r)}(g_i^{-1})\right]_{\nu(r),\mu(r)}\left[AM\right]_{i,s}$$

$$= \sum_{j=1}^{n} \sigma(g_j) \sum_{t=1}^{n_{\lambda(s)}} \left[F_{\lambda(s)}(g_j)\right]_{t,\,\nu(s)} \frac{n_{\lambda(r)}}{n} \sum_{i=1}^{n} \left[F_{\lambda(r)}(g_i^{-1})\right]_{\nu(r),\mu(r)}\left[F_{\lambda(s)}(g_i)\right]_{\mu(s),t}$$

$$= \sum_{j=1}^{n} \sigma(g_j) \sum_{t=1}^{n_{\lambda(s)}} \left[F_{\lambda(s)}(g_j)\right]_{t,\,\nu(s)} \delta\big(\lambda(r), \lambda(s)\big)\, \delta\big(\mu(r), \mu(s)\big)\, \delta\big(\nu(r), t\big)$$

$$= \sum_{j=1}^{n} \sigma(g_j) \sum_{t=1}^{n_{\lambda(r)}} \left[F_{\lambda(s)}(g_j)\right]_{t,\,\nu(s)} \delta\big(\lambda(r), \lambda(s)\big)\, \delta\big(\mu(r), \mu(s)\big)\, \delta\big(\nu(r), t\big).$$

The preceding line was obtained from its predecessor by replacing $n_{\lambda(s)}$ above the second summation sign with $n_{\lambda(r)}$. Namely, if $\lambda(r) \neq \lambda(s)$, then $\delta\big(\lambda(r), \lambda(s)\big) = 0$ and both lines are equal to 0. For the integers t such that $1 \leq t \leq n_{\lambda(r)}$, we note that $\delta\big(\nu(r), t\big)$ is nonzero if and only if $t = \nu(r)$. Thus, the preceding formula yields

$$(13.13) \qquad \big[B\big]_{r,s} = \delta\big(\lambda(r), \lambda(s)\big)\, \delta\big(\mu(r), \mu(s)\big) \sum_{j=1}^{n} \sigma(g_j) \Big[F_{\lambda(s)}(g_j)\Big]_{\nu(r),\,\nu(s)},$$
$$\text{for } r,\, s = 1,\, 2,\, \ldots,\, n.$$

Consequently, B satisfies (13.4) and therefore its submatrices

$$\overbrace{\mathscr{B}_{1,1}}^{1\times 1}, \ldots, \overbrace{\mathscr{B}_{\ell,1}, \mathscr{B}_{\ell,2}, \ldots, \mathscr{B}_{\ell,n_\ell}}^{n_\ell \text{ blocks of size } n_\ell \times n_\ell}, \ldots, \overbrace{\mathscr{B}_{q,1}, \mathscr{B}_{q,2}, \ldots, \mathscr{B}_{q,n_q}}^{n_q \text{ blocks of size } n_q \times n_q}$$

from (13.2) are such that the components of B outside them are zero.

To establish that B also satisfies (13.5), suppose that α, β_1, β_2 are integers subject to $1 \leq \alpha \leq q$ and $1 \leq \beta_1, \beta_2 \leq n_\alpha$. Then, for any integers i, j that satisfy $1 \leq i, j \leq n_\alpha$, we use the abbreviation $\kappa = n_1^2 + n_2^2 + \cdots + n_{\alpha-1}^2$ to obtain

$$\big[\mathscr{B}_{\alpha,\beta_1}\big]_{i,j} = \big[B\big]_{r_1,s_1}, \text{ with } r_1 = \kappa + (\beta_1 - 1)n_\alpha + i \text{ and } s_1 = \kappa + (\beta_1 - 1)n_\alpha + j,$$

as well as

$$\big[\mathscr{B}_{\alpha,\beta_2}\big]_{i,j} = \big[B\big]_{r_2,s_2}, \text{ with } r_2 = \kappa + (\beta_2 - 1)n_\alpha + i \text{ and } s_2 = \kappa + (\beta_2 - 1)n_\alpha + j.$$

With $\lambda(r_1) = \lambda(s_1) = \lambda(r_2) = \lambda(s_2) = \alpha$, $\mu(r_1) = \mu(s_1) = \beta_1$, $\mu(r_2) = \mu(s_2) = \beta_2$ as well as $\nu(r_1) = \nu(r_2) = i$ and $\nu(s_1) = \nu(s_2) = j$, we see that (13.13) yields

$$\big[B\big]_{r_1,s_1} = \sum_{k=1}^{n} \sigma(g_k) \Big[F_\alpha(g_k)\Big]_{i,j} = \big[B\big]_{r_2,s_2}.$$

Consequently, we have $\big[\mathscr{B}_{\alpha\beta_1}\big]_{i,j} = \big[\mathscr{B}_{\alpha\beta_2}\big]_{i,j}$, for $1 \leq i, j \leq n_\alpha$. We use this to conclude that $\mathscr{B}_{\alpha,\beta_1} = \mathscr{B}_{\alpha,\beta_2}$ is valid and B therefore satisfies (13.5). Since both (13.4) and (13.5) are satisfied, B is an element of $\mathfrak{B}$.

(ii) For an $n \times n$ matrix A having components in $\mathbb{F}$, suppose that $M^{-1}AM$ is an element of $\mathfrak{B}$ and set $B = M^{-1}AM$. Then, B satisfies (13.4) and (13.5).

For $\ell = 1, 2, \ldots, q$, let P_ℓ be the set of pairs (a, b) of integers a, b defined by

$$P_\ell = \left\{ (a, b) : \left(\sum_{k=1}^{\ell-1} n_k^2 \right) < a,\, b \leq \left(\sum_{k=1}^{\ell-1} n_k^2 \right) + n_l \right\}.$$

For $\ell = 1, 2, \ldots, q$ and each (a, b) in P_ℓ, let $Q_{\ell,a,b}$ be the set of ordered pairs (i, j) of integers i, j defined by

$$Q_{\ell,a,b} = \Big\{ (i, j) : i = a + (m-1)n_\ell,\ j = b + (m-1)n_\ell,\ (a, b) \in P_\ell,\ 1 \leq m \leq n_\ell \Big\}.$$

For $1 \leq \ell \leq q$, (a, b) in P_ℓ, and (i, j) in $Q_{\ell,a,b}$, we see that

$$(13.14) \qquad \lambda(i) = \lambda(j) = \ell, \quad \mu(i) = \mu(j) = m, \quad \nu(i) = \nu(a), \quad \nu(j) = \nu(b),$$

and

$$(13.15) \qquad\qquad\qquad \big[B\big]_{i,j} = \big[B\big]_{a,b}.$$

For r, $s = 1, 2, \ldots, n$, we use $A = MBM^{-1}$, (13.4), the preceding notation about $Q_{\ell,a,b}$, (13.15), (13.6), (13.7), and (13.14) to obtain

$$
\begin{aligned}
\left[A\right]_{r,s} = \left[MBM^{-1}\right]_{r,s} &= \sum_{i=1}^{n}\sum_{j=1}^{n}\left[M\right]_{r,i}\left[B\right]_{i,j}\left[L\right]_{j,s} \\
&= \sum_{i=1}^{n}\sum_{j=1}^{n}\left[M\right]_{r,i}\delta\big(\lambda(i),\lambda(j)\big)\,\delta\big(\mu(i),\mu(j)\big)\left[B\right]_{i,j}\left[L\right]_{j,s} \\
&= \sum_{\ell=1}^{q}\sum_{(a,b)\in P_\ell}\sum_{(i,j)\in Q_{\ell,a,b}}\left[M\right]_{r,i}\left[B\right]_{i,j}\left[L\right]_{j,s} \\
&= \sum_{\ell=1}^{q}\sum_{(a,b)\in P_\ell}\sum_{(i,j)\in Q_{\ell,a,b}}\left[M\right]_{r,i}\left[B\right]_{a,b}\left[L\right]_{j,s} \\
&= \sum_{\ell=1}^{q}\sum_{(a,b)\in P_\ell}\left[B\right]_{a,b}\sum_{(i,j)\in Q_{\ell,a,b}}\left[M\right]_{r,i}\left[L\right]_{j,s} \\
&= \sum_{\ell=1}^{q}\sum_{(a,b)\in P_\ell}\left[B\right]_{a,b}\sum_{(i,j)\in Q_{\ell,a,b}}\left[F_{\lambda(i)}(g_r)\right]_{\mu(i),\nu(i)}\frac{n_{\lambda(j)}}{n}\left[F_{\lambda(j)}(g_s^{-1})\right]_{\nu(j),\mu(j)} \\
&= \sum_{\ell=1}^{q}\frac{n_\ell}{n}\sum_{(a,b)\in P_\ell}\left[B\right]_{a,b}\sum_{m=1}^{n_\ell}\left[F_\ell(g_r)\right]_{m,\nu(a)}\left[F_\ell(g_s^{-1})\right]_{\nu(b),m} \\
&= \sum_{\ell=1}^{q}\frac{n_\ell}{n}\sum_{(a,b)\in P_\ell}\left[B\right]_{a,b}\sum_{m=1}^{n_\ell}\left[F_\ell(g_s^{-1})\right]_{\nu(b),m}\left[F_\ell(g_r)\right]_{m,\nu(a)}.
\end{aligned}
$$

We use the property $F_\ell\big(g_s^{-1}g_r\big) = F_\ell(g_s^{-1})\,F_\ell(g_r)$ that F_ℓ possesses as a matrix representation for G of degree n_ℓ to deduce

$$
\left[F_\ell\big((g_r^{-1}g_s)^{-1}\big)\right]_{\nu(b),\nu(a)} = \left[F_\ell(g_s^{-1}g_r)\right]_{\nu(b),\nu(a)} = \sum_{m=1}^{n_\ell}\left[F_\ell(g_s^{-1})\right]_{\nu(b),m}\left[F_\ell(g_r)\right]_{m,\nu(a)}
$$

and obtain

$$
(13.16) \qquad \left[A\right]_{r,s} = \sum_{\ell=1}^{q}\frac{n_\ell}{n}\sum_{(a,b)\in P_\ell}\left[B\right]_{a,b}\left[F_\ell\big((g_r^{-1}g_s)^{-1}\big)\right]_{\nu(b),\nu(a)}.
$$

Therefore, in terms of the function τ from G to $\mathbb{F}$ defined by

$$
\tau(x) = \sum_{\ell=1}^{q}\frac{n_\ell}{n}\sum_{(a,b)\in P_\ell}\left[B\right]_{a,b}\left[F_\ell(x^{-1})\right]_{\nu(b),\nu(a)}, \qquad \text{for each } x \text{ in } G,
$$

we use (13.16) to see that A is given by

$$
\left[A\right]_{r,s} = \tau\big(g_r^{-1}g_s\big), \quad \text{for } r,\, s = 1, 2, \ldots, n.
$$

Thus, A is a group-pattern matrix for G and $\mathscr{L}$. Therefore, A is an element of $\mathfrak{A}$.

(iii) Suppose that A is an element of $\mathfrak{A}$ and the blocks of the corresponding matrix $B = M^{-1}AM$ in $\mathfrak{B}$ are designated as in (13.8) when viewed consecutively from upper-left to lower-right. For integers ℓ, i, j that satisfy $1 \leq \ell \leq q$ and $1 \leq i, j \leq n_\ell$, we set

$$(13.17) \qquad \kappa = \sum_{k=1}^{\ell-1} n_k^2, \quad r = \kappa + i, \quad s = \kappa + j,$$

and find that (11.7)–(11.8) on page 105 give

$$(13.18) \qquad \lambda(r) = \lambda(s) = \ell, \quad \mu(r) = \mu(s) = 1, \quad \nu(r) = i, \quad \text{and} \quad \nu(s) = j.$$

We use (13.17), (13.18), and (13.13) to see that the leftmost of the n_ℓ blocks in (13.8) labeled $\mathscr{B}_\ell$ has the component $[\mathscr{B}_\ell]_{i,j}$ in its ith row and jth column given by

$$[\mathscr{B}_\ell]_{i,j} = [B]_{\kappa+i,\kappa+j} = [B]_{r,s} = \sum_{k=1}^{n} \sigma(g_k) \Big[F_{\lambda(s)}(g_k)\Big]_{\nu(r),\nu(s)}$$

$$= \sum_{k=1}^{n} \sigma(g_k) \Big[F_\ell(g_k)\Big]_{i,j} = \left[\sum_{k=1}^{n} \sigma(g_k) F_\ell(g_k)\right]_{i,j}.$$

Thus, we have

$$\mathscr{B}_\ell = \sum_{k=1}^{n} \sigma(g_k) F_\ell(g_k).$$

This shows that (13.9) is valid and completes the proof of Theorem 13.3. $\square$

13.3. Immediate consequences of Theorem 13.3

13.3.1. Factorizations for determinants of group-pattern matrices.
Theorem 13.3 shows that a factorization for the determinant of a group-pattern matrix A for G and $\mathscr{L}$ can be obtained whenever a complete set $\{F_1, F_2, \ldots, F_q\}$ of pairwise-inequivalent irreducible matrix representations for G is known. Namely, after M is constructed from $F_1, F_2, \ldots, F_q$, the blocks for $B = M^{-1}AM$ indicated in (13.8) yield the factorization

$$(13.19) \qquad \det(A) = (\det M)^{-1} \det(A) \det(M) = \det(M^{-1}AM) = \det(B)$$

$$= \big(\det(\mathscr{B}_1)\big)^{n_1} \big(\det(\mathscr{B}_2)\big)^{n_2} \cdots \big(\det(\mathscr{B}_q)\big)^{n_q}.$$

A context where these factors are irreducible in a unique-factorization domain is presented for Theorem 14.2 on page 137.

13.3.2. $\mathfrak{A}$ and $\mathfrak{B}$ are isomorphic as rings and vector spaces over $\mathbb{F}$.

THEOREM 13.4. *With respect to matrix addition, matrix multiplication, and scalar multiplication by elements of $\mathbb{F}$, the sets $\mathfrak{A}$ and $\mathfrak{B}$ form isomorphic rings as well as isomorphic vector spaces and algebras over $\mathbb{F}$. Moreover, in terms of the nonsingular matrix M defined by (13.6), the function Φ from $\mathfrak{A}$ to $\mathfrak{B}$ having*

$$(13.20) \qquad \Phi(A) = M^{-1}AM, \quad \text{for each } A \text{ in } \mathfrak{A},$$

is a ring isomorphism and vector-space isomorphism of $\mathfrak{A}$ onto $\mathfrak{B}$.

PROOF. The notation for Theorem 4.2 on page 42 shows that $\mathfrak{A} = \mathcal{M}[\mathbb{F}, G, \mathscr{L}]$. We use Theorem 4.2 to see that: with respect to matrix addition, multiplication, and scalar multiplication, $\mathfrak{A}$ is a ring and a vector space over $\mathbb{F}$.

For the same matrix operations, it is clear that the set $\mathfrak{B}$ of block-diagonal matrices over $\mathbb{F}$ specified by $n_1, n_2, \ldots, n_q$ forms a ring and a vector space over $\mathbb{F}$.

We use both parts of the *if-and-only-if* statement for Theorem 13.3 to see that a one-to-one function Φ from $\mathfrak{A}$ onto $\mathfrak{B}$ is defined by (13.20). In particular, if B is in $\mathfrak{B}$, then $A = MBM^{-1}$ is an $n \times n$ matrix over $\mathbb{F}$, $M^{-1}AM$ is in $\mathfrak{B}$, A is in $\mathfrak{A}$, and Φ is onto. For A, B in $\mathfrak{A}$ and c in $\mathbb{F}$, we have

$$\Phi(A + B) = M^{-1}(A + B)M = M^{-1}AM + M^{-1}BM = \Phi(A) + \Phi(B),$$

$$\Phi(AB) = M^{-1}(AB)M = (M^{-1}AM)(M^{-1}BM) = \Phi(A)\,\Phi(B),$$

$$\Phi(cA) = M^{-1}(cA)M = c\,(M^{-1}AM) = c\,\Phi(A).$$

Thus, Φ is a ring isomorphism of $\mathfrak{A}$ onto $\mathfrak{B}$ and Φ is a vector-space isomorphism of $\mathfrak{A}$ onto $\mathfrak{B}$. We also have (4.14) on page 43. This completes the proof. $\qquad \square$

13.3.3. A natural basis for $\mathfrak{B}$ as a vector space over $\mathbb{F}$. The context for Theorem 2.16 on page 22 and Theorem 4.2 on page 42 shows that a basis for $\mathfrak{A}$ as a vector space over $\mathbb{F}$ is given by

$$(13.21) \qquad \qquad \{P_1, P_2, \ldots, P_n\},$$

where, for $1 \le k \le n$, P_k is the $n \times n$ permutation matrix in $\mathfrak{A}$ that is given by

$$(13.22) \qquad [P_k]_{r,s} = \begin{cases} 1, & \text{if } g_r^{-1}g_s = g_k, \\ 0, & \text{if } g_r^{-1}g_s \ne g_k, \end{cases} \quad \text{for } k, r, s = 1, 2, \ldots, n.$$

We use the isomorphism Φ of Theorem 13.4 to see that $\mathfrak{B}$ has the basis

$$(13.23) \qquad \qquad \{B_1, B_2, \ldots, B_n\},$$

where $B_k = \Phi(P_k)$, for $k = 1, 2, \ldots, n$.

Since each element of $\mathfrak{A}$ has a unique representation

$$(13.24) \qquad a_1 P_1 + a_2 P_2 + \cdots + a_n P_n, \quad \text{with } a_1, a_2, \ldots, a_n \text{ in } \mathbb{F},$$

we observe that the isomorphism Φ of Theorem 13.4 is given by

$$(13.25) \qquad \Phi\big(a_1 P_1 + a_2 P_2 + \cdots + a_n P_n\big) = a_1 B_1 + a_2 B_2 + \cdots + a_n B_n.$$

THEOREM 13.5. *For the context about* (13.20), (13.21), *and* (13.23), *let the successive blocks of the Frobenius block-diagonal matrix B_k in $\mathfrak{B}$ be represented by*

$$\overbrace{\mathcal{B}_{k,1}}^{1 \times 1}, \ldots, \overbrace{\mathcal{B}_{k,\ell}, \mathcal{B}_{k,\ell}, \ldots, \mathcal{B}_{k,\ell}}^{n_\ell \text{ blocks of size } n_\ell \times n_\ell}, \ldots, \overbrace{\mathcal{B}_{k,q}, \mathcal{B}_{k,q}, \ldots, \mathcal{B}_{k,q}}^{n_q \text{ blocks of size } n_q \times n_q}, \quad \text{for } k = 1, 2, \ldots, n.$$

Then, these blocks are given by

$$(13.26) \qquad \mathcal{B}_{k,\ell} = F_\ell(g_k), \quad \text{for } 1 \le k \le n \text{ and } 1 \le \ell \le q.$$

PROOF. For $k = 1, 2, \ldots, n$, let ϕ_k be the function from G to $\mathbb{F}$ defined by

$$(13.27) \qquad \phi_k(x) = \begin{cases} 1, & \text{if } x = g_k, \\ 0, & \text{if } x \ne g_k. \end{cases}$$

Since the definition of P_k in (13.22) for (13.21) is equivalent to

(13.28) $$\left[P_k\right]_{r,s} = \phi_k\left(g_r^{-1}g_s\right), \quad \text{for } k,\, r,\, s = 1,\, 2,\, \dots,\, n,$$

we use $B_k = M^{-1}P_kM$, for $1 \le k \le n$, (13.9), (13.28) and (13.27) to obtain

$$\mathcal{B}_{k,\ell} = \sum_{\nu=1}^{n} \phi_k(g_\nu)\,F_\ell(g_\nu) = F_\ell(g_k), \quad \text{for } 1 \le k \le n \text{ and } 1 \le \ell \le q.$$

Thus, (13.26) is valid. This completes the proof.

$\square$

OBSERVATION 13.6. We employ Theorem 2.16 on page 22 to see that the assignment $g_k \mapsto P_k$, for each g_k in G, is a matrix representation for G. Thus, an application of Theorem 13.4 shows that the assignment $g_k \mapsto B_k$, for each g_k in G, is a matrix representation for G whose nature is made explicit by Theorem 13.5.

Among several ways that a suitable matrix M can be defined to satisfy the first assertion of Theorem 13.3, the simplicity of (13.9) and (13.26) motivated our selection of M as defined by (13.6).

EXAMPLE 13.7. An application of Theorem 13.4 to the block-diagonalization of (11.21) shows that: the ring and vector space $\mathfrak{A}$ of matrices A having the form

$$A = \begin{bmatrix} a_1 & a_2 & a_3 & a_4 & a_5 & a_6 \\ a_3 & a_1 & a_2 & a_5 & a_6 & a_4 \\ a_2 & a_3 & a_1 & a_6 & a_4 & a_5 \\ a_4 & a_5 & a_6 & a_1 & a_2 & a_3 \\ a_5 & a_6 & a_4 & a_3 & a_1 & a_2 \\ a_6 & a_4 & a_5 & a_2 & a_3 & a_1 \end{bmatrix}, \quad \text{with } a_1,\, a_2,\, \dots,\, a_6 \text{ in } \mathbb{F},$$

is isomorphic to the ring and vector space $\mathfrak{B}$ of matrices having the form

$$B = \left[\begin{array}{c|c|cc|cc} d_1 & 0 & 0 & 0 & 0 & 0 \\ \hline 0 & d_2 & 0 & 0 & 0 & 0 \\ \hline 0 & 0 & d_3 & d_4 & 0 & 0 \\ 0 & 0 & d_5 & d_6 & 0 & 0 \\ \hline 0 & 0 & 0 & 0 & d_3 & d_4 \\ 0 & 0 & 0 & 0 & d_5 & d_6 \end{array}\right], \quad \text{with } d_1,\, d_2,\, \dots,\, d_6 \text{ in } \mathbb{F}.$$

The basis (13.21) for $\mathfrak{A}$ is given by

$$P_1 = \begin{bmatrix} 1 & 0 & 0 & 0 & 0 & 0 \\ 0 & 1 & 0 & 0 & 0 & 0 \\ 0 & 0 & 1 & 0 & 0 & 0 \\ 0 & 0 & 0 & 1 & 0 & 0 \\ 0 & 0 & 0 & 0 & 1 & 0 \\ 0 & 0 & 0 & 0 & 0 & 1 \end{bmatrix}, \quad P_2 = \begin{bmatrix} 0 & 1 & 0 & 0 & 0 & 0 \\ 0 & 0 & 1 & 0 & 0 & 0 \\ 1 & 0 & 0 & 0 & 0 & 0 \\ 0 & 0 & 0 & 0 & 1 & 0 \\ 0 & 0 & 0 & 0 & 0 & 1 \\ 0 & 0 & 0 & 1 & 0 & 0 \end{bmatrix}, \quad P_3 = \begin{bmatrix} 0 & 0 & 1 & 0 & 0 & 0 \\ 1 & 0 & 0 & 0 & 0 & 0 \\ 0 & 1 & 0 & 0 & 0 & 0 \\ 0 & 0 & 0 & 0 & 0 & 1 \\ 0 & 0 & 0 & 1 & 0 & 0 \\ 0 & 0 & 0 & 0 & 1 & 0 \end{bmatrix},$$

$$P_4 = \begin{bmatrix} 0 & 0 & 0 & 1 & 0 & 0 \\ 0 & 0 & 0 & 0 & 0 & 1 \\ 0 & 0 & 0 & 0 & 1 & 0 \\ 1 & 0 & 0 & 0 & 0 & 0 \\ 0 & 0 & 1 & 0 & 0 & 0 \\ 0 & 1 & 0 & 0 & 0 & 0 \end{bmatrix}, \quad P_5 = \begin{bmatrix} 0 & 0 & 0 & 0 & 1 & 0 \\ 0 & 0 & 0 & 1 & 0 & 0 \\ 0 & 0 & 0 & 0 & 0 & 1 \\ 0 & 1 & 0 & 0 & 0 & 0 \\ 1 & 0 & 0 & 0 & 0 & 0 \\ 0 & 0 & 1 & 0 & 0 & 0 \end{bmatrix}, \quad P_6 = \begin{bmatrix} 0 & 0 & 0 & 0 & 0 & 1 \\ 0 & 0 & 0 & 0 & 1 & 0 \\ 0 & 0 & 0 & 1 & 0 & 0 \\ 0 & 0 & 1 & 0 & 0 & 0 \\ 0 & 1 & 0 & 0 & 0 & 0 \\ 1 & 0 & 0 & 0 & 0 & 0 \end{bmatrix}.$$

We apply Theorem 13.5 and see that the complete set $\{F_1, F_2, F_3\}$ of pairwise-inequivalent irreducible matrix representations for the symmetric group $\mathfrak{S}_3$, given in (11.17), yields the basis elements of (13.23) for $\mathfrak{B}$ as

$$
B_1 = \begin{bmatrix} 1 & 0 & 0 & 0 & 0 & 0 \\ 0 & 1 & 0 & 0 & 0 & 0 \\ 0 & 0 & 1 & 0 & 0 & 0 \\ 0 & 0 & 0 & 1 & 0 & 0 \\ 0 & 0 & 0 & 0 & 1 & 0 \\ 0 & 0 & 0 & 0 & 0 & 1 \end{bmatrix}, \quad
B_2 = \begin{bmatrix} 1 & 0 & 0 & 0 & 0 & 0 \\ 0 & 1 & 0 & 0 & 0 & 0 \\ 0 & 0 & \omega & 0 & 0 & 0 \\ 0 & 0 & 0 & \omega^2 & 0 & 0 \\ 0 & 0 & 0 & 0 & \omega & 0 \\ 0 & 0 & 0 & 0 & 0 & \omega^2 \end{bmatrix}, \quad
B_3 = \begin{bmatrix} 1 & 0 & 0 & 0 & 0 & 0 \\ 0 & 1 & 0 & 0 & 0 & 0 \\ 0 & 0 & \omega^2 & 0 & 0 & 0 \\ 0 & 0 & 0 & \omega & 0 & 0 \\ 0 & 0 & 0 & 0 & \omega^2 & 0 \\ 0 & 0 & 0 & 0 & 0 & \omega \end{bmatrix},
$$

$$
B_4 = \begin{bmatrix} 1 & 0 & 0 & 0 & 0 & 0 \\ 0 & -1 & 0 & 0 & 0 & 0 \\ 0 & 0 & 0 & 1 & 0 & 0 \\ 0 & 0 & 1 & 0 & 0 & 0 \\ 0 & 0 & 0 & 0 & 0 & 1 \\ 0 & 0 & 0 & 0 & 1 & 0 \end{bmatrix}, \quad
B_5 = \begin{bmatrix} 1 & 0 & 0 & 0 & 0 & 0 \\ 0 & -1 & 0 & 0 & 0 & 0 \\ 0 & 0 & 0 & \omega^2 & 0 & 0 \\ 0 & 0 & \omega & 0 & 0 & 0 \\ 0 & 0 & 0 & 0 & 0 & \omega^2 \\ 0 & 0 & 0 & 0 & \omega & 0 \end{bmatrix}, \quad
B_6 = \begin{bmatrix} 1 & 0 & 0 & 0 & 0 & 0 \\ 0 & -1 & 0 & 0 & 0 & 0 \\ 0 & 0 & 0 & \omega & 0 & 0 \\ 0 & 0 & \omega^2 & 0 & 0 & 0 \\ 0 & 0 & 0 & 0 & 0 & \omega \\ 0 & 0 & 0 & 0 & \omega^2 & 0 \end{bmatrix}.
$$

For this context, the isomorphism Φ of (13.25) for Theorem 13.4 specializes to be

$$
\Phi(A) = \Phi\left(\sum_{k=1}^{6} a_k P_k\right) = \sum_{k=1}^{6} a_k B_k, \quad \text{for each } A \text{ in } \mathfrak{A}.
$$

13.4. Particular situations for Theorem 13.3

13.4.1. The condition that each B in $\mathfrak{B}$ is a diagonal matrix.

PROPOSITION 13.8. *Each matrix in $\mathfrak{B}$ is a diagonal matrix if and only if G is abelian.*

PROOF. Each matrix in $\mathfrak{B}$ is a diagonal matrix if and only if

$$(13.29) \qquad n_k = 1, \quad \text{for } 1 \le k \le q.$$

In view of $n_1^2 + n_2^2 + \cdots + n_q^2 = n$, we see that (13.29) is equivalent to $q = n$. However, Definition 10.15 on page 99 shows that a group G of order n has $q = n$ conjugacy classes if and only if G is abelian. This completes the proof. $\qquad\square$

13.4.2. Theorem 8.3 in relation to Theorem 13.3.

PROPOSITION 13.9. *Suppose that G is abelian and $\mathcal{F} = \mathbb{F}$. Then, $q = n$ and a complete set $\mathcal{C} = \{F_1, F_2, \ldots, F_n\}$ of pairwise-inequivalent irreducible matrix representations for G, each of degree 1 with components in $\mathbb{C}$, is given by*

$$(13.30) \qquad F_\ell(g_k) = [\chi_\ell(g_k)], \quad \text{for } 1 \le \ell \le n \text{ and } 1 \le k \le n,$$

with respect to the group characters $\chi_1, \chi_2, \ldots, \chi_n$ for G of page 79. Moreover, when $\mathcal{C}$ is employed for the construction, Theorem 13.3 reduces to Theorem 8.3.

PROOF. We use Theorem 8.1 on page 79 to see that n matrix representations of degree 1 for G are defined by (13.30). Clearly, these representations are pairwise-inequivalent and irreducible. In view of $n_\ell = 1$, for $1 \le \ell \le n$, Theorem 10.17 on page 99 shows that the set $\mathcal{C}$ of these representations is complete.

With $q = n$ as well as $\lambda(s) = s$, $\mu(s) = 1$ and $\nu(s) = 1$, for $1 \leq s \leq n$, we use $\chi_s(g_r) = \chi_r(g_s)$, for $1 \leq r, s \leq n$, from (8.8) to see that (13.6) and $\mathscr{C}$ yield

$$\left[M\right]_{r,s} = \left[F_{\lambda(s)}(g_r)\right]_{\mu(s),\nu(s)} = \left[F_s(g_r)\right]_{1,1} = \chi_s(g_r) = \chi_r(g_s),$$
$$\text{for } r, s = 1, 2, \ldots, n.$$

Thus, the matrix M constructed for Theorem 13.3 with (13.6) and $\mathscr{C}$ is equal to the matrix M for Theorem 8.3 based on (8.6). Then, the assertions of Theorem 13.3 are those of Theorem 8.3 with (8.13). This completes the proof. □

13.5. Use of a block-diagonalization to deduce others

For the context of Theorem 13.3, suppose that two lists

(13.31) $\mathscr{L} = (g_1, g_2, \ldots, g_n)$ and $\mathscr{L}_* = (h_1, h_2, \ldots, h_n)$

are given for the n elements of G. Then, in addition to $\mathfrak{A}$, $\mathfrak{B}$, and M for G with $\mathscr{L}$, there are also $\mathfrak{A}_*$, $\mathfrak{B}$, and M_* for G with $\mathscr{L}_*$. Let π be the permutation of $\{1, 2, \ldots, n\}$ such that

(13.32) $h_k = g_{\pi(k)},$ for $k = 1, 2, \ldots, n.$

PROPOSITION 13.10. *The matrix M_* is given by $M_* = P_* M$, where P_* is the $n \times n$ permutation matrix having*

(13.33) $[P_*]_{r,s} = \begin{cases} 1, & \text{if } s = \pi(r), \\ 0, & \text{if } s \neq \pi(r), \end{cases}$ for $r, s = 1, 2, \ldots, n.$

PROOF. For $r, s = 1, 2, \ldots, n$, we use (13.6), (13.33), and (13.32) to obtain

$$\left[P_* M\right]_{r,s} = \sum_{k=1}^{n} \left[P_*\right]_{r,k} \left[M\right]_{k,s}$$

$$= \sum_{k=1}^{n} \left[P_*\right]_{r,k} \left[F_{\lambda(s)}(g_k)\right]_{\mu(s),\nu(s)}$$

$$= \left[F_{\lambda(s)}(g_{\pi(r)})\right]_{\mu(s),\nu(s)} = \left[F_{\lambda(s)}(h_r)\right]_{\mu(s),\nu(s)} = \left[M_*\right]_{r,s}.$$

This yields $M_* = P_* M$ and completes the proof. □

PROPOSITION 13.11. *Suppose that $\mathbb{F}$ contains elements $x_1, x_2, \ldots, x_n$ that are algebraically independent over $\mathbb{C}$. Let σ and σ_* be functions from G to $\mathbb{F}$ such that $\sigma(g_k) = x_k$ and $\sigma_*(h_k) = y_k$, for $1 \leq k \leq n$. Then, for the $n \times n$ group-pattern matrices A in $\mathfrak{A}$ and A_* in $\mathfrak{A}_*$ defined by*

(13.34) $[A]_{r,s} = \sigma(g_r^{-1} g_s)$ and $[A_*]_{r,s} = \sigma_*(h_r^{-1} h_s),$ *for $r, s = 1, 2, \ldots, n,$*

the matrices $B = M^{-1} A M$ and $B_* = M_*^{-1} A_* M_*$ in $\mathfrak{B}$ are related in the sense that B_* is obtained from B by replacing x_k in B with $y_{\pi^{-1}(k)}$, for $1 \leq k \leq n$.

PROOF. Theorem 13.3 shows that the blocks in (13.8) for $B = M^{-1} A M$ are given by (13.9) as

(13.35) $\mathscr{B}_\ell = \sum_{k=1}^{n} \sigma(g_k) F_\ell(g_k) = \sum_{k=1}^{n} x_k F_\ell(g_k),$ for $\ell = 1, 2, \ldots, q.$

We use Theorem 13.3 to see that the blocks

$$\overbrace{\mathscr{B}_1^*}^{1 \times 1}, \ldots, \quad \overbrace{\mathscr{B}_\ell^*, \mathscr{B}_\ell^*, \ldots, \mathscr{B}_\ell^*}^{n_\ell \text{ blocks of size } n_\ell \times n_\ell}, \ldots, \quad \overbrace{\mathscr{B}_q^*, \mathscr{B}_q^*, \ldots, \mathscr{B}_q^*}^{n_q \text{ blocks of size } n_q \times n_q}$$

for $B_* = M_*^{-1} A_* M_*$ are provided by (13.9) as

$$
\begin{aligned}
(13.36) \qquad \mathscr{B}_\ell^* &= \sum_{\nu=1}^{n} \sigma_*(h_\nu) \, F_\ell(h_\nu) \\
&= \sum_{\nu=1}^{n} y_\nu \, F_\ell(h_\nu) \\
&= \sum_{\nu=1}^{n} y_{\pi^{-1}\left(\pi(\nu)\right)} F_\ell(g_{\pi(\nu)}) \\
&= \sum_{k=1}^{n} y_{\pi^{-1}(k)} F_\ell(g_k), \quad \text{for } \ell = 1, 2, \ldots, q.
\end{aligned}
$$

A comparison of (13.36) with (13.35) shows that B_* is obtained from B by replacing each x_k in B with $y_{\pi^{-1}(k)}$, for $k = 1, 2, \ldots, n$. This completes the proof. $\qquad \square$

Propositions 13.10 and 13.11 are illustrated in Section 23.2 on pages 203–204.

CHAPTER 14

A in $\mathfrak{A}$ and B in $\mathfrak{B}$ when $B = M^{-1}AM$

14.1. Introduction

We use the context of the preceding chapter with the additional assumption that Proposition 2.7 of page 13 has been applied, if necessary, to ensure that the first element of the list $\mathscr{L} = (g_1, g_2, \ldots, g_n)$ for G is $g_1 = e$. Then, each group-pattern matrix in $\mathfrak{A}$ is conveniently specified by its first row. Namely, if the first row of A in $\mathfrak{A}$ is $(a_1, a_2, \ldots, a_n)$, then we have

$$(14.1) \qquad \left[A\right]_{r,s} = \sigma\left(g_r^{-1}g_s\right), \quad \text{for } r, s = 1, 2, \ldots, n,$$

where σ is the function from G to $\mathbb{F}$ having $\sigma(g_k) = \left[A\right]_{1,k} = a_k$, for $1 \leq k \leq n$. In particular, $\mathfrak{A}$ is uniquely specified by $\mathbb{F}$, G, and $\mathscr{L}$.

We note that $\mathfrak{B}$ is uniquely specified by $\mathbb{F}$ and any complete set $\mathscr{C}$ of pairwise-inequivalent irreducible matrix representations for G. Namely, the number of elements in $\mathscr{C}$ is equal to the number q of conjugacy classes for G; and the q elements of $\mathscr{C}$ can be named $F_1, F_2, \ldots, F_q$ in such a manner that

$$n_1 = \text{degree}(F_1) \leq n_2 = \text{degree}(F_2) \leq \cdots \leq n_q = \text{degree}(F_q).$$

Moreover, if $\mathscr{C}_1$ is also a complete set of pairwise-inequivalent irreducible matrix representations for G, then the q elements of $\mathscr{C}_1$ can be named $\Phi_1, \Phi_2, \ldots, \Phi_q$ in such a manner that Φ_k is equivalent to F_k, for $k = 1, 2, \ldots, q$. This gives

$$n_1 = \text{degree}(\Phi_1) \leq n_2 = \text{degree}(\Phi_2) \leq \cdots \leq n_q = \text{degree}(\Phi_q).$$

Thus, G uniquely specifies the integers q, n_1, n_2, $\ldots n_q$ that define the pattern of the submatrices for each B in $\mathfrak{B}$ and the functions $\lambda(s)$, $\mu(s)$, $\nu(s)$, for $1 \leq s \leq n$.

14.2. Use of n components $d_1, d_2, \ldots, d_n$ to specify any B in $\mathfrak{B}$

The submatrices of B in $\mathfrak{B}$ are indicated by

$$(14.2) \qquad \overbrace{\mathscr{B}_1}^{1 \times 1}, \ldots, \overbrace{\mathscr{B}_\ell, \mathscr{B}_\ell, \ldots, \mathscr{B}_\ell}^{n_\ell \text{ blocks of size } n_\ell \times n_\ell}, \ldots, \overbrace{\mathscr{B}_q, \mathscr{B}_q, \ldots, \mathscr{B}_q}^{n_q \text{ blocks of size } n_q \times n_q};$$

and the components of B outside these submatrices are zero. Hence, to specify B, it is sufficient to specify the $n_1^2 + n_2^2 + \cdots + n_q^2 = n$ components of the submatrices

$$(14.3) \qquad \mathscr{B}_1, \mathscr{B}_2, \ldots, \mathscr{B}_k, \ldots, \mathscr{B}_q.$$

This we do with the corresponding $\lambda(s)$, $\mu(s)$, $\nu(s)$ for (11.7)–(11.8) by introducing

$$(14.4) \qquad d_t = \left[\mathscr{B}_{\lambda(t)}\right]_{\mu(t),\nu(t)}, \quad \text{for } 1 \leq t \leq n.$$

In particular, we have $n_1 = 1$ and $\mathscr{B}_1 = [d_1]$. Moreover, for any integer k such that $1 < k \leq q$ and $n_k = 1$, we have $\mathscr{B}_1 = [d_1]$, $\mathscr{B}_2 = [d_2]$, $\ldots$, $\mathscr{B}_k = [d_k]$.

For $1 \leq k \leq q$, we apply (11.7) and (11.8) to see that (14.4) yields

$$(14.5) \quad \mathscr{B}_k = \begin{bmatrix} d_{\ell+1} & d_{\ell+2} & \cdots & d_{\ell+n_k} \\ d_{\ell+n_k+1} & d_{\ell+n_k+2} & \cdots & d_{\ell+2n_k} \\ d_{\ell+2n_k+1} & d_{\ell+2n_k+2} & \cdots & d_{\ell+3n_k} \\ \vdots & \vdots & \vdots & \vdots \\ d_{\ell+(n_k-1)n_k+1} & d_{\ell+(n_k-1)n_k+2} & \cdots & d_{\ell+n_k^2} \end{bmatrix},$$

$$\text{where } \ell = n_1^2 + n_2^2 + \cdots + n_{k-1}^2.$$

If G is abelian, then $q = n$. That requires $n_k = 1$, for $1 \leq k \leq n$, and $\mathscr{B}_k = [d_k]$, for $1 \leq k \leq n$. If G is nonabelian, then $1 < q < n$ and $n_q \geq 2$.

14.3. Supplement for Theorem 13.3

By representing each matrix A in $\mathfrak{A}$ with its first row $(a_1, a_2, \ldots, a_n)$ and by representing each matrix B in $\mathfrak{B}$ with the $1 \times n$ matrix $[d_1, d_2, \ldots, d_n]$ whose components enable (14.3) and (14.2) to specify B via (14.4), we simplify the relation between A and B given in Theorem 13.3.

THEOREM 14.1. *If A is in $\mathfrak{A}$, then $M^{-1}AM$ is the matrix B in $\mathfrak{B}$ given by*

$$(14.6) \qquad [d_1, d_2, \ldots, d_n] = [a_1, a_2, \ldots, a_n]M.$$

Moreover, for any elements $d_1, d_2, \ldots, d_n$ of $\mathbb{F}$, if B is the matrix in $\mathfrak{B}$ specified by $[d_1, d_2, \ldots, d_n]$, then MBM^{-1} is the matrix in $\mathfrak{A}$ whose first row corresponds to

$$(14.7) \qquad [a_1, a_2, \ldots, a_n] = [d_1, d_2, \ldots, d_n]M^{-1}.$$

PROOF. (i) Suppose that A is in $\mathfrak{A}$ and let its first row be $(a_1, a_2, \ldots, a_n)$. Theorem 13.3 shows that the matrix $B = M^{-1}AM$ is an element of $\mathfrak{B}$. Then, (14.2) and (14.3) for B specify $[d_1, d_2, \ldots, d_n]$ through (14.4). With A given by (14.1) and $\sigma(g_k) = a_k$, for $1 \leq k \leq n$, we see that (13.9) of Theorem 13.3 yields

$$(14.8) \qquad \mathscr{B}_\ell = \sum_{k=1}^{n} \sigma(g_k) F_\ell(g_k) = \sum_{k=1}^{n} a_k F_\ell(g_k), \quad \text{for } \ell = 1, 2, \ldots, q.$$

For $1 \leq t \leq n$, we apply (14.4), (14.8), and (13.6) on page 124 to obtain

$$d_t = \left[\mathscr{B}_{\lambda(t)} \right]_{\mu(t),\nu(t)} = \left[\sum_{k=1}^{n} a_k F_{\lambda(t)}(g_k) \right]_{\mu(t),\nu(t)}$$

$$= \sum_{k=1}^{n} a_k \left[F_{\lambda(t)}(g_k) \right]_{\mu(t),\nu(t)} = \sum_{k=1}^{n} a_k [M]_{k,t}$$

Consequently, (14.6) is valid.

(ii) Suppose that B is a matrix in $\mathfrak{B}$ specified by $[d_1, d_2, \ldots, d_n]$. Then, Theorem 13.3 shows that the matrix A defined by $A = MBM^{-1}$ is an element of $\mathfrak{A}$. Since the matrix $M^{-1}AM$ is the matrix B in $\mathfrak{B}$, the specification of B is given by Part (i) of this proof as $[d_1, d_2, \ldots, d_n] = [a_1, a_2, \ldots, a_n]M$. We multiply both sides of this last relation on the right by M^{-1} to obtain (14.7) and compete the proof. $\square$

14.4. Complete factorizations for the determinants
of standard group-pattern matrices

Let X_1, X_2, ..., X_n denote algebraically independent variables over $\mathbb{C}$. Then, the ring $\mathfrak{R} = \mathbb{C}[X_1, X_2, \ldots, X_n]$ of polynomials in those variables over $\mathbb{C}$ is a unique-factorization domain; e.g., see [**31**, page 199]. For a group G of order n and a list $\mathscr{L} = (g_1, g_2, \ldots, g_n)$ of its elements with $g_1 = e$, the standard group-pattern matrix A for G and $\mathscr{L}$ is given by

$$(14.9) \qquad [A]_{r,s} = \sigma(g_r^{-1} g_s), \quad \text{for } r, s = 1, 2, \ldots, n,$$

where σ is the function from G to $\mathfrak{R}$ that has $\sigma(g_k) = X_k$, for $k = 1, 2, \ldots, n$. The determinant of A is a homogeneous polynomial in $\mathfrak{R}$ of degree n and therefore it has a representation as a product of irreducible polynomials in $\mathfrak{R}$.

A factorization for $\det(A)$ *can be obtained from the construction of* M *for* Theorem 13.3. Namely, in terms of $\mathbb{F}$ as a field that contains $\mathfrak{R}$ as a subring, set $\mathfrak{A} = \mathcal{M}[\mathbb{F}, G, \mathscr{L}]$, let $\mathfrak{B}$ denote the corresponding set of block-diagonal matrices over $\mathbb{F}$, and let M be an $n \times n$ matrix with components in $\mathbb{C}$ that relates the matrices of $\mathfrak{A}$ and $\mathfrak{B}$ according to Theorem 13.3 on page 125. Then, the standard group-pattern matrix A of (14.9) specifies $B = M^{-1}AM$ as the block-diagonal matrix whose blocks are represented by (14.2) with components in $\mathfrak{R}$. Thus, we have

$$(14.10) \qquad \det(A) = \det(MBM^{-1}) = \det B = \prod_{k=1}^{q} \big(\det(\mathscr{B}_k)\big)^{n_k}.$$

The next result shows that (14.10) provides a suitable factorization.

THEOREM 14.2. *The polynomial* $\det(\mathscr{B}_k)$ *is irreducible in* $\mathfrak{R}$, *for* $1 \leq k \leq q$.

PROOF. The first row of A is $(X_1, X_2, , \ldots, X_n)$ and the blocks in (14.3) for B have their components given by $d_1, d_2, \ldots, d_n$. We use (14.6) and (14.7) to obtain

$$(14.11) \qquad d_i = \sum_{j=1}^{n} \alpha_{i,j} X_j, \quad \text{for } i = 1, 2, \ldots, n,$$

and

$$(14.12) \qquad X_i = \sum_{j=1}^{n} \beta_{i,j} d_j, \quad \text{for } i = 1, 2, \ldots, n,$$

where the coefficients $\alpha_{i,j}$ and $\beta_{i,j}$ are elements in $\mathbb{C}$ because they are components of M and M^{-1} in some order. Therefore, formulas (14.11) and (14.12) show that $d_1, d_2, \ldots, d_n$ are algebraically independent over $\mathbb{C}$. Thus, for $1 \leq k \leq q$, the $(n_k)^2$ components of the matrix $\mathscr{B}_k$ in (14.5) are $(n_k)^2$ independent variables over $\mathbb{C}$. We apply [**3**, page 176, Theorem 1] or Exercise 1 on page 138 to establish that there is no nontrivial factorization for $\det(\mathscr{B}_k)$ as a polynomial combination of the variables $d_{\ell+1}, d_{\ell+2}, \ldots, d_{\ell+n_k^2}$ over $\mathbb{C}$ where $\ell = n_1^2 + n_2^2 + \cdots + n_{k-1}^2$.

For a fixed integer k that satisfies $1 \leq k \leq n$, let $\phi(X_1, X_2, \ldots, X_n)$ and $\psi(X_1, X_2, \ldots, X_n)$ be polynomials in $\mathfrak{R}$ such that

$$(14.13) \qquad \det(\mathscr{B}_k) = \phi(X_1, X_2, \ldots, X_n)\, \psi(X_1, X_2, \ldots, X_n).$$

The left member of (14.13) is a polynomial combination of $d_{\ell+1}, d_{\ell+2}, \ldots, d_{\ell+n_k^2}$ over $\mathbb{C}$. Therefore, when X_i in the right member of (14.13) is replaced by the right member of (14.12), for $1 \leq i \leq n$, we obtain a factorization

$$(14.14) \qquad \det(\mathscr{B}_k) = \Phi\big(d_{\ell+1}, d_{\ell+2}, \ldots, d_{\ell+n_k^2}\big)\, \Psi\big(d_{\ell+1}, d_{\ell+2}, \ldots, d_{\ell+n_k^2}\big)$$

for $\det(\mathscr{B}_k)$ as a product of two polynomial combinations of $d_{\ell+1}, d_{\ell+2}, \ldots, d_{\ell+n_k^2}$ over $\mathbb{C}$. Since $\det(\mathscr{B}_k)$ has only trivial factorizations of this kind, one of the factors in both (14.14) and (14.13) is a nonzero element of $\mathbb{C}$. This shows that $\det(\mathscr{B}_k)$ is an irreducible polynomial in $\mathfrak{R}$ and completes the proof. $\qquad\square$

OBSERVATION 14.3. If names such as A, B, C, ... are assigned to the elements of G in some order, then there is a permutation π of $\{1, 2, \ldots, n\}$ such that $g_{\pi(1)} = A$, $g_{\pi(2)} = B$, $g_{\pi(3)} = C$, ... Then, the complete factorization for the Gruppendeterminante polynomial $\mathfrak{P}(G)$ for G in Frobenius notation is obtained by replacing the variables of $X_{\pi(1)}$, $X_{\pi(2)}$, $X_{\pi(3)}$, ... in the right-most member of (14.10) by the corresponding variables of X_A, X_B, X_C, ...

EXERCISE

1. Supposes that the k^2 components $b_{i,j}$, for $1 \leq i, j \leq k$, of the $k \times k$ matrix

$$(14.15) \qquad B = \begin{bmatrix} b_{1,1} & b_{1,2} & \cdots & b_{1,k} \\ b_{2,1} & b_{2,2} & \cdots & b_{2,k} \\ \vdots & \vdots & \ddots & \vdots \\ b_{k,1} & b_{k,2} & \cdots & b_{k,k} \end{bmatrix}$$

are independent variables over $\mathbb{C}$ and suppose that $\det(B) = P_1 P_2$, where P_1 and P_2 are polynomials in the components of B over $\mathbb{C}$.

    A: Show $b_{1,1}$ appears in only one of P_1 and P_2. Assume $b_{1,1}$ is not in P_2.

    B: Show that no component of B in its first column appears in P_2.

    C: Show that no component of B in any of its rows appears in P_2.

    D: Conclude that P_2 is a nonzero element of $\mathbb{C}$.

Thus, $\det(B)$ is an irreducible polynomial over $\mathbb{C}$.

CHAPTER 15

The Dihedral Group $\mathcal{D}_n$

15.1. Introduction

The elements of the dihedral group $\mathcal{D}_n$ of order $2n$, for $n \geq 3$, can be identified as the $2n$ symmetries of a regular polygon having n sides. There are clearly n distinct rotational symmetries and there are n distinct reflectional symmetries specified by the n lines that pass through the center of the polygon and either a vertex or a midpoint of a side. For instance, the 6 symmetries of an equilateral triangle are exhibited for (1.4) and the 8 symmetries of a square are shown for (1.8).

To represent the elements of $\mathcal{D}_n$ conveniently with respect to a rectangular Cartesian coordinate system, we first recall that the transformation formulas for a rotation through θ radians about the origin are given by

$$(15.1) \qquad \begin{cases} x' = (\cos\theta)\,x - (\sin\theta)\,y, \\ y' = (\sin\theta)\,x + (\cos\theta)\,y, \end{cases} \quad \text{or} \quad \begin{bmatrix} x' \\ y' \end{bmatrix} = \begin{bmatrix} \cos\theta & -\sin\theta \\ \sin\theta & \cos\theta \end{bmatrix} \begin{bmatrix} x \\ y \end{bmatrix}.$$

Also, transformation formulas that represent a reflection in the x-axis are given by

$$(15.2) \qquad \begin{cases} x' = 1 \cdot x + 0 \cdot y, \\ y' = 0 \cdot x - 1 \cdot y, \end{cases} \quad \text{or} \quad \begin{bmatrix} x' \\ y' \end{bmatrix} = \begin{bmatrix} 1 & 0 \\ 0 & -1 \end{bmatrix} \begin{bmatrix} x \\ y \end{bmatrix}.$$

Therefore, we position a regular polygon in the plane of the coordinate system so that the center of the polygon is at the origin and one of its vertices is on the positive x-axis. Thus, for an equilateral triangle, the position is

(15.3)

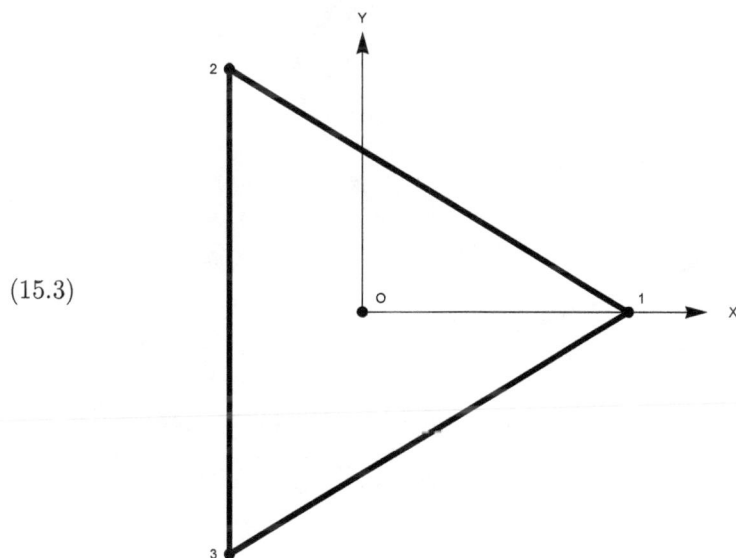

while, for a square, the position is

(15.4)

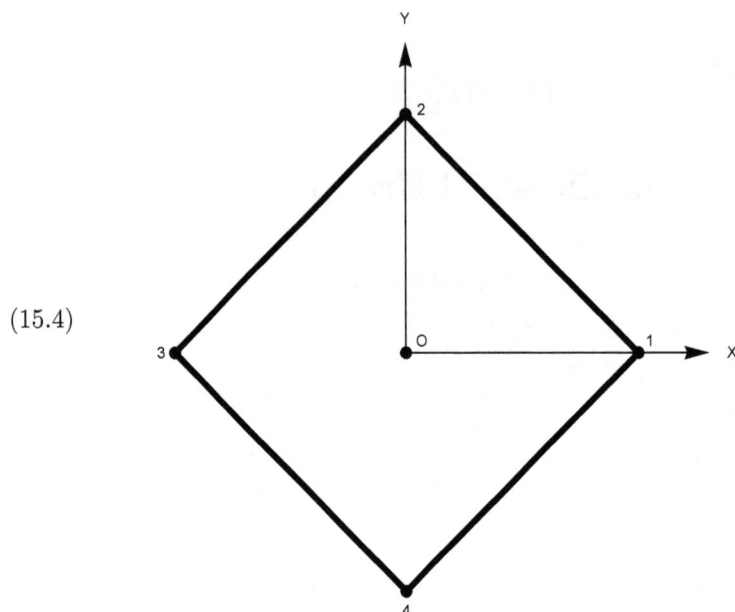

and the position is

(15.5)

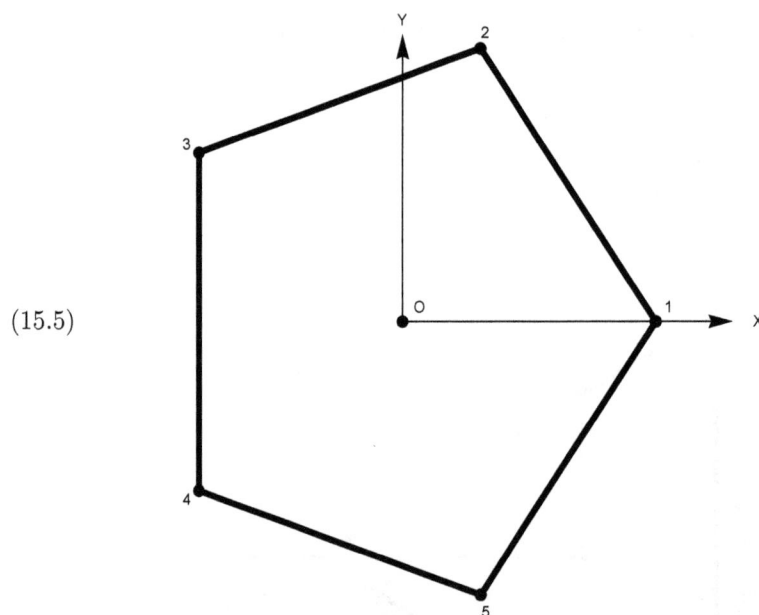

for a regular pentagon. Thus, as elements in $\mathcal{D}_n$, the n rotations are given by

$$(15.6) \quad g_k = \begin{bmatrix} \cos\left(\dfrac{2\pi}{n}(k-1)\right) & -\sin\left(\dfrac{2\pi}{n}(k-1)\right) \\[2mm] \sin\left(\dfrac{2\pi}{n}(k-1)\right) & \cos\left(\dfrac{2\pi}{n}(k-1)\right) \end{bmatrix}, \quad \text{for } k = 1, 2, \ldots, n,$$

and, by first performing a rotation followed by a reflection in the x-axis, we see that the n reflections of $\mathcal{D}_n$ are given by

$$(15.7) \quad g_k = \begin{bmatrix} 1 & 0 \\ 0 & -1 \end{bmatrix} \begin{bmatrix} \cos\left(\dfrac{2\pi}{n}(k-n-1)\right) & -\sin\left(\dfrac{2\pi}{n}(k-n-1)\right) \\ \sin\left(\dfrac{2\pi}{n}(k-n-1)\right) & \cos\left(\dfrac{2\pi}{n}(k-n-1)\right) \end{bmatrix}$$

$$= \begin{bmatrix} \cos\left(\dfrac{2\pi}{n}(k-n-1)\right) & -\sin\left(\dfrac{2\pi}{n}(k-n-1)\right) \\ -\sin\left(\dfrac{2\pi}{n}(k-n-1)\right) & -\cos\left(\dfrac{2\pi}{n}(k-n-1)\right) \end{bmatrix},$$

$$\text{for } k = n+1, \, n+2, \, \ldots, \, 2n.$$

Alternatively, after defining $\mathcal{D}_n$ as the set of the $2n$ matrices of size 2×2 given by (15.6) and (15.7), one could then use trigonometric identities to show that $\mathcal{D}_n$ is closed under matrix multiplication and forms a group. For convenience, we set

$$(15.8) \quad e = g_1 = \begin{bmatrix} 1 & 0 \\ 0 & 1 \end{bmatrix}, \quad \alpha = g_2 = \begin{bmatrix} \cos\left(\frac{2\pi}{n}\right) & -\sin\left(\frac{2\pi}{n}\right) \\ \sin\left(\frac{2\pi}{n}\right) & \cos\left(\frac{2\pi}{n}\right) \end{bmatrix}, \quad \beta = \begin{bmatrix} 1 & 0 \\ 0 & -1 \end{bmatrix}.$$

Thus, with $\alpha^n = e$, the list

$$(15.9) \qquad \mathcal{L} = \left(g_1, g_2, g_3, \ldots, g_n, g_{n+1}, g_{n+2}, g_{n+3}, \ldots, g_{2n} \right)$$

of elements for $\mathcal{D}_n$ is given by

$$(15.10) \qquad \mathcal{L} = \left(e, \alpha, \alpha^2, \ldots, \alpha^{n-1}, \beta, \beta\alpha, \beta\alpha^2, \ldots, \beta\alpha^{n-1} \right).$$

In particular, we have

$$(15.11) \quad g_k = \alpha^{k-1}, \quad \text{for } 1 \le k \le n, \quad \text{and} \quad g_{n+k} = \beta\alpha^{k-1}, \quad \text{for } 1 \le k \le n.$$

PROPOSITION 15.1. *The inverses of $g_1, g_2, \ldots, g_{2n}$ in $\mathcal{D}_n$ are given by*

$$(15.12) \quad g_k^{-1} = \beta g_k \beta, \quad \text{for } 1 \le k \le n, \quad \text{and} \quad g_{n+k}^{-1} = g_{n+k}, \quad \text{for } 1 \le k \le n,$$

along with $g_k^{-1} = g_{n+2-k}$, for $2 \le k \le n$.

PROOF. For $1 \le k \le n$, we set $\theta_k = \left(\frac{2\pi}{n}\right)(k-1)$. Then, (15.6) yields

$$(15.13) \quad g_k = \begin{bmatrix} \cos(\theta_k) & -\sin(\theta_k) \\ \sin(\theta_k) & \cos(\theta_k) \end{bmatrix} \quad \text{and} \quad g_k^{-1} = \begin{bmatrix} \cos(-\theta_k) & -\sin(-\theta_k) \\ \sin(-\theta_k) & \cos(-\theta_k) \end{bmatrix}.$$

We use (15.13) and β in (15.8) to verify, for $1 \le k \le n$, that

$$\beta g_k \beta = \begin{bmatrix} 1 & 0 \\ 0 & -1 \end{bmatrix} \begin{bmatrix} \cos(\theta_k) & -\sin(\theta_k) \\ \sin(\theta_k) & \cos(\theta_k) \end{bmatrix} \begin{bmatrix} 1 & 0 \\ 0 & -1 \end{bmatrix}$$

$$= \begin{bmatrix} \cos(\theta_k) & \sin(\theta_k) \\ -\sin(\theta_k) & \cos(\theta_k) \end{bmatrix} = \begin{bmatrix} \cos(-\theta_k) & -\sin(-\theta_k) \\ \sin(-\theta_k) & \cos(-\theta_k) \end{bmatrix} = g_k^{-1}$$

and $g_{n+k}\, g_{n+k} = (\beta g_k)(\beta g_k) = (\beta g_k \beta)\, g_k = g_k^{-1} g_k = e$. Hence, (15.12) is valid. For $2 \le k \le n$ and therefore $2 \le n+2-k \le n$, we have

$$g_k = \alpha^{k-1}, \quad g_{n+2-k} = \alpha^{n+1-k}, \quad g_k\, g_{n+2-k} = \alpha^n = e, \quad \text{and} \quad g_k^{-1} = g_{n+2-k}.$$

This completes the proof. $\qquad\qquad\qquad\qquad\qquad\qquad\qquad\qquad\qquad\square$

15.2. Primary group-pattern matrices for $\mathcal{D}_n$

When $n \geq 3$, a *primary group-pattern matrix* A_{2n} for $\mathcal{D}_n$ is defined with respect to the list $\mathscr{L}$ in (15.9) and a function σ from $\mathcal{D}_n$ to a set S by

(15.14) $$[A_{2n}]_{r,s} = \sigma(g_r^{-1}g_s), \quad \text{for } 1 \leq r,\, s \leq 2n.$$

PROPOSITION 15.2. *When σ for (15.14) yields $\sigma(g_k) = x_k$, for $1 \leq k \leq 2n$, the $2n \times 2n$ primary group-pattern matrix A_{2n} for $\mathcal{D}_n$ and $\mathscr{L}$ has the form*

(15.15) $$A_{2n} = \left[\begin{array}{c|c} C & L \\ \hline L & C \end{array}\right],$$

where C is the $n \times n$ circulant matrix whose first row is $(x_1, x_2, \ldots, x_n)$ and L is the $n \times n$ leftward-circulant matrix whose first row is $(x_{n+1}, x_{n+2}, \ldots, x_{2n})$.

PROOF. In terms of the $2n \times 2n$ group-pattern matrix A_{2n}, we define four $n \times n$ submatrices C_1, L_1 L_2, and C_2 of A by

(15.16) $$[C_1]_{r,s} = [A_{2n}]_{r,s}, \qquad \text{for } 1 \leq r,\, s \leq n,$$

(15.17) $$[L_1]_{r,s} = [A_{2n}]_{r,n+s}, \qquad \text{for } 1 \leq r,\, s \leq n,$$

(15.18) $$[L_2]_{r,s} = [A_{2n}]_{n+r,s}, \qquad \text{for } 1 \leq r,\, s \leq n,$$

(15.19) $$[C_2]_{r,s} = [A_{2n}]_{n+r,n+s}, \qquad \text{for } 1 \leq r,\, s \leq n,$$

in order to express A_{2n} as

(15.20) $$A_{2n} = \left[\begin{array}{c|c} C_1 & L_1 \\ \hline L_2 & C_2 \end{array}\right].$$

(i) For (15.16) and $1 \leq r,\, s \leq n$, we have $[C_1]_{r,s} = \sigma(g_r^{-1}g_s)$ where (15.11) yields $g_r^{-1}g_s = \alpha^{1-r}\alpha^{s-1} = \alpha^{s-r}$. The first row of C_1 is $(x_1, x_2, x_3, \ldots, x_n)$ and the multiplication table

$\cdot$	$g_1 = e$	$g_2 = \alpha$	$g_3 = \alpha^2$	$g_4 = \alpha^3$	$\cdots$	$g_n = \alpha^{n-1}$
$g_1^{-1} = e$	e	α	α^2	α^3	$\cdots$	α^{n-1}
$g_2^{-1} = \alpha^{n-1}$	α^{n-1}	e	α	α^2	$\cdots$	α^{n-2}
$g_3^{-1} = \alpha^{n-2}$	α^{n-2}	α^{n-1}	e	α	$\cdots$	α^{n-3}
$g_4^{-1} = \alpha^{n-3}$	α^{n-3}	α^{n-2}	α^{n-1}	e	$\cdots$	α^{n-4}
$\vdots$	$\vdots$	$\vdots$	$\vdots$	$\vdots$	$\ddots$	$\vdots$
$g_n^{-1} = \alpha$	α	α^2	α^3	α^4	$\cdots$	e

shows that C_1 is equal to the $n \times n$ circulant matrix

(15.21) $$C = \begin{bmatrix} x_1 & x_2 & x_3 & x_4 & \cdots & x_n \\ x_n & x_1 & x_2 & x_3 & \cdots & x_{n-1} \\ x_{n-1} & x_n & x_1 & x_2 & \cdots & x_{n-2} \\ x_{n-2} & x_{n-1} & x_n & x_1 & \cdots & x_{n-3} \\ \vdots & \vdots & \vdots & \vdots & \ddots & \vdots \\ x_2 & x_3 & x_4 & x_5 & \cdots & x_1 \end{bmatrix}.$$

(ii) For (15.17) and $1 \leq r,\ s \leq n$, we have $\left[L_1\right]_{r,s} = \sigma(g_r^{-1} g_{n+s})$, where (15.11) yields

$$g_r^{-1} g_{n+s} = (\beta g_r \beta)\beta \alpha^{s-1} = \beta g_r \alpha^{s-1} = \beta \alpha^{r-1} \alpha^{s-1} = \beta \alpha^{r+s-2}.$$

Thus, the first row of L_1 is

$$\big(\sigma(\beta),\ \sigma(\beta\alpha),\ \sigma(\beta\alpha^2),\ \ldots,\ \sigma(\beta\alpha^{n-1})\big) = \big(x_{n+1},\ x_{n+2},\ x_{n+3},\ \ldots,\ x_{2n}\big)$$

and, with (15.12) for its construction, the multiplication table

$\cdot$	$g_{n+1} = \beta$	$g_{n+2} = \beta\alpha$	$g_{n+3} = \beta\alpha^2$	$\cdots$	$g_{2n} = \beta\alpha^{n-1}$
$g_1^{-1} = e$	β	$\beta\alpha$	$\beta\alpha^2$	$\cdots$	$\beta\alpha^{n-1}$
$g_2^{-1} = \alpha^{n-1}$	$\beta\alpha$	$\beta\alpha^2$	$\beta\alpha^3$	$\cdots$	β
$g_3^{-1} = \alpha^{n-2}$	$\beta\alpha^2$	$\beta\alpha^3$	$\beta\alpha^4$	$\cdots$	$\beta\alpha$
$\vdots$	$\vdots$	$\vdots$	$\vdots$	$\ddots$	$\vdots$
$g_n^{-1} = \alpha$	$\beta\alpha^{n-1}$	β	$\beta\alpha$	$\cdots$	$\beta\alpha^{n-2}$

shows that L_1 is equal to the $n \times n$ leftward-circulant matrix

$$(15.22) \qquad L = \begin{bmatrix} x_{n+1} & x_{n+2} & x_{n+3} & \cdots & x_{2n} \\ x_{n+2} & x_{n+3} & x_{n+4} & \cdots & x_{n+1} \\ x_{n+3} & x_{n+4} & x_{n+5} & \cdots & x_{n+2} \\ \vdots & \vdots & \vdots & \ddots & \vdots \\ x_{2n} & x_{n+1} & x_{n+2} & \cdots & x_{2n-1} \end{bmatrix}.$$

(iii) For (15.18) and $1 \leq r,\ s \leq n$, we have $\left[L_2\right]_{r,s} = \sigma(g_{n+r}^{-1} g_s)$ where (15.11) yields

$$g_{n+r}^{-1} g_s = g_{n+r} g_s = \beta \alpha^{r-1} \alpha^{s-1} = \beta \alpha^{r+s-2}.$$

Thus, the first row of L_2 is

$$\big(\sigma(\beta),\ \sigma(\beta\alpha),\ \sigma(\beta\alpha^2),\ \ldots,\ \sigma(\beta\alpha^{n-1})\big) = \big(x_{n+1},\ x_{n+2},\ x_{n+3},\ \ldots,\ x_{2n}\big)$$

and the multiplication table

$\cdot$	$g_1 = e$	$g_2 = \alpha$	$g_3 = \alpha^2$	$\cdots$	$g_n = \alpha^{n-1}$
$g_{n+1}^{-1} = \beta$	β	$\beta\alpha$	$\beta\alpha^2$	$\cdots$	$\beta\alpha^{n-1}$
$g_{n+2}^{-1} = \beta\alpha$	$\beta\alpha$	$\beta\alpha^2$	$\beta\alpha^3$	$\cdots$	β
$g_{n+3}^{-1} = \beta\alpha^2$	$\beta\alpha^2$	$\beta\alpha^3$	$\beta\alpha^4$	$\cdots$	$\beta\alpha$
$\vdots$	$\vdots$	$\vdots$	$\vdots$	$\ddots$	$\vdots$
$g_{n+n}^{-1} = \beta\alpha^{n-1}$	$\beta\alpha^{n-1}$	β	$\beta\alpha$	$\cdots$	$\beta\alpha^{n-2}$

shows that L_2 is equal to L in (15.22).

(iv) For (15.19) and $1 \leq r$, $s \leq n$, we have $\left[C_2\right]_{r,s} = \sigma(g_{n+r}^{-1} \, g_{n+s})$ where (15.11) yields

$$g_{n+r}^{-1} \, g_{n+s} = g_{n+r} \, g_{n+s} = \beta \alpha^{r-1} \beta \alpha^{s-1} = \alpha^{1-r} \alpha^{s-1} = \alpha^{s-r}.$$

Thus, the first row of C_2 is

$$\big(\sigma(e), \, \sigma(\alpha), \, \sigma(\alpha^2), \, \ldots, \, \sigma(\alpha^{n-1})\big) = \big(x_1, \, x_2, \, x_3, \, \ldots, \, x_n\big)$$

and, with (15.12) for its construction, the multiplication table

$\cdot$	$g_{n+1} = \beta$	$g_{n+2} = \beta\alpha$	$g_{n+3} = \beta\alpha^2$	$\cdots$	$g_{2n} = \beta\alpha^{n-1}$
$g_{n+1}^{-1} = \beta$	e	α	α^2	$\cdots$	α^{n-1}
$g_{n+2}^{-1} = \beta\alpha$	α^{n-1}	e	α	$\cdots$	α^{n-2}
$g_{n+3}^{-1} = \beta\alpha^2$	α^{n-2}	α^{n-1}	e	$\cdots$	α^{n-3}
$\vdots$	$\vdots$	$\vdots$	$\vdots$	$\ddots$	$\vdots$
$g_{2n}^{-1} = \beta\alpha^{n-1}$	α	α^2	α^3	$\cdots$	e

shows that C_2 is the circulant matrix C in (15.21).

Since the matrix A_{2n} in (15.20) is given by (15.15), this completes the proof. $\quad\square$

OBSERVATION 15.3. Let C_1, C_2 denote $n \times n$ circulant matrices over a field $\mathfrak{F}$ and let L_1, L_2 denote $n \times n$ leftward-circulant matrices over $\mathfrak{F}$. Then, the matrices

$$A_1 = \left[\begin{array}{c|c} C_1 & L_1 \\ \hline L_1 & C_1 \end{array}\right] \quad \text{and} \quad A_2 = \left[\begin{array}{c|c} C_2 & L_2 \\ \hline L_2 & C_2 \end{array}\right]$$

are $2n \times 2n$ group-pattern matrices of the type (15.15). Therefore, their product

$$A_1 A_2 = \left[\begin{array}{c|c} C_1 C_2 + L_1 L_2 & C_1 L_2 + L_1 C_2 \\ \hline L_1 C_2 + C_1 L_2 & L_1 L_2 + C_1 C_2 \end{array}\right]$$

is also a group-pattern matrix of the type (15.15).

Theorem 6.10 on page 69 provides more detail. Namely, it establishes that $C_1 C_2$, $L_1 L_2$, $C_1 C_2 + L_1 L_2$ are circulant matrices and $C_1 L_2$, $L_1 C_2$, $C_1 L_2 + L_1 C_2$ are leftward-circulant matrices.

Natural generalizations of these multiplicative properties of circulant matrices and leftward circulant matrices are presented in Theorem 17.7 on page 167.

OBSERVATION 15.4. When the elements of $\mathfrak{S}_3$ are regarded as the permutations of the vertices of an equilateral triangle, we can regard $\mathcal{D}_3$ as $\mathfrak{S}_3$ and see that the visual appearance of (15.15) for $n = 3$ is similar to that of (1.7) on page 5, or (2.17) on page 17, or (11.15) on page 107.

For $\mathcal{D}_4$, see (1.10) on page 6. For $\mathcal{D}_5$, see Z_1 in (2.36) on page 25.

15.3. Other group-pattern matrices for $\mathcal{D}_3$, $\mathcal{D}_4$, and $\mathcal{D}_5$

Exercises 1, 2, and 3 on pages 39-40 show that the number of standard group-pattern matrices for $\mathcal{D}_3$, $\mathcal{D}_4$, and $\mathcal{D}_5$ are given by $\mathfrak{N}_{\mathfrak{G}}(\mathcal{D}_3) = 20$, $\mathfrak{N}_{\mathfrak{G}}(\mathcal{D}_4) = 630$, and $\mathfrak{N}_{\mathfrak{G}}(\mathcal{D}_5) = 18,144$. In fact, we have the following result.

PROPOSITION 15.5. *The number of standard group-pattern matrices for $\mathcal{D}_n$ is*

$$(15.23) \qquad \mathfrak{N}_{\mathfrak{G}}(\mathcal{D}_n) = \frac{(2n-1)!}{n\,\varphi(n)}, \quad \text{for } n \geq 3,$$

where φ is the Euler totient function.

COMMENT. A method to compute $\varphi(n)$ is presented in (3.31) on page 40.

PROOF. We note that the order of $\mathcal{D}_n$ is $2n$, Thus, Theorem 3.7 on page 30 yields $\mathfrak{N}_{\mathfrak{G}}(\mathcal{D}_n) = (2n-1)!/m$, where m is the number of automorphisms of $\mathcal{D}_n$. Since Proposition 15.9 on page 150 shows that m is given by $m = n\,\varphi(n)$, we see that (15.23) is valid. This completes the proof. $\qquad\square$

OBSERVATION 15.6. When the technique of Section 3.6 is applied to display the standard group-pattern matrices for $\mathcal{D}_3$, $\mathcal{D}_4$, and $\mathcal{D}_5$, the ones that catch our attention as having interesting patterns are those of Proposition 15.2 along with

$$H_6 = \left[\begin{array}{ccc|ccc} y_1 & y_2 & y_3 & y_4 & y_5 & y_6 \\ y_3 & y_1 & y_2 & y_6 & y_4 & y_5 \\ y_2 & y_3 & y_1 & y_5 & y_6 & y_4 \\ \hline y_4 & y_6 & y_5 & y_1 & y_3 & y_2 \\ y_5 & y_4 & y_6 & y_2 & y_1 & y_3 \\ y_6 & y_5 & y_4 & y_3 & y_2 & y_1 \end{array}\right], \quad \text{for } \mathcal{D}_3,$$

$$H_8 = \left[\begin{array}{cccc|cccc} y_1 & y_2 & y_3 & y_4 & y_5 & y_6 & y_7 & y_8 \\ y_4 & y_1 & y_2 & y_3 & y_8 & y_5 & y_6 & y_7 \\ y_3 & y_4 & y_1 & y_2 & y_7 & y_8 & y_5 & y_6 \\ y_2 & y_3 & y_4 & y_1 & y_6 & y_7 & y_8 & y_5 \\ \hline y_5 & y_8 & y_7 & y_6 & y_1 & y_4 & y_3 & y_2 \\ y_6 & y_5 & y_8 & y_7 & y_2 & y_1 & y_4 & y_3 \\ y_7 & y_6 & y_5 & y_8 & y_3 & y_2 & y_1 & y_4 \\ y_8 & y_7 & y_6 & y_5 & y_4 & y_3 & y_2 & y_1 \end{array}\right], \quad \text{for } \mathcal{D}_4,$$

$$H_{10} = \left[\begin{array}{ccccc|ccccc} y_1 & y_2 & y_3 & y_4 & y_5 & y_6 & y_7 & y_8 & y_9 & y_{10} \\ y_5 & y_1 & y_2 & y_3 & y_4 & y_{10} & y_6 & y_7 & y_8 & y_9 \\ y_4 & y_5 & y_1 & y_2 & y_3 & y_9 & y_{10} & y_6 & y_7 & y_8 \\ y_3 & y_4 & y_5 & y_1 & y_2 & y_8 & y_9 & y_{10} & y_6 & y_7 \\ y_2 & y_3 & y_4 & y_5 & y_1 & y_7 & y_8 & y_9 & y_{10} & y_6 \\ \hline y_6 & y_{10} & y_9 & y_8 & y_7 & y_1 & y_5 & y_4 & y_3 & y_2 \\ y_7 & y_6 & y_{10} & y_9 & y_8 & y_2 & y_1 & y_5 & y_4 & y_3 \\ y_8 & y_7 & y_6 & y_{10} & y_9 & y_3 & y_2 & y_1 & y_5 & y_4 \\ y_9 & y_8 & y_7 & y_6 & y_{10} & y_4 & y_3 & y_2 & y_1 & y_5 \\ y_{10} & y_9 & y_8 & y_7 & y_6 & y_5 & y_4 & y_3 & y_2 & y_1 \end{array}\right], \quad \text{for } \mathcal{D}_5,$$

as secondary group-pattern matrices. These matrices have symmetry about their centers and each consists of four blocks of circulant matrices.

15.4. Secondary group-pattern matrices for $\mathcal{D}_n$

We introduce $\mathscr{L}_H = \big(h_1, h_2, \ldots, h_n, h_{n+1}, h_{n+2}, \ldots, h_{2n}\big)$ as the list obtained from $\mathscr{L} = \big(g_1, g_2, \ldots, g_n, g_{n+1}, g_{n+2}, g_{n+3}, \ldots, g_{2n-1}, g_{2n}\big)$ of (15.9) on page 141 by writing the last $n-1$ elements $g_{n+2}, g_{n+3}, \ldots, g_{2n-1}, g_{2n}$ of $\mathscr{L}$ in the reverse order $g_{2n}, g_{2n-1}, \ldots, g_{n+3}, g_{n+2}$. Thus, $\mathscr{L}_H$ is defined directly in terms of (15.6) and (15.7) by $h_k = g_k$, for $1 \leq k \leq n+1$, and $h_k = g_{3n+2-k}$, for $n+2 \leq k \leq 2n$.

When $n \geq 3$, a *secondary group-pattern matrix* H_{2n} for $\mathcal{D}_n$ is defined with respect to the list $\mathscr{L}_H$ and a function τ from $\mathcal{D}_n$ to a set S by

$$(15.24) \qquad \big[H_{2n}\big]_{r,s} = \tau\big(h_r^{-1}h_s\big), \quad \text{for } 1 \leq r, s \leq 2n.$$

PROPOSITION 15.7. *When τ for (15.24) yields $\tau(h_k) = y_k$, for $1 \leq k \leq 2n$, the $2n \times 2n$ secondary group-pattern matrix H_{2n} for $\mathcal{D}_n$ and $\mathscr{L}_H$ has the form*

$$(15.25) \qquad H_{2n} = \left[\begin{array}{c|c} C_1 & C_2 \\ \hline C_3 & C_4 \end{array}\right],$$

where C_1, C_2, C_3 and C_4 are the $n \times n$ circulant matrices such that:

(1) *the first row of C_1 is $\big(y_1, y_2, y_3, \ldots, y_{n-1}, y_n\big)$,*

(2) *the first row of C_2 is $\big(y_{n+1}, y_{n+2}, y_{n+3}, \ldots, y_{2n-1}, y_{2n}\big)$,*

(3) *the first row of C_3 is $\big(y_{n+1}, y_{2n}, y_{2n-1}, \ldots, y_{n+3}, y_{n+2}\big)$, and*

(4) *the first row of C_4 is $\big(y_1, y_n, y_{n-1}, \ldots, y_3, y_2\big)$.*

PROOF. In terms of the $2n \times 2n$ group-pattern matrix H_{2n}, we define four $n \times n$ submatrices B_1, B_2, B_3, and B_4 by

$$(15.26) \quad \big[B_1\big]_{r,s} = \big[H_{2n}\big]_{r,s}, \qquad \text{for } 1 \leq r, s \leq n,$$

$$(15.27) \quad \big[B_2\big]_{r,s} = \big[H_{2n}\big]_{r,n+s}, \qquad \text{for } 1 \leq r, s \leq n,$$

$$(15.28) \quad \big[B_3\big]_{r,s} = \big[H_{2n}\big]_{n+r,s}, \qquad \text{for } 1 \leq r, s \leq n,$$

$$(15.29) \quad \big[B_4\big]_{r,s} = \big[H_{2n}\big]_{n+r,n+s}, \qquad \text{for } 1 \leq r, s \leq n,$$

in order to express H_{2n} as

$$(15.30) \qquad H_{2n} = \left[\begin{array}{c|c} B_1 & B_2 \\ \hline B_3 & B_4 \end{array}\right].$$

We use (15.9)–(15.11) to rewrite $\mathscr{L}_H$ as

$$\mathscr{L}_H = (e, \alpha, \alpha^2, \ldots, \alpha^{n-1}, \beta, \beta\alpha^{n-1}, \beta\alpha^{n-2}, \ldots, \beta\alpha).$$

With $\alpha^n = e$, we observe that

$$(15.31) \quad h_k = \alpha^{k-1}, \quad \text{for } 1 \leq k \leq n, \quad \text{and} \quad h_{n+k} = \beta\alpha^{n+1-k}, \quad \text{for } 1 \leq k \leq n.$$

(i) For (15.26) and $1 \leq r, s \leq n$, we have $\big[B_1\big]_{r,s} = \tau(h_r^{-1}h_s)$ where (15.31) yields

$$h_r^{-1}h_s = \big(\alpha^{r-1}\big)^{-1}\alpha^{s-1} = \alpha^{s-r}.$$

The first row of B_1 is

$$\big(\tau(e), \tau(\alpha), \tau(\alpha^2), \ldots, \tau(\alpha^{n-1})\big) = \big(y_1, y_2, y_3, \ldots, y_n\big).$$

and the multiplication table

$\cdot$	$h_1 = e$	$h_2 = \alpha$	$h_3 = \alpha^2$	$h_4 = \alpha^3$	$\cdots$	$h_n = \alpha^{n-1}$
$h_1^{-1} = e$	e	α	α^2	α^3	$\cdots$	α^{n-1}
$h_2^{-1} = \alpha^{n-1}$	α^{n-1}	e	α	α^2	$\cdots$	α^{n-2}
$h_3^{-1} = \alpha^{n-2}$	α^{n-2}	α^{n-1}	e	α	$\cdots$	α^{n-3}
$h_4^{-1} = \alpha^{n-3}$	α^{n-3}	α^{n-2}	α^{n-1}	e	$\cdots$	α^{n-4}
$\vdots$	$\vdots$	$\vdots$	$\vdots$	$\vdots$	$\ddots$	$\vdots$
$h_n^{-1} = \alpha$	α	α^2	α^3	α^4	$\cdots$	e

shows that B_1 is equal to the $n \times n$ circulant matrix

$$C_1 = \begin{bmatrix} y_1 & y_2 & y_3 & y_4 & \cdots & y_n \\ y_n & y_1 & y_2 & y_3 & \cdots & y_{n-1} \\ y_{n-1} & y_n & y_1 & y_2 & \cdots & y_{n-2} \\ y_{n-2} & y_{n-1} & y_n & y_1 & \cdots & y_{n-3} \\ \vdots & \vdots & \vdots & \vdots & \ddots & \vdots \\ y_2 & y_3 & y_4 & y_5 & \cdots & y_1 \end{bmatrix}.$$

(ii) For (15.27) and $1 \leq r, s \leq n$, we have $\left[B_2\right]_{r,s} = \tau(h_r^{-1} h_{n+s})$ where (15.31) yields

$$h_r^{-1} h_{n+s} = \alpha^{1-r} \beta \alpha^{n+1-s} = (\beta \alpha^{r-1} \beta) \beta \alpha^{n+1-s} = \beta \alpha^{n+r-s}.$$

The first row of B_2 is

$$\left(\tau(\beta), \, \tau(\beta \alpha^{n-1}), \, \tau(\beta \alpha^{n-2}), \, \ldots, \, \tau(\beta \alpha) \right)$$
$$= \left(\tau(h_{n+1}), \, \tau(h_{n+2}), \, \tau(h_{n+3}), \, \ldots, \, \tau(h_{2n}) \right)$$
$$= \left(y_{n+1}, \, y_{n+2}, \, y_{n+3}, \, \ldots, \, y_{2n} \right)$$

and, with (15.12) for its construction, the multiplication table

$\cdot$	$h_{n+1} = \beta$	$h_{n+2} = \beta \alpha^{n-1}$	$h_{n+3} = \beta \alpha^{n-2}$	$\cdots$	$h_{2n} = \beta \alpha$
$h_1^{-1} = e$	β	$\beta \alpha^{n-1}$	$\beta \alpha^{n-2}$	$\cdots$	$\beta \alpha$
$h_2^{-1} = \alpha^{n-1}$	$\beta \alpha$	β	$\beta \alpha^{n-1}$	$\cdots$	$\beta \alpha^2$
$h_3^{-1} = \alpha^{n-2}$	$\beta \alpha^2$	$\beta \alpha$	β	$\cdots$	$\beta \alpha^3$
$\vdots$	$\vdots$	$\vdots$	$\vdots$	$\ddots$	$\vdots$
$h_n^{-1} = \alpha$	$\beta \alpha^{n-1}$	$\beta \alpha^{n-2}$	$\beta \alpha^{n-3}$	$\cdots$	β

shows that B_2 is equal to the $n \times n$ circulant matrix

$$C_2 = \begin{bmatrix} y_{n+1} & y_{n+2} & y_{n+3} & \cdots & y_{2n} \\ y_{2n} & y_{n+1} & y_{n+2} & \cdots & y_{2n-1} \\ y_{2n-1} & y_{2n} & y_{n+1} & \cdots & y_{2n-2} \\ \vdots & \vdots & \vdots & \ddots & \vdots \\ y_{n+2} & y_{n+3} & y_{n+4} & \cdots & y_{n+1} \end{bmatrix}.$$

(iii) For (15.28) and $1 \le r,\, s \le n$, we have $\left[B_3\right]_{r,s} = \tau(h_{n+r}^{-1}\, h_s)$ where (15.31) yields

$$h_{n+r}^{-1}\, h_s = h_{n+r}\, h_s = \beta\alpha^{n+1-r}\alpha^{s-1} = \beta\alpha^{s-r}.$$

Thus, the first row of B_3 is

$$\big(\tau(\beta),\, \tau(\beta\alpha),\, \tau(\beta\alpha^2),\, \ldots,\, \beta\alpha^{n-1}\big)$$
$$= \big(\tau(h_{n+1}),\, \tau(h_{2n}),\, \tau(h_{2n-1}),\, \ldots,\, \tau(h_{n+2})\big)$$
$$= \big(y_{n+1},\, y_{2n},\, y_{2n-1},\, \ldots,\, y_{n+2}\big)$$

and the multiplication table

$\cdot$	$h_1 = e$	$h_2 = \alpha$	$h_3 = \alpha^2$	$\cdots$	$h_n = \alpha^{n-1}$
$h_{n+1}^{-1} = \beta$	β	$\beta\alpha$	$\beta\alpha^2$	$\cdots$	$\beta\alpha^{n-1}$
$h_{n+2}^{-1} = \beta\alpha^{n-1}$	$\beta\alpha^{n-1}$	β	$\beta\alpha$	$\cdots$	$\beta\alpha^{n-2}$
$h_{n+3}^{-1} = \beta\alpha^{n-2}$	$\beta\alpha^{n-2}$	$\beta\alpha^{n-1}$	β	$\cdots$	$\beta\alpha^{n-3}$
$\vdots$	$\vdots$	$\vdots$	$\vdots$	$\ddots$	$\vdots$
$h_{n+n}^{-1} = \beta\alpha$	$\beta\alpha$	$\beta\alpha^2$	$\beta\alpha^3$	$\cdots$	β

shows that B_3 is equal to the $n \times n$ circulant matrix

$$C_3 = \begin{bmatrix} y_{n+1} & y_{2n} & y_{2n-1} & \cdots & y_{n+2} \\ y_{n+2} & y_{n+1} & y_{2n} & \cdots & y_{n+3} \\ y_{n+3} & y_{n+2} & y_{n+1} & \cdots & y_{n+4} \\ \vdots & \vdots & \vdots & \ddots & \vdots \\ y_{2n} & y_{2n-1} & y_{2n-2} & \cdots & y_{n+1} \end{bmatrix}.$$

(iv) For (15.29) and $1 \le r,\, s \le n$, we have $\left[B_4\right]_{r,s} = \tau(h_{n+r}^{-1}\, h_{n+s})$ where (15.31) yields

$$h_{n+r}^{-1}\, h_{n+s} = h_{n+r}\, h_{n+s} = \beta\alpha^{n+1-r}\beta\alpha^{n+1-s} = \alpha^{r-1}\alpha^{n+1-s} = \alpha^{n+r-s}.$$

The first row of B_4 is

$$\big(\tau(e),\, \tau(\alpha^{n-1}),\, \tau(\alpha^{n-2}),\, \ldots,\, \tau(\alpha)\big)$$
$$= \big(\tau(h_1),\, \tau(h_n),\, \tau(h_{n-1}),\, \ldots,\, \tau(h_2)\big)$$
$$= \big(y_1,\, y_n,\, y_{n-1},\, \ldots,\, y_2\big)$$

and, with (15.12) for the construction, the multiplication table

$\cdot$	$h_{n+1} = \beta$	$h_{n+2} = \beta\alpha^{n-1}$	$h_{n+3} = \beta\alpha^{n-2}$	$\cdots$	$h_{2n} = \beta\alpha$
$h_{n+1}^{-1} = \beta$	e	α^{n-1}	α^{n-2}	$\cdots$	α
$h_{n+2}^{-1} = \beta\alpha^{n-1}$	α	e	α^{n-1}	$\cdots$	α^2
$h_{n+3}^{-1} = \beta\alpha^{n-2}$	α^2	α	e	$\cdots$	α^3
$\vdots$	$\vdots$	$\vdots$	$\vdots$	$\ddots$	$\vdots$
$h_{2n}^{-1} = \beta\alpha$	α^{n-1}	α^{n-2}	α^{n-3}	$\cdots$	e

shows that B_4 is equal to the $n \times n$ circulant matrix

$$C_4 = \begin{bmatrix} y_1 & y_n & y_{n-1} & \cdots & y_2 \\ y_2 & y_1 & y_n & \cdots & y_3 \\ y_3 & y_2 & y_1 & \cdots & y_4 \\ \vdots & \vdots & \vdots & \ddots & \vdots \\ y_n & y_{n-1} & y_{n-2} & \cdots & y_1 \end{bmatrix}.$$

Since the matrix H_{2n} in (15.30) is given by (15.25), this completes the proof. $\square$

15.5. The automorphisms for $\mathcal{D}_n$

Definition 3.9 on page 30 shows that an automorphism χ of $\mathcal{D}_n$ is a one-to-one function from $\mathcal{D}_n$ onto $\mathcal{D}_n$ such that $\chi(xy) = \chi(x)\chi(y)$, for each x, y in $\mathcal{D}_n$. The elements of $\mathcal{D}_n$ belong to its two disjoint subsets E_1 and E_2 defined by

$$E_1 = \{e, \alpha, \alpha^2, \ldots, \alpha^{n-1}\} \quad \text{and} \quad E_2 = \{\beta, \beta\alpha, \beta\alpha^2, \ldots, \beta\alpha^{n-1}\}.$$

PROPOSITION 15.8. *A function χ from G to G is an automorphism of G if and only if integers b, c exist such that: $1 \le b, c \le n$, b is relatively prime to n, and*

$$(15.32) \qquad \chi(x) = \begin{cases} \alpha^{bk}, & \text{if } x = \alpha^k \text{ and } 0 \le k \le n-1, \\ \beta\alpha^{bk+c-1}, & \text{if } x = \beta\alpha^k \text{ and } 0 \le k \le n-1. \end{cases}$$

PROOF. (i) Suppose that χ is an automorphism of $\mathcal{D}_n$. Since E_1 forms a cyclic subgroup of $\mathcal{D}_n$, the restriction of χ to E_1 is an automorphism of E_1. Exercise 10 of page 40 shows that there is an integer b such that $1 \le b \le n$, b is relatively prime to n, and $\chi(\alpha^k) = \alpha^{bk}$, for $0 \le k \le n-1$. Since χ is one-to-one, the element $\chi(\beta)$ is in E_2. Thus, there is an integer c such that $1 \le c \le n$, $\chi(\beta) = \beta\alpha^{c-1}$, and

$$\chi(\beta\alpha^k) = \chi(\beta)\chi(\alpha^k) = (\beta\alpha^{c-1})(\alpha^{bk}) = \beta\alpha^{bk+c-1}, \quad \text{for } 0 \le k \le n-1.$$

Hence, χ is given by (15.32) where $1 \le b, c \le n$ and b is relatively prime to n.

(ii) In terms of integers b, c for which $1 \le b, c \le n$ and b is relatively prime to n, let χ be the function from G to G defined by (15.32).

To show that χ is one-to-one, suppose that x_1, x_2 are elements of G for which $\chi(x_1) = \chi(x_2)$. Then, we can not have either x_1 in E_1 and x_2 in E_2 or x_1 in E_2 and x_2 in E_1. If $x_1 = \alpha^{k_1}$, $x_2 = \alpha^{k_2}$, $0 \le k_1, k_2 \le n-1$, and $\chi(x_1) = \chi(x_2)$, then $\alpha^{bk_1} = \alpha^{bk_2}$, $\alpha^{b(k_1-k_2)} = e$, the period of α is n, n divides $b(k_1 - k_2)$, n is relatively prime to b, n divides $k_1 - k_2$, $|k_1 - k_2| < n$, and $x_1 = x_2$. If $x_1 = \beta\alpha^{k_1}$, $x_2 = \beta\alpha^{k_2}$, $0 \le k_1, k_2 \le n-1$, and $\chi(x_1) = \chi(x_2)$, then $\beta\alpha^{bk_1+c-1} = \beta\alpha^{bk_2+c-1}$, $\alpha^{bk_1} = \alpha^{bk_2}$, $\alpha^{b(k_1-k_2)} = e$, etc., and $x_1 = x_2$. Thus, χ is a one-to-one function from G into G. Since G is finite, χ is therefore a one-to-one function from G onto G.

Let k denote any integer. Then, there are unique integers q and r such that $k = qn + r$, where $0 \le r < n$. With $\alpha^n = e$, we observe that

$$\chi(\alpha^k) = \chi(\alpha^{qn+r}) = \chi(\alpha^r) = \alpha^{br} = \alpha^{b(k-qn)} = \alpha^{hk}$$

and

$$\chi(\beta\alpha^k) = \chi(\beta\alpha^{qn+r}) = \chi(\beta\alpha^r) = \beta\alpha^{br+c-1} = \beta\alpha^{b(k-qn)+c-1} = \beta\alpha^{bk+c-1}.$$

Hence, (15.32) remains valid when k is any integer. There are four cases.

(1) If $x = \alpha^{k_1}$ and $y = \alpha^{k_2}$, then $xy = \alpha^{k_1+k_2}$ and
$$\chi(xy) = \alpha^{b(k_1+k_2)} = \alpha^{bk_1}\alpha^{bk_2} = \chi(x)\chi(y).$$

(2) If $x = \alpha^{k_1}$ and $y = \beta\alpha^{k_2}$, then $xy = \beta(\beta\alpha^{k_1}\beta)\alpha^{k_2} = \beta\alpha^{-k_1+k_2}$ and
$$\chi(xy) = \beta\alpha^{b(-k_1+k_2)+c-1} = \left(\beta\alpha^{b(-k_1)}\beta\right)\beta\alpha^{bk_2+c-1} = \left(\alpha^{bk_1}\right)\chi(y) = \chi(x)\chi(y).$$

(3) If $x = \beta\alpha^{k_1}$ and $y = \alpha^{k_2}$, then $xy = \beta\alpha^{k_1+k_2}$ and
$$\chi(xy) = \beta\alpha^{b(k_1+k_2)+c-1} = \left(\beta\alpha^{bk_1+c-1}\right)\alpha^{bk_2} = \chi(x)\chi(y).$$

(4) If $x = \beta\alpha^{k_1}$ and $y = \beta\alpha^{k_2}$, then $xy = \left(\beta\alpha^{k_1}\beta\right)\alpha^{k_2} = \alpha^{-k_1+k_2}$ and
$$\chi(xy) = \alpha^{b(-k_1+k_2)} = \alpha^{-bk_1-c+1}\alpha^{bk_2+c-1} = \beta\alpha^{bk_1+c-1}\beta\alpha^{bk_2+c-1} = \chi(x)\chi(y).$$

Consequently, χ is an automorphism for G. This completes the proof. $\square$

Properties of the Euler totient function φ are given in Exercise 7 on page 40.

PROPOSITION 15.9. *There are $n\varphi(n)$ automorphisms for $\mathcal{D}_n$; they are given by*

(15.33) $$\chi_{b,c}(x) = \begin{cases} \alpha^{bk}, & \text{if } x = \alpha^k, \quad \text{for } 0 \leq k \leq n-1, \\ \beta\alpha^{bk+c-1}, & \text{if } x = \beta\alpha^k, \text{ for } 0 \leq k \leq n-1, \end{cases}$$

as the pair (b, c) of integers b, c ranges through those that satisfy $1 \leq b, c \leq n$ with b relatively prime to n.

PROOF. Proposition 15.8 shows that each automorphism for $\mathcal{D}_n$ is obtained as described for (15.33).

For integers b_1, c_1, b_2, c_2 that satisfy $1 \leq b_1, c_1, b_2, c_2 \leq n$ with b_1, b_2 relatively prime to n, suppose that $(b_1, c_1) \neq (b_2, c_2)$. If $b_1 \neq b_2$, then $\alpha^{b_1} \neq \alpha^{b_2}$ and
$$\chi_{b_1,c_1}(\alpha) = \alpha^{b_1} \neq \alpha^{b_2} = \chi_{b_2,c_2}(\alpha).$$
If $c_1 \neq c_2$, then $\alpha^{c_1-1} \neq \alpha^{c_2-1}$ and
$$\chi_{b_1,c_1}(\beta) = \beta\alpha^{c_1-1} \neq \beta\alpha^{c_2-1} = \chi_{b_2,c_2}(\beta).$$
This shows that $\chi_{b_1,c_1} \neq \chi_{b_2,c_2}$. Thus, there is a one-to-one correspondence between the automorphisms of G and the ordered couples (b, c) with $1 \leq b, c \leq n$ and b relatively prime to n. Since there are n selections for c and $\varphi(n)$ selections for b, this shows that $\mathcal{D}_n$ has $n\varphi(n)$ automorphisms and completes the proof. $\square$

15.6. Matrix representations of degree 1 for $\mathcal{D}_n$

We recall from (15.9) and (15.10) of page 141 that the list $\mathscr{L}$ of the $2n$ elements in $\mathcal{D}_n$ is given by

(15.34) $$\mathscr{L} = \left(e, \alpha, \alpha^2, \ldots, \alpha^{n-1}, \beta, \beta\alpha, \beta\alpha^2, \ldots, \beta\alpha^{n-1}\right),$$

where $\alpha^n = e$, $\beta^2 = e$, and $\beta\alpha^k\beta = \alpha^{-k}$, for any integer k. In terms of the two 1×1 matrices $[1]$ and $[-1]$, we introduce the set $\mathcal{T} = \left\{[1], [-1]\right\}$.

PROPOSITION 15.10. *Let F be a matrix representation of degree 1 for $\mathcal{D}_n$. Then, F is a function from $\mathcal{D}_n$ to $\mathcal{T}$ such that*

(15.35) $F(\alpha^k) = \left(F(\alpha)\right)^k$ *and* $F(\beta\alpha^k) = F(\beta)\left(F(\alpha)\right)^k$,   *for any integer k.*

PROOF. In view of $F(e) = F(ee) = F(e)\,F(e)$ and $F(e) \neq [\,0\,]$, we observe that $F(e) = [\,1\,]$. The relation $\big(F(\beta)\big)^2 = F\big(\beta^2\big) = F(e) = [\,1\,]$ shows that $F(\beta)$ is an element of $\mathcal{T}$. Due to $(\beta\alpha)^2 = (\beta\alpha\beta)\alpha = \alpha^{-1}\alpha = e$, we have

$$\big(F(\beta\alpha)\big)^2 = F\big((\beta\alpha)^2\big) = F(e) = [\,1\,]$$

and see that $F(\beta\alpha)$ is an element of $\mathcal{T}$. Since $F(\beta)$ is also an element of $\mathcal{T}$ and $\mathcal{T}$ is closed under matrix multiplication, the product $F(\beta)\,F(\beta\alpha) = F(\beta\beta\alpha) = F(\alpha)$ is an element of $\mathcal{T}$. Consequently, both $F(\alpha)$ and $F(\beta)$ are elements of $\mathcal{T}$.

Because F is a matrix representation for $\mathcal{D}_n$, (15.35) is satisfied.

Each element x of $\mathcal{D}_n$ is expressible either as α^k or as $\beta\alpha^k$, for some k. Thus, an application of (15.35) shows that, for each x in $\mathcal{D}_n$, $F(x)$ is an element of $\mathcal{T}$. This completes the proof. $\qquad\square$

PROPOSITION 15.11. *For any given function F from $\mathcal{D}_n$ to $\mathcal{T}$, F is a matrix representation for $\mathcal{D}_n$ if and only if the subsets*

$$S_1 = \big\{\; x\colon x \in \mathcal{D}_n \text{ and } F(x) = [\,1\,]\big\} \quad\text{and}\quad S_2 = \big\{\; x\colon x \in \mathcal{D}_n \text{ and } F(x) = [-1]\big\},$$

of $\mathcal{D}_n$ satisfy each of the four conditions:

(1) *if $x \in S_1$ and $y \in S_1$, then $xy \in S_1$;*
(2) *if $x \in S_1$ and $y \in S_2$, then $xy \in S_2$;*
(3) *if $x \in S_2$ and $y \in S_1$, then $xy \in S_2$;*
(4) *if $x \in S_2$ and $y \in S_2$, then $xy \in S_1$.*

PROOF. Suppose that F is a matrix representation of $\mathcal{D}_n$.
(a) If $x \in S_1$ and $y \in S_1$, then $F(xy) = F(x)\,F(y) = [\,1\,][\,1\,] = [\,1\,]$ and $xy \in S_1$.
(b) If $x \in S_1$ and $y \in S_2$, then $F(xy) = F(x)\,F(y) = [\,1\,][-1] = [-1]$ and $xy \in S_2$.
(c) If $x \in S_2$ and $y \in S_1$, then $F(xy) = F(x)\,F(y) = [-1][\,1\,] = [-1]$ and $xy \in S_2$.
(d) If $x \in S_2$ and $y \in S_2$, then $F(xy) = F(x)\,F(y) = [-1][-1] = [\,1\,]$ and $xy \in S_1$.
Thus, conditions (1), (2), (3), and (4) are satisfied.

Suppose that (1), (2), (3), (4) are satisfied and let x, y be any elements in $\mathcal{D}_n$.
(1) If $x \in S_1$ and $y \in S_1$, then $xy \in S_1$ and $F(xy) = [\,1\,] = [\,1\,][\,1\,] = F(x)\,F(y)$.
(2) If $x \in S_1$ and $y \in S_2$, then $xy \in S_2$ and $F(xy) = [-1] = [\,1\,][-1] = F(x)\,F(y)$.
(3) If $x \in S_2$ and $y \in S_1$, then $xy \in S_2$ and $F(xy) = [-1] = [-1][\,1\,] = F(x)\,F(y)$.
(4) If $x \in S_2$ and $y \in S_2$, then $xy \in S_1$ and $F(xy) = [\,1\,] = [-1][-1] = F(x)\,F(y)$.
Thus, we have $F(xy) = F(x)\,F(y)$, for each x, y in $\mathcal{D}_n$, and F is therefore a matrix representation for $\mathcal{D}_n$. This completes the proof. $\qquad\square$

PROPOSITION 15.12. *Let n be an integer that satisfies $n \geq 3$. When n is odd, there are precisely two matrix representations of degree 1 for $\mathcal{D}_n$ and they are*

$$(15.36) \qquad\qquad F_1(g_k) = [\,1\,], \quad \text{for } 1 \leq k \leq 2n,$$

and

$$(15.37) \qquad\qquad F_2(g_k) = \begin{cases} [\,1\,], & \text{for } 1 \leq k \leq n, \\ [-1], & \text{for } n+1 \leq k \leq 2n. \end{cases}$$

Moreover, when n is an even integer $n = 2m$, there are precisely four matrix representations of degree 1 for $\mathcal{D}_n$ and they are given by (15.36), (15.37) as well as

$$(15.38) \qquad F_3(g_k) = \begin{cases} [\,1\,], & \text{for } k = 2\ell - 1 \text{ and } 1 \leq \ell \leq n, \\ [-1], & \text{for } k = 2\ell \text{ and } 1 \leq \ell \leq n, \end{cases}$$

and

$$(15.39) \qquad F_4(g_k) = \begin{cases} [\,1\,], & \text{for } k = 2\ell - 1 \text{ and } 1 \leq \ell \leq m, \\ [-1], & \text{for } k = 2\ell \text{ and } 1 \leq \ell \leq m, \\ [-1], & \text{for } k = 2\ell - 1 \text{ and } m + 1 \leq \ell \leq n, \\ [\,1\,], & \text{for } k = 2\ell \text{ and } m + 1 \leq \ell \leq n. \end{cases}$$

PROOF. Let F be a function from $\mathcal{D}_n$ to the set $\mathcal{T} = \{[\,1\,], [-1]\}$ whose values are defined in terms of $F(\alpha)$ and $F(\beta)$ in $\mathcal{T}$ by

$$(15.40) \quad F(\alpha^k) = \left(F(\alpha)\right)^k \quad \text{and} \quad F(\beta\alpha^k) = F(\beta)\left(F(\alpha)\right)^k, \quad \text{for any integer } k.$$

Proposition 15.10 shows that any matrix representation of degree 1 for $\mathcal{D}_n$ is a function of that kind.

(i) Suppose that $F(\alpha) = [\,1\,]$ and $F(\beta) = [\,1\,]$. Then, we use (15.40) to deduce that $F(x) = [\,1\,]$, for each x in $\mathcal{D}_n$. Hence, for the subsets S_1 and S_2 of $\mathcal{D}_n$ defined in Proposition 15.11, we see that $S_1 = \mathcal{D}_n$ and S_2 is the vacuous set. Since they satisfy the conditions of Proposition 15.11, we conclude that F, as F_1 in (15.36), is a matrix representation for $\mathcal{D}_n$.

(ii) Suppose that $F(\alpha) = [\,1\,]$ and $F(\beta) = [-1]$. Then, we use (15.40) to obtain

$$S_1 = \left\{\alpha^{k-1} : \text{ for } 1 \leq k \leq n\right\} \quad \text{and} \quad S_2 = \left\{\beta\alpha^{k-1} : \text{ for } 1 \leq k \leq n\right\}.$$

Since S_1 and S_2 satisfy the conditions of Proposition 15.11, we see that F, as F_2 in (15.37), is a matrix representation for $\mathcal{D}_n$.

(iii) To show that there does not exist a matrix representation F of degree 1 for $\mathcal{D}_n$ when $F(\alpha) = [-1]$ and n is an odd integer, we suppose that such an F exists and see that leads to $F(\alpha^{n-1}) = [\,1\,]$ and the contradiction

$$[\,1\,] = F(e) = F(\alpha^n) = F(\alpha)\,F(\alpha^{n-1}) = [-1]\,[\,1\,] = [-1].$$

(iv) Suppose that $F(\alpha) = [-1]$, $F(\beta) = [\,1\,]$, and n is an even integer $n = 2m$. We use

$$(15.41) \qquad g_k = \begin{cases} \alpha^{k-1}, & \text{for } 1 \leq k \leq 2m, \\ \beta\alpha^{k-2m-1}, & \text{for } 2m + 1 \leq k \leq 2n, \end{cases}$$

to verify that

$$F(g_k) = \begin{cases} [\,1\,], & \text{for } k = 1, 3, \ldots, 2m - 1, 2m + 1, 2m + 3, \ldots, 2n - 1, \\ [-1], & \text{for } k = 2, 4, \ldots, 2m, 2m + 2, 2m + 4, \ldots, 2n, \end{cases}$$

and obtain

$$S_1 = \left\{g_k : k = 2\ell - 1 \text{ and } 1 \leq \ell \leq n\right\} \quad \text{and} \quad S_2 = \left\{g_k : k = 2\ell \text{ and } 1 \leq \ell \leq n\right\}.$$

Thus, F is the function F_3 defined by (15.38). By rewriting S_1 and S_2 as

$$(15.42) \qquad S_1 = \left\{e, \alpha^2, \alpha^4, \ldots, \alpha^{2m-2}, \beta, \beta\alpha^2, \beta\alpha^4, \ldots, \beta\alpha^{2m-2}\right\}$$

and

$$(15.43) \qquad S_2 = \left\{\alpha, \alpha^3, \alpha^5, \ldots, \alpha^{2m-1}, \beta\alpha, \beta\alpha^3, \beta\alpha^5, \ldots, \beta\alpha^{2m-1}\right\},$$

it becomes easy to check (with n even) that the conditions for Proposition 15.11 are satisfied and therefore F, as F_3 in (15.38), is a matrix representation for $\mathcal{D}_n$.

(v) Suppose that $F(\alpha) = [-1]$, $F(\beta) = [-1]$, and n is an even integer $n = 2m$. We use (15.41) to deduce that

$$
F(g_k) = \begin{cases}
[\,1\,], & \text{for } k = 1,\,3,\,\ldots,\,2m-1, \\
[-1], & \text{for } k = 2,\,4,\,\ldots,\,2m, \\
[-1], & \text{for } k = 2m+1,\,2m+3,\,\ldots,\,2n-1, \\
[\,1\,], & \text{for } k = 2m+2,\,2m+4,\,\ldots,\,2n.
\end{cases}
$$

Thus, F is the function F_4 in (15.39). For the corresponding S_1 and S_2, we have

(15.44) $\qquad S_1 = \left\{ e,\, \alpha^2,\, \alpha^4,\, \ldots,\, \alpha^{2m-2},\, \beta\alpha,\, \beta\alpha^3,\, \beta\alpha^5,\, \ldots,\, \beta\alpha^{2m-1} \right\}$

and

(15.45) $\qquad S_2 = \left\{ \alpha,\, \alpha^3,\, \alpha^5,\, \ldots,\, \alpha^{2m-1},\, \beta,\, \beta\alpha^2,\, \beta\alpha^4,\, \ldots,\, \beta\alpha^{2m-2} \right\}.$

It is easy to check (with n even) that the conditions for Proposition 15.11 are satisfied and therefore F, as F_4 in (15.39), is a matrix representation for $\mathcal{D}_n$.

The results of Parts (i)–(v) yield Proposition 15.12 and complete its proof. $\quad\square$

EXERCISES

1. When n is an even integer, check that the sets S_1 and S_2 in (15.42) and (15.43) satisfy the conditions (1), (2), (3), (4) of Proposition 15.11.

2. When n is an even integer, check that the sets S_1 and S_2 in (15.44) and (15.45) satisfy the conditions (1), (2), (3), (4) of Proposition 15.11.

3. For $n \geq 3$, show that the $2n \times 2n$ matrix H_{2n} defined by (15.24) on page 146 is symmetric with respect to its center.

CHAPTER 16

Frobenius Block-Diagonalizations
of Primary Group-Pattern Matrices for $\mathcal{D}_n$

The primary group-pattern matrices A_6 for $\mathcal{D}_3$, A_8 for $\mathcal{D}_4$, and A_{10} for $\mathcal{D}_5$ are characterized in Proposition 15.2 on page 142. To obtain Frobenius block-diagonalizations for them in this chapter by means of Theorem 13.3, emphasis will be placed on the matrix representation of degree 2 for each $\mathcal{D}_n$ in Section 15.1 and the matrix representations of degree 1 for each $\mathcal{D}_n$ in Proposition 15.12.

16.1. Frobenius block-diagonalization of A_6 for $\mathcal{D}_3$

The six elements of the dihedral group $\mathcal{D}_3$ of order 6 specify the symmetries for an equilateral triangle positioned as indicated in (15.3). For convenience, we set

(16.1) $\qquad c_1 = \cos(\tfrac{2\pi}{3}), \quad s_1 = \sin(\tfrac{2\pi}{3}), \quad c_2 = \cos(\tfrac{4\pi}{3}), \quad s_2 = \sin(\tfrac{4\pi}{3}).$

The three elements

(16.2) $\quad g_1 = \begin{bmatrix} 1 & 0 \\ 0 & +1 \end{bmatrix}, \qquad g_2 = \begin{bmatrix} c_1 & -s_1 \\ +s_1 & c_1 \end{bmatrix}, \qquad g_3 = \begin{bmatrix} c_2 & -s_2 \\ +s_2 & c_2 \end{bmatrix}$

of $\mathcal{D}_3$ are given by (15.6) for $n = 3$. They represent rotations about the center of the triangle through angles of 0, $2\pi/3$, and $4\pi/3$ radians. The other three elements

(16.3) $\quad g_4 = \begin{bmatrix} 1 & 0 \\ 0 & -1 \end{bmatrix}, \qquad g_5 = \begin{bmatrix} c_1 & -s_1 \\ -s_1 & -c_1 \end{bmatrix}, \qquad g_6 = \begin{bmatrix} c_2 & -s_2 \\ -s_2 & -c_2 \end{bmatrix}$

of $\mathcal{D}_3$ are obtained from (15.7) for $n = 3$. Each of them represents a reflection in one of the three lines through the center and a vertex of the equilateral triangle.

For a function σ from $\mathcal{D}_3$ to $\mathbb{F}$ having $\sigma(g_k) = x_k$, for $1 \le k \le 6$, the primary group-pattern matrix for $\mathcal{D}_3$ is given by Proposition 15.2 on page 142 as

(16.4) $\qquad A_6 = \left[\begin{array}{ccc|ccc} x_1 & x_2 & x_3 & x_4 & x_5 & x_6 \\ x_3 & x_1 & x_2 & x_5 & x_6 & x_4 \\ x_2 & x_3 & x_1 & x_6 & x_4 & x_5 \\ \hline x_4 & x_5 & x_6 & x_1 & x_2 & x_3 \\ x_5 & x_6 & x_4 & x_3 & x_1 & x_2 \\ x_6 & x_4 & x_5 & x_2 & x_3 & x_1 \end{array} \right].$

As a check, the multiplication table for $\mathcal{D}_3$ and the list $\mathscr{L}_3 = (g_1, g_2, g_3, g_4, g_5, g_6)$ of its six elements can be verified to be the one in (11.14) on page 107. Its pattern is transferred to (16.4).

Since $n = 3$ is an odd integer, Proposition 15.12 on page 151 shows that the matrix representations of degree 1 for $\mathcal{D}_3$ are given by (15.36) and (15.37) as

(16.5) $\qquad\qquad F_1(g_k) = [1], \quad \text{for } k = 1, 2, 3, 4, 5, 6,$

along with

(16.6) $F_2(g_k) = [1]$,   for $k = 1, 2, 3$,   and $F_2(g_k) = [-1]$,   for $k = 4, 5, 6$,

A matrix representation F_3 of degree 2 for $\mathcal{D}_3$ is given by (16.2) and (16.3) as

(16.7) $F_3(g_k) = g_k$,   for $k = 1, 2, 3, 4, 5, 6$.

Since $\mathcal{D}_3$ is nonabelian and the degree of F_3 is 2, Theorem 10.11 on page 98 shows that the matrix representation F_3 is irreducible. With $n_1 = 1$, $n_2 = 1$, $n_3 = 2$, and $n_1^2 + n_2^2 + n_3^2 = 6$, we use Theorem 10.17 on page 99 to conclude that $\{F_1, F_2, F_3\}$ is a complete set of pairwise-inequivalent irreducible matrix representations for $\mathcal{D}_3$.

We use (11.18) on page 108 and (13.6) on page 124 to obtain

$$
(16.8) \qquad M = \begin{bmatrix}
1 & 1 & 1 & 0 & 0 & 1 \\
1 & 1 & c_1 & -s_1 & s_1 & c_1 \\
1 & 1 & c_2 & -s_2 & s_2 & c_2 \\
1 & -1 & 1 & 0 & 0 & -1 \\
1 & -1 & c_1 & -s_1 & -s_1 & -c_1 \\
1 & -1 & c_2 & -s_2 & -s_2 & -c_2
\end{bmatrix} .
$$

For the permutation π of $\{1, 2, 3, 4, 5, 6\}$ defined by $\pi(2) = 3$, $\pi(3) = 2$, and $\pi(k) = k$, for $k = 1, 4, 5, 6$, we have $g_k^{-1} = g_{\pi(k)}$, for $1 \le k \le 6$. Thus, we can replace g_s^{-1} with $g_{\pi(s)}$ in (13.7), for $1 \le s \le 6$, to obtain the inverse of M as

$$
(16.9) \qquad M^{-1} = L = \frac{1}{6}\begin{bmatrix}
1 & 1 & 1 & 1 & 1 & 1 \\
1 & 1 & 1 & -1 & -1 & -1 \\
2 & +2c_2 & +2c_1 & 2 & +2c_1 & +2c_2 \\
0 & +2s_2 & +2s_1 & 0 & -2s_1 & -2s_2 \\
0 & -2s_2 & -2s_1 & 0 & -2s_1 & -2s_2 \\
2 & +2c_2 & +2c_1 & -2 & -2c_1 & -2c_2
\end{bmatrix} .
$$

For the corresponding *Mathematica* notebook, see page 235. Theorem 13.3 yields

$$
(16.10) \qquad M^{-1}A_6 M = \left[\begin{array}{c|c|cc|cc}
d_1 & 0 & 0 & 0 & 0 & 0 \\ \hline
0 & d_2 & 0 & 0 & 0 & 0 \\ \hline
0 & 0 & d_3 & d_4 & 0 & 0 \\
0 & 0 & d_5 & d_6 & 0 & 0 \\ \hline
0 & 0 & 0 & 0 & d_3 & d_4 \\
0 & 0 & 0 & 0 & d_5 & d_6
\end{array}\right] ,
$$

where the components $d_1, d_2, \ldots, d_6$ of this Frobenius block-diagonalization for A_6 are given by (14.6), (16.4), (16.8), and (16.1) as

$$
(16.11) \qquad \begin{cases}
d_1 = x_1 + x_2 + x_3 + x_4 + x_5 + x_6, \\
d_2 = x_1 + x_2 + x_3 - x_4 - x_5 - x_6, \\
d_3 = x_1 + x_4 - \frac{1}{2}\left(x_2 + x_3 + x_5 + x_6\right), \\
d_4 = -\frac{\sqrt{3}}{2}\left(x_2 - x_3 + x_5 - x_6\right), \\
d_5 = \frac{\sqrt{3}}{2}\left(x_2 - x_3 - x_5 + x_6\right), \\
d_6 = x_1 - x_4 - \frac{1}{2}\left(x_2 + x_3 - x_5 - x_6.\right).
\end{cases}
$$

A direct evaluation of $M^{-1}A_6 M$ also yields (16.10) with (16.11). In fact, (16.10) with (16.11) is valid for any field $\mathfrak{F}$ in which $1/(6 \cdot 1)$ and $\sqrt{3}$ have meaning.

OBSERVATION 16.1. We note that, for A_6 in (16.4), the factorization

$$\det(A_6) = d_1 d_2 \big(d_3 d_6 - d_4 d_5\big)^2 = d_1 d_2 \begin{pmatrix} + x_1^2 + x_2^2 + x_3^2 - x_1 x_2 - x_1 x_3 - x_2 x_3 \\ - x_4^2 - x_5^2 - x_6^2 + x_4 x_5 + x_4 x_6 + x_5 x_6 \end{pmatrix}^2$$

based on (16.10)–(16.11) is consistent with that of (11.22) and (10.7).

16.2. Frobenius block-diagonalization of A_8 for $\mathcal{D}_4$

The eight elements of the dihedral group $\mathcal{D}_4$ of order 8 specify the symmetries for a square positioned as indicated in (15.4). The four elements

$$g_1 = \begin{bmatrix} 1 & 0 \\ 0 & 1 \end{bmatrix}, \quad g_2 = \begin{bmatrix} 0 & -1 \\ 1 & 0 \end{bmatrix}, \quad g_3 = \begin{bmatrix} -1 & 0 \\ 0 & -1 \end{bmatrix}, \quad g_4 = \begin{bmatrix} 0 & 1 \\ -1 & 0 \end{bmatrix}$$

of $\mathcal{D}_4$ are given by (15.6) for $n = 4$. They represent rotations about the center of the square through angles of 0, $\pi/2$, π, and $3\pi/2$ radians. The other four elements

$$g_5 = \begin{bmatrix} 1 & 0 \\ 0 & -1 \end{bmatrix}, \quad g_6 = \begin{bmatrix} 0 & -1 \\ -1 & 0 \end{bmatrix}, \quad g_7 = \begin{bmatrix} -1 & 0 \\ 0 & 1 \end{bmatrix}, \quad g_8 = \begin{bmatrix} 0 & 1 \\ 1 & 0 \end{bmatrix}$$

of $\mathcal{D}_4$ are obtained from (15.7) for $n = 4$. Each of them represents a reflection in one of the four lines through the center of the square and either a vertex or a midpoint of an edge. For a function σ from $\mathcal{D}_4$ to $\mathbb{F}$ having $\sigma(g_k) = x_k$, for $1 \leq k \leq 8$, Proposition 15.2 on page 142 yields

$$(16.12) \qquad A_8 = \left[\begin{array}{cccc|cccc} x_1 & x_2 & x_3 & x_4 & x_5 & x_6 & x_7 & x_8 \\ x_4 & x_1 & x_2 & x_3 & x_6 & x_7 & x_8 & x_5 \\ x_3 & x_4 & x_1 & x_2 & x_7 & x_8 & x_5 & x_6 \\ x_2 & x_3 & x_4 & x_1 & x_8 & x_5 & x_6 & x_7 \\ \hline x_5 & x_6 & x_7 & x_8 & x_1 & x_2 & x_3 & x_4 \\ x_6 & x_7 & x_8 & x_5 & x_4 & x_1 & x_2 & x_3 \\ x_7 & x_8 & x_5 & x_6 & x_3 & x_4 & x_1 & x_2 \\ x_8 & x_5 & x_6 & x_7 & x_2 & x_3 & x_4 & x_1 \end{array} \right]$$

as the corresponding primary group-pattern matrix for $\mathcal{D}_4$. As a check, we obtain the multiplication table

$$(16.13)$$

$\cdot$	g_1	g_2	g_3	g_4	g_5	g_6	g_7	g_8
g_1	g_1	g_2	g_3	g_4	g_5	g_6	g_7	g_8
g_4	g_4	g_1	g_2	g_3	g_6	g_7	g_8	g_5
g_3	g_3	g_4	g_1	g_2	g_7	g_8	g_5	g_6
g_2	g_2	g_3	g_4	g_1	g_8	g_5	g_6	g_7
g_5	g_5	g_6	g_7	g_8	g_1	g_2	g_3	g_4
g_6	g_6	g_7	g_8	g_5	g_4	g_1	g_2	g_3
g_7	g_7	g_8	g_5	g_6	g_3	g_4	g_1	g_2
g_8	g_8	g_5	g_6	g_7	g_2	g_3	g_4	g_1

as the one specified by the list $\mathscr{L} = \big(g_1, g_2, g_3, g_4, g_5, g_6, g_7, g_8\big)$.

Since $n = 4$ is an even integer, Proposition 15.12 on page 151 shows that $\mathcal{D}_4$ has precisely four matrix representations of degree 1 and they are given by

$$(16.14) \quad \begin{cases} F_1(g_k) = [1], \text{ for } 1 \le k \le 8, \\ F_2(g_k) = [1], \text{ for } 1 \le k \le 4, \quad \text{ and } F_2(g_k) = [-1], \text{ for } 5 \le k \le 8, \\ F_3(g_k) = [1], \text{ for } k = 1, 3, 5, 7, \text{ and } F_3(g_k) = [-1], \text{ for } k = 2, 4, 6, 8, \\ F_4(g_k) = [1], \text{ for } k = 1, 3, 6, 8, \text{ and } F_3(g_k) = [-1], \text{ for } k = 2, 4, 5, 7. \end{cases}$$

The function F_5 from $\mathcal{D}_4$ to $\mathcal{D}_4$ defined by

$$(16.15) \qquad\qquad F_5(g_k) = g_k, \quad \text{ for } 1 \le k \le 8,$$

is a matrix representation for $\mathcal{D}_4$ of degree 2. Moreover, since $\mathcal{D}_4$ is nonabelian, Theorem 10.11 on page 98 establishes that F_5 is irreducible. Since we have

$$n_1 = 1, \ n_2 = 1, \ n_3 = 1, \ n_4 = 1, \ n_5 = 2, \text{ and } n_1^2 + n_2^2 + n_3^2 + n_4^2 + n_5^2 = 8,$$

Theorem 10.17 on page 99 shows that $\{F_1, F_2, F_3, F_4, F_5\}$ is a complete set of pairwise-inequivalent irreducible matrix representations for $\mathcal{D}_4$. In terms of the abbreviation $\kappa_s = (\lambda(s), \mu(s), \nu(s))$, for $1 \le s \le 8$, Section 11.2 yields

$$(16.16) \quad \begin{cases} \kappa_1 = (1, 1, 1), \quad \kappa_2 = (2, 1, 1), \quad \kappa_3 = (3, 1, 1), \quad \kappa_4 = (4, 1, 1), \\ \kappa_5 = (5, 1, 1), \quad \kappa_6 = (5, 1, 2), \quad \kappa_7 = (5, 2, 1), \quad \kappa_8 = (5, 2, 2), \end{cases}$$

We use (13.6) on page 124 with (16.14)–(16.16) to obtain

$$(16.17) \qquad M = \begin{bmatrix} 1 & 1 & 1 & 1 & 1 & 0 & 0 & 1 \\ 1 & 1 & -1 & -1 & 0 & -1 & 1 & 0 \\ 1 & 1 & 1 & 1 & -1 & 0 & 0 & -1 \\ 1 & 1 & -1 & -1 & 0 & 1 & -1 & 0 \\ 1 & -1 & 1 & -1 & 1 & 0 & 0 & -1 \\ 1 & -1 & -1 & 1 & 0 & -1 & -1 & 0 \\ 1 & -1 & 1 & -1 & -1 & 0 & 0 & 1 \\ 1 & -1 & -1 & 1 & 0 & 1 & 1 & 0 \end{bmatrix}.$$

In view of (16.13), we see that the permutation π of $\{1, 2, \ldots, 8\}$ having $\pi(2) = 4$, $\pi(4) = 2$, and $\pi(k) = k$, for $k = 1, 3, 5, 6, 7, 8$, yields $g_k^{-1} = g_{\pi(k)}$, for $1 \le k \le 8$. We employ this with (16.16) and (13.7) on page 124 to obtain

$$(16.18) \qquad L = \frac{1}{8} \begin{bmatrix} 1 & 1 & 1 & 1 & 1 & 1 & 1 & 1 \\ 1 & 1 & 1 & 1 & -1 & -1 & -1 & -1 \\ 1 & -1 & 1 & -1 & 1 & -1 & 1 & -1 \\ 1 & -1 & 1 & -1 & -1 & 1 & -1 & 1 \\ 2 & 0 & -2 & 0 & 2 & 0 & -2 & 0 \\ 0 & -2 & 0 & 2 & 0 & -2 & 0 & 2 \\ 0 & 2 & 0 & -2 & 0 & -2 & 0 & 2 \\ 2 & 0 & -2 & 0 & -2 & 0 & 2 & 0 \end{bmatrix}.$$

Here, for the Frobenius block-diagonal matrix $M^{-1}A_8M = LA_8M$, we have

$$(16.19) \qquad M^{-1}A_8M = \left[\begin{array}{c|c|c|c|c|c|c|c} d_1 & 0 & 0 & 0 & 0 & 0 & 0 & 0 \\ \hline 0 & d_2 & 0 & 0 & 0 & 0 & 0 & 0 \\ \hline 0 & 0 & d_3 & 0 & 0 & 0 & 0 & 0 \\ \hline 0 & 0 & 0 & d_4 & 0 & 0 & 0 & 0 \\ \hline 0 & 0 & 0 & 0 & d_5 & d_6 & 0 & 0 \\ 0 & 0 & 0 & 0 & d_7 & d_8 & 0 & 0 \\ \hline 0 & 0 & 0 & 0 & 0 & 0 & d_5 & d_6 \\ 0 & 0 & 0 & 0 & 0 & 0 & d_7 & d_8 \end{array}\right],$$

where we illustrate (13.9) of Theorem 13.3 by using it with (16.14)–(16.15) to obtain

$$[d_1] = \sum_{k=1}^{6} x_k F_1(g_k) = [x_1 + x_2 + x_3 + x_4 + x_5 + x_6 + x_7 + x_8],$$

$$[d_2] = \sum_{k=1}^{6} x_k F_2(g_k) = [x_1 + x_2 + x_3 + x_4 - x_5 - x_6 - x_7 - x_8],$$

$$[d_3] = \sum_{k=1}^{6} x_k F_3(g_k) = [x_1 - x_2 + x_3 - x_4 + x_5 - x_6 + x_7 - x_8],$$

$$[d_4] = \sum_{k=1}^{6} x_k F_4(g_k) = [x_1 - x_2 + x_3 - x_4 - x_5 + x_6 - x_7 + x_8],$$

$$\left[\begin{array}{cc} d_5 & d_6 \\ d_7 & d_8 \end{array}\right] = \sum_{k=1}^{6} x_k F_5(g_k) = \left[\begin{array}{cc} x_1 - x_3 + x_5 - x_7 & -x_2 + x_4 - x_6 + x_8 \\ x_2 - x_4 - x_6 + x_8 & +x_1 - x_3 - x_5 + x_7 \end{array}\right].$$

However, it is simpler to compute $d_1, d_2, \ldots, d_8$ directly via Theorem 14.1 as

$$[d_1, d_2, d_3, d_4, d_5, d_6, d_7, d_8] = [x_1, x_2, x_3, x_4, x_5, x_6, x_7, x_8]M.$$

One can view (16.19) as an identity that is valid when the components of A in (16.12), M in (16.17), and L in (16.18) are elements of a field $\mathfrak{F}$ in which $1/(8 \cdot 1)$ is meaningful.

16.3. Frobenius block-diagonalization of A_{10} for $\mathcal{D}_5$

The ten elements of the dihedral group $\mathcal{D}_5$ of order 10 specify the symmetries for a regular pentagon positioned as indicated in (15.5). As a convenience, we set

$$(16.20) \qquad c_k = \cos(k\tfrac{2\pi}{5}) \quad \text{and} \quad s_k = \sin(k\tfrac{2\pi}{5}), \quad \text{for } k = 1, 2, 3, 4.$$

For $n = 5$, we use (15.6) and (15.7) to obtain $g_1 = \begin{bmatrix} 1 & 0 \\ 0 & 1 \end{bmatrix}$, $g_6 = \begin{bmatrix} 1 & 0 \\ 0 & -1 \end{bmatrix}$,

$$g_2 = \begin{bmatrix} c_1 & -s_1 \\ s_1 & c_1 \end{bmatrix}, \quad g_3 = \begin{bmatrix} c_2 & -s_2 \\ s_2 & c_2 \end{bmatrix}, \quad g_4 = \begin{bmatrix} c_3 & -s_3 \\ s_3 & c_3 \end{bmatrix}, \quad g_5 = \begin{bmatrix} c_4 & -s_4 \\ s_4 & c_4 \end{bmatrix},$$

$$g_7 = \begin{bmatrix} c_1 & -s_1 \\ -s_1 & -c_1 \end{bmatrix}, \quad g_8 = \begin{bmatrix} c_2 & -s_2 \\ -s_2 & -c_2 \end{bmatrix}, \quad g_9 = \begin{bmatrix} c_3 & -s_3 \\ -s_3 & -c_3 \end{bmatrix}, \quad g_{10} = \begin{bmatrix} c_4 & -s_4 \\ -s_4 & -c_4 \end{bmatrix}.$$

The elements g_1, g_2, g_3, g_4, g_5 represent rotations about the center of the pentagon through angles of $0, 2\pi/5, 4\pi/5, 6\pi/5, 8\pi/5$ radians. Each of $g_6, g_7, g_8, g_9, g_{10}$ represents a reflection in a line through the center of the pentagon and one of its five vertices.

For the function σ from $\mathcal{D}_5$ to $\mathbb{F}$ having $\sigma(g_k) = x_k$, when $k = 1, 2, \ldots, 10$, we see that the corresponding primary group-pattern matrix for $\mathcal{D}_5$ is specified by Proposition 15.2 on page 142 as

$$(16.21) \qquad A_{10} = \left[\begin{array}{ccccc|ccccc} x_1 & x_2 & x_3 & x_4 & x_5 & x_6 & x_7 & x_8 & x_9 & x_{10} \\ x_5 & x_1 & x_2 & x_3 & x_4 & x_7 & x_8 & x_9 & x_{10} & x_6 \\ x_4 & x_5 & x_1 & x_2 & x_3 & x_8 & x_9 & x_{10} & x_6 & x_7 \\ x_3 & x_4 & x_5 & x_1 & x_2 & x_9 & x_{10} & x_6 & x_7 & x_8 \\ x_2 & x_3 & x_4 & x_5 & x_1 & x_{10} & x_6 & x_7 & x_8 & x_9 \\ \hline x_6 & x_7 & x_8 & x_9 & x_{10} & x_1 & x_2 & x_3 & x_4 & x_5 \\ x_7 & x_8 & x_9 & x_{10} & x_6 & x_5 & x_1 & x_2 & x_3 & x_4 \\ x_8 & x_9 & x_{10} & x_6 & x_7 & x_4 & x_5 & x_1 & x_2 & x_3 \\ x_9 & x_{10} & x_6 & x_7 & x_8 & x_3 & x_4 & x_5 & x_1 & x_2 \\ x_{10} & x_6 & x_7 & x_8 & x_9 & x_2 & x_3 & x_4 & x_5 & x_1 \end{array} \right].$$

To provide a check, we note that the multiplication table

$$(16.22)$$

$\cdot$	g_1	g_2	g_3	g_4	g_5	g_6	g_7	g_8	g_9	g_{10}
g_1	g_1	g_2	g_3	g_4	g_5	g_6	g_7	g_8	g_9	g_{10}
g_5	g_5	g_1	g_2	g_3	g_4	g_7	g_8	g_9	g_{10}	g_6
g_4	g_4	g_5	g_1	g_2	g_3	g_8	g_9	g_{10}	g_6	g_7
g_3	g_3	g_4	g_5	g_1	g_2	g_9	g_{10}	g_6	g_7	g_8
g_2	g_2	g_3	g_4	g_5	g_1	g_{10}	g_6	g_7	g_8	g_9
g_6	g_6	g_7	g_8	g_9	g_{10}	g_1	g_2	g_3	g_4	g_5
g_7	g_7	g_8	g_9	g_{10}	g_6	g_5	g_1	g_2	g_3	g_4
g_8	g_8	g_9	g_{10}	g_6	g_7	g_4	g_5	g_1	g_2	g_3
g_9	g_9	g_{10}	g_6	g_7	g_8	g_3	g_4	g_5	g_1	g_2
g_{10}	g_{10}	g_6	g_7	g_8	g_9	g_2	g_3	g_4	g_5	g_1

is the one for $\mathcal{D}_5$ provided by the list

$$(16.23) \qquad \begin{aligned} \mathscr{L}_5 &= \big(g_1, g_2, g_3, g_4, g_5, g_6, g_7, g_8, g_9, g_{10} \big) \\ &= \big(e, \alpha, \alpha^2, \alpha^3, \alpha^4, \beta, \beta\alpha, \beta\alpha^2, \beta\alpha^3, \beta\alpha^4 \big). \end{aligned}$$

Since $n = 5$ is an odd integer, Proposition 15.12 on page 151 shows that $\mathcal{D}_5$ has precisely two matrix representations of degree 1 and they are given by

$$(16.24) \qquad \begin{cases} F_1(g_k) = [1], \text{ for } 1 \le k \le 10, \\ F_2(g_k) = [1], \text{ for } 1 \le k \le 5, \text{ and } F_2(g_k) = [-1], \text{ for } 6 \le k \le 10. \end{cases}$$

The function F_3 from $\mathcal{D}_5$ to $\mathcal{D}_5$ defined by

$$(16.25) \qquad F_3(g_k) = g_k, \quad \text{for } 1 \le k \le 10,$$

is a matrix representation for $\mathcal{D}_5$ of degree 2. Since $\mathcal{D}_5$ is nonabelian, Theorem 10.11 on page 98 establishes that F_3 is irreducible.

For another representation of degree 2, we note that an automorphism ϕ of $\mathcal{D}_5$ is defined by setting $n = 5$, $b = 2$, and $c = 1$ in (15.33) on page 150 to obtain:

$$(16.26) \quad \begin{cases} \phi(g_1) = g_1, & \phi(g_2) = g_3, & \phi(g_3) = g_5, & \phi(g_4) = g_2, & \phi(g_5) = g_4, \\ \phi(g_6) = g_6, & \phi(g_7) = g_8, & \phi(g_8) = g_{10}, & \phi(g_9) = g_7, & \phi(g_{10}) = g_9. \end{cases}$$

Thus, a matrix representation F_4 of degree 2 for $\mathcal{D}_5$ is given by $F_4(g_k) = F_3\big(\phi(g_k)\big)$, for $1 \le k \le 10$. Therefore, in terms of the permutation ψ of $\{1, 2, \ldots, 10\}$ having $\psi(1) = 1$, $\psi(2) = 3$, $\psi(3) = 5$, $\psi(4) = 2$, $\psi(5) = 4$, $\psi(6) = 6$, $\psi(7) = 8$, $\psi(8) = 10$, $\psi(9) = 7$, and $\psi(10) = 9$, we note that

$$(16.27) \qquad F_4(g_k) = g_{\psi(k)}, \quad \text{for } 1 \le k \le 10.$$

Section 16.4 shows that F_4 is not equivalent to F_3. Therefore, with $n_1 = 1$, $n_2 = 1$, $n_3 = 2$, $n_4 = 2$, and $n_1^2 + n_2^2 + n_3^2 + n_4^2 = 10$, we use Theorem 10.17 on page 99 to see that $\{F_1, F_2, F_3, F_4\}$ is a complete set of pairwise-inequivalent irreducible matrix representations for $\mathcal{D}_5$. For our context here, Section 11.2 shows the formula

$$\kappa_s = \big(\lambda(s), \mu(s), \nu(s)\big), \quad \text{for } 1 \le s \le 10,$$

yields

$$\begin{cases} \kappa_1 = (1, 1, 1), & \kappa_2 = (2, 1, 1), & \kappa_3 = (3, 1, 1), & \kappa_4 = (3, 1, 2), & \kappa_5 = (3, 2, 1), \\ \kappa_6 = (3, 2, 2), & \kappa_7 = (4, 1, 1), & \kappa_8 = (4, 1, 2), & \kappa_9 = (4, 2, 1), & \kappa_{10} = (4, 2, 2). \end{cases}$$

We employ this with (13.6) on page 124, the definitions of $g_1, g_2, \ldots, g_{10}$, and the formulas (16.24)–(16.27) for F_1, F_2, F_3, F_4 to obtain

$$(16.28) \quad M = \begin{bmatrix} 1 & 1 & 1 & 0 & 0 & 1 & 1 & 0 & 0 & 1 \\ 1 & 1 & c_1 & -s_1 & s_1 & c_1 & c_2 & -s_2 & s_2 & c_2 \\ 1 & 1 & c_2 & -s_2 & s_2 & c_2 & c_4 & -s_4 & s_4 & c_4 \\ 1 & 1 & c_3 & -s_3 & s_3 & c_3 & c_1 & -s_1 & s_1 & c_1 \\ 1 & 1 & c_4 & -s_4 & s_4 & c_4 & c_3 & -s_3 & s_3 & c_3 \\ 1 & -1 & 1 & 0 & 0 & -1 & 1 & 0 & 0 & -1 \\ 1 & -1 & c_1 & -s_1 & -s_1 & -c_1 & c_2 & -s_2 & -s_2 & -c_2 \\ 1 & -1 & c_2 & -s_2 & -s_2 & -c_2 & c_4 & -s_4 & -s_4 & -c_4 \\ 1 & -1 & c_3 & -s_3 & -s_3 & -c_3 & c_1 & -s_1 & -s_1 & -c_1 \\ 1 & -1 & c_4 & -s_4 & -s_4 & -c_4 & c_3 & -s_3 & -s_3 & -c_3 \end{bmatrix}.$$

The multiplication table (16.22) shows that $g_k^{-1} = g_{\pi(k)}$, for $1 \le k \le 10$, when π is the permutation of $\{1, 2, \ldots, 10\}$ such that $\pi(2) = 5$, $\pi(3) = 4$, $\pi(4) = 3$, $\pi(5) = 2$, and $\pi(k) = k$, for $k = 1, 6, 7, 8, 9, 10$. We use this with (13.7) and the formulas for F_1, F_2, F_3, F_4 to find that the inverse L of M is given by

$$L = \frac{1}{10} \begin{bmatrix} 1 & 1 & 1 & 1 & 1 & 1 & 1 & 1 & 1 & 1 \\ 1 & 1 & 1 & 1 & 1 & -1 & -1 & -1 & -1 & -1 \\ 2 & 2c_4 & 2c_3 & 2c_2 & 2c_1 & 2 & 2c_1 & 2c_2 & 2c_3 & 2c_4 \\ 0 & 2s_4 & 2s_3 & 2s_2 & 2s_1 & 0 & -2s_1 & -2s_2 & -2s_3 & -2s_4 \\ 0 & -2s_4 & -2s_3 & -2s_2 & -2s_1 & 0 & -2s_1 & -2s_2 & -2s_3 & -2s_4 \\ 2 & 2c_4 & 2c_3 & 2c_2 & 2c_1 & -2 & -2c_1 & -2c_2 & -2c_3 & -2c_4 \\ 2 & 2c_3 & 2c_1 & 2c_4 & 2c_2 & 2 & 2c_2 & 2c_4 & 2c_1 & 2c_3 \\ 0 & 2s_3 & 2s_1 & 2s_4 & 2s_2 & 0 & -2s_2 & -2s_4 & -2s_1 & -2s_3 \\ 0 & -2s_3 & -2s_1 & -2s_4 & -2s_2 & 0 & -2s_2 & -2s_4 & -2s_1 & -2s_3 \\ 2 & 2c_3 & 2c_1 & 2c_4 & 2c_2 & -2 & -2c_2 & -2c_4 & -2c_1 & -2c_3 \end{bmatrix}.$$

At this point, we use (16.20) to introduce

$$(16.29) \quad \begin{cases} c_1 = \dfrac{\sqrt{5}-1}{4}, \quad c_2 = \dfrac{-\sqrt{5}-1}{4}, \quad c_3 = \dfrac{-\sqrt{5}-1}{4}, \quad c_4 = \dfrac{\sqrt{5}-1}{4}, \\[2mm] s_1 = \sqrt{\dfrac{5+\sqrt{5}}{8}}, \quad s_2 = \sqrt{\dfrac{5-\sqrt{5}}{8}}, \quad s_3 = -\sqrt{\dfrac{5-\sqrt{5}}{8}}, \quad s_4 = -\sqrt{\dfrac{5+\sqrt{5}}{8}}. \end{cases}$$

Then, for A_{10} in (16.21), we employ machine computations to obtain

$$(16.30) \quad M^{-1}A_{10}M = \left[\begin{array}{c|c|cc|cc|cc|cc} d_1 & 0 & 0 & 0 & 0 & 0 & 0 & 0 & 0 & 0 \\ \hline 0 & d_2 & 0 & 0 & 0 & 0 & 0 & 0 & 0 & 0 \\ \hline 0 & 0 & d_3 & d_4 & 0 & 0 & 0 & 0 & 0 & 0 \\ 0 & 0 & d_5 & d_6 & 0 & 0 & 0 & 0 & 0 & 0 \\ \hline 0 & 0 & 0 & 0 & d_3 & d_4 & 0 & 0 & 0 & 0 \\ 0 & 0 & 0 & 0 & d_5 & d_6 & 0 & 0 & 0 & 0 \\ \hline 0 & 0 & 0 & 0 & 0 & 0 & d_7 & d_8 & 0 & 0 \\ 0 & 0 & 0 & 0 & 0 & 0 & d_9 & d_{10} & 0 & 0 \\ \hline 0 & 0 & 0 & 0 & 0 & 0 & 0 & 0 & d_7 & d_8 \\ 0 & 0 & 0 & 0 & 0 & 0 & 0 & 0 & d_9 & d_{10} \end{array} \right],$$

where the expressions for $d_1, d_2, \ldots, d_{10}$ involve the right-hand members of (16.29). With $c_3 = c_2$, $c_4 = c_1$, $s_3 = -s_2$, and $s_4 = -s_1$, we use (16.28) and (14.6) of Theorem 14.1 on page 136 to verify that

$$\begin{aligned} d_1 &= x_1 + x_2 + x_3 + x_4 + x_5 + x_6 + x_7 + x_8 + x_9 + x_{10}, \\ d_2 &= x_1 + x_2 + x_3 + x_4 + x_5 - x_6 - x_7 - x_8 - x_9 - x_{10}, \\ d_3 &= x_1 + x_6 + c_1(x_2 + x_5 + x_7 + x_{10}) + c_2(x_3 + x_4 + x_8 + x_9), \\ d_4 &= s_1(-x_2 + x_5 - x_7 + x_{10}) + s_2(-x_3 + x_4 - x_8 + x_9), \\ d_5 &= s_1(x_2 - x_5 - x_7 + x_{10}) + s_2(x_3 - x_4 - x_8 + x_9), \\ d_6 &= x_1 - x_6 + c_1(x_2 + x_5 - x_7 - x_{10}) + c_2(x_3 + x_4 - x_8 - x_9), \\ d_7 &= x_1 + x_6 + c_1(x_3 + x_4 + x_8 + x_9) + c_2(x_2 + x_5 + x_7 + x_{10}), \\ d_8 &= s_1(x_3 - x_4 + x_8 - x_9) + s_2(-x_2 + x_5 - x_7 + x_{10}), \\ d_9 &= s_1(-x_3 + x_4 + x_8 - x_9) + s_2(x_2 - x_5 - x_7 + x_{10}), \\ d_{10} &= x_1 - x_6 + c_1(x_3 + x_4 - x_8 - x_9) + c_2(x_2 + x_5 - x_7 - x_{10}). \end{aligned}$$

For the corresponding *Mathematica* notebook, see page 235.

16.4. Details about the matrix representations used for $\mathcal{D}_5$

PROPOSITION 16.2. *The matrix representations F_3 and F_4 for $\mathcal{D}_5$ in (16.25) and (16.27) are not equivalent.*

PROOF. For an indirect argument, suppose that F_3 and F_4 are equivalent. Then, there is a nonsingular matrix $S = \begin{bmatrix} a & b \\ c & d \end{bmatrix}$, with a, b, c, d in $\mathbb{C}$, such that

$$(16.31) \qquad S^{-1}F_3(g_k)S = F_4(g_k), \quad \text{for } 1 \le k \le 10.$$

For $k = 6$, this requires

$$\begin{bmatrix} a & b \\ -c & -d \end{bmatrix} = \begin{bmatrix} 1 & 0 \\ 0 & -1 \end{bmatrix}\begin{bmatrix} a & b \\ c & d \end{bmatrix} = g_6\,S = F_3(g_6)\,S = S\,F_4(g_6) = S\,g_6$$

$$= \begin{bmatrix} a & b \\ c & d \end{bmatrix}\begin{bmatrix} 1 & 0 \\ 0 & -1 \end{bmatrix} = \begin{bmatrix} a & -b \\ c & -d \end{bmatrix}$$

and therefore yields $b = c = 0$. We use this with (16.31), for $k = 2$, to obtain

$$\begin{bmatrix} ac_1 & -ds_1 \\ as_1 & dc_1 \end{bmatrix} = \begin{bmatrix} c_1 & -s_1 \\ s_1 & c_1 \end{bmatrix}\begin{bmatrix} a & 0 \\ 0 & d \end{bmatrix} = g_2\,S = F_3(g_2)\,S = S\,F_4(g_2) = S\,g_3$$

$$= \begin{bmatrix} a & 0 \\ 0 & d \end{bmatrix}\begin{bmatrix} c_2 & -s_2 \\ s_2 & c_2 \end{bmatrix} = \begin{bmatrix} ac_2 & -as_2 \\ ds_2 & dc_2 \end{bmatrix}.$$

With $c_1 \neq c_2$, this requires $a = d = 0$ and yields the contradiction that S is the 2×2 zero matrix. Thus, F_3 and F_4 are not equivalent. This completes the proof. $\square$

OBSERVATION 16.3. In terms of the automorphism ϕ for $\mathcal{D}_5$ defined by (16.26), we recall that F_4 is given by $F_4(x) = F_3\big(\phi(x)\big) = \phi(x)$, for each x in $\mathcal{D}_5$. Since $\phi\circ\phi$ and $\phi\circ\phi\circ\phi$ are also automorphisms of $\mathcal{D}_5$, we note that the formulas

$$F_5(x) = F_3\big((\phi\circ\phi)(x)\big) = (\phi\circ\phi)(x), \qquad \text{for each } x \text{ in } \mathcal{D}_5,$$

$$F_6(x) = F_3\big((\phi\circ\phi\circ\phi)(x)\big) = (\phi\circ\phi\circ\phi)(x), \quad \text{for each } x \text{ in } \mathcal{D}_5,$$

define group representations F_5 and F_6 for $\mathcal{D}_5$. Because $\{F_1, F_2, F_3, F_4\}$ is a complete set of pairwise-inequivalent irreducible matrix representations for $\mathcal{D}_5$ and the representations F_5 and F_6 are clearly irreducible, Definition 10.13 on page 99 shows that: F_5 is equivalent to F_3 or F_4; and, F_6 is equivalent to F_3 or F_4.

We use (16.26) and (16.22) to construct the table

x	g_1	g_2	g_3	g_4	g_5	g_6	g_7	g_8	g_9	g_{10}
$F_3(x) = x$	g_1	g_2	g_3	g_4	g_5	g_6	g_7	g_8	g_9	g_{10}
$F_5(x) = (\phi\circ\phi)(x)$	g_1	g_5	g_4	g_3	g_2	g_6	g_{10}	g_9	g_8	g_7
$S^{-1}x\,S = g_6\,x\,g_6$	g_1	g_5	g_4	g_3	g_2	g_6	g_{10}	g_9	g_8	g_7

and therefore conclude that

$$F_5(x) = S^{-1}F_3(x)S, \quad \text{for each } x \text{ in } \mathcal{D}_5, \text{ where} \quad S = g_6 = \begin{bmatrix} 1 & 0 \\ 0 & -1 \end{bmatrix}.$$

Thus, F_5 and F_3 are equivalent. In this formula, we replace x with $\phi(x)$ to obtain

$$F_6(x) = S^{-1}F_4(x)S, \quad \text{for each } x \text{ in } \mathcal{D}_5, \text{ where} \quad S = g_6 = \begin{bmatrix} 1 & 0 \\ 0 & -1 \end{bmatrix}.$$

Consequently, F_6 and F_4 are equivalent. Next, we have $\phi\circ\phi\circ\phi\circ\phi = id$.

CHAPTER 17

Auxiliary-Pattern Matrices

17.1. Introduction

Let G be a group of order n, let $\mathscr{L} = \left(g_1, g_2, \ldots, g_n\right)$ be a list of the elements in G, and let $\mathcal{S}$ be a set. Just as G and $\mathscr{L}$ specify the multiplication table

(17.1)

$\cdot$	g_1	g_2	g_3	$\cdots$	g_s	$\cdots$	g_{n-1}	g_n
g_1^{-1}	e	$g_1^{-1}g_2$	$g_1^{-1}g_3$	$\cdots$	$g_1^{-1}g_s$	$\cdots$	$g_1^{-1}g_{n-1}$	$g_1^{-1}g_n$
g_2^{-1}	$g_2^{-1}g_1$	e	$g_2^{-1}g_3$	$\cdots$	$g_2^{-1}g_s$	$\cdots$	$g_2^{-1}g_{n-1}$	$g_2^{-1}g_n$
g_3^{-1}	$g_3^{-1}g_1$	$g_3^{-1}g_2$	e	$\cdots$	$g_3^{-1}g_s$	$\cdots$	$g_3^{-1}g_{n-1}$	$g_3^{-1}g_n$
$\vdots$	$\vdots$	$\vdots$	$\vdots$	$\vdots$	$\vdots$	$\vdots$	$\vdots$	$\vdots$
g_r^{-1}	$g_r^{-1}g_1$	$g_r^{-1}g_2$	$g_r^{-1}g_3$	$\cdots$	$g_r^{-1}g_s$	$\cdots$	$g_r^{-1}g_{n-1}$	$g_r^{-1}g_n$
$\vdots$	$\vdots$	$\vdots$	$\vdots$	$\vdots$	$\vdots$	$\vdots$	$\vdots$	$\vdots$
g_{n-1}^{-1}	$g_{n-1}^{-1}g_1$	$g_{n-1}^{-1}g_2$	$g_{n-1}^{-1}g_3$	$\cdots$	$g_{n-1}^{-1}g_s$	$\cdots$	e	$g_{n-1}^{-1}g_n$
g_n^{-1}	$g_n^{-1}g_1$	$g_n^{-1}g_2$	$g_n^{-1}g_3$	$\cdots$	$g_n^{-1}g_s$	$\cdots$	$g_n^{-1}g_{n-1}$	e

from (2.2) on page 11 that provides the group-pattern for Definition 2.2 of a group-pattern matrix, they also specify the multiplication table

(17.2)

$\cdot$	g_1	g_2	g_3	$\cdots$	g_s	$\cdots$	g_{n-1}	g_n
g_1	$g_1\,g_1$	$g_1\,g_2$	$g_1\,g_3$	$\cdots$	$g_1\,g_s$	$\cdots$	$g_1\,g_{n-1}$	$g_1\,g_n$
g_2	$g_2\,g_1$	$g_2\,g_2$	$g_2\,g_3$	$\cdots$	$g_2\,g_s$	$\cdots$	$g_2\,g_{n-1}$	$g_2\,g_n$
g_3	$g_3\,g_1$	$g_3\,g_2$	$g_3\,g_3$	$\cdots$	$g_3\,g_s$	$\cdots$	$g_3\,g_{n-1}$	$g_3\,g_n$
$\vdots$	$\vdots$	$\vdots$	$\vdots$	$\vdots$	$\vdots$	$\vdots$	$\vdots$	$\vdots$
g_r	$g_r\,g_1$	$g_r\,g_2$	$g_r\,g_3$	$\cdots$	$g_r\,g_s$	$\cdots$	$g_r\,g_{n-1}$	$g_r\,g_n$
$\vdots$	$\vdots$	$\vdots$	$\vdots$	$\vdots$	$\vdots$	$\vdots$	$\vdots$	$\vdots$
g_{n-1}	$g_{n-1}\,g_1$	$g_{n-1}\,g_2$	$g_{n-1}\,g_3$	$\cdots$	$g_{n-1}\,g_s$	$\cdots$	$g_{n-1}\,g_{n-1}$	$g_{n-1}\,g_n$
g_n	$g_n\,g_1$	$g_n\,g_2$	$g_n\,g_3$	$\cdots$	$g_n\,g_s$	$\cdots$	$g_n\,g_{n-1}$	$g_n\,g_n$

to be considered here. We note that the rows of (17.2) are a permutation of the rows for (17.1). In this regard, see the permutation matrix N in (17.5) for Proposition 17.4 on the next page. Also, (17.2) is symmetric with respect to its principal diagonal if and only is G is abelian.

EXAMPLE 17.1. When G is a cyclic group of order n with h as a generator such that $g_k = h^{k-1}$, when $1 \leq k \leq n$, and $g_1 = h^0 = h^n = e$, the tables (17.1) and (17.2) are the tables (6.16) and (6.17) of page 67 that serve as patterns to introduce $n \times n$ circulant matrices and $n \times n$ leftward-circulant matrices.

17.2. Auxiliary-pattern matrices and immediate deductions

DEFINITION 17.2. An $n \times n$ matrix B having components in a set S is an
auxiliary-pattern matrix for a group G of order n and a list $\mathscr{L} = (g_1, g_2, \ldots, g_n)$
of the elements in G when there is a function τ from G to S such that

$$(17.3) \qquad [B]_{r,s} = \tau(g_r g_s), \quad \text{for } r, s = 1, 2, \ldots, n.$$

The auxiliary-pattern matrix B of (17.3) is **injective** if and only if τ is one-to-one.

EXAMPLE 17.3. Let ϕ be the function from G to a field $\mathfrak{F}$ that is defined by

$$(17.4) \qquad \phi(x) = \begin{cases} 1, & \text{if } x = e, \\ 0, & \text{if } x \neq e, \end{cases}, \quad \text{for each } x \text{ in } G,$$

and let N be the $n \times n$ auxiliary-pattern matrix for G and $\mathscr{L}$ having

$$(17.5) \qquad [N]_{r,s} = \phi(g_r g_s), \quad \text{for } r, s = 1, 2, \ldots, n.$$

We use (17.5) and (17.4) to see that N is a symmetric permutation matrix for which
$N^{-1} = N^T = N$. It is a generalization of N in (6.19) and (8.18).

To formulate Proposition 17.4, let σ and τ be functions from G to a field $\mathfrak{F}$.
Then, a group-pattern matrix A and an auxiliary-pattern matrix B are defined by

$$(17.6) \quad [A]_{r,s} = \sigma(g_r^{-1} g_s), \quad \text{and} \quad [B]_{r,s} = \tau(g_r g_s), \quad \text{for } r, s = 1, 2, \ldots, n.$$

PROPOSITION 17.4. *The products BA and NA are auxiliary-pattern matrices
for G and $\mathscr{L}$ while the product NB is a group-pattern matrix for G and $\mathscr{L}$.
Moreover, when $\sigma = \tau$, the relations $NA = B$ and $NB = A$ are valid.*

PROOF. We use (17.6) to obtain

$$[BA]_{r,s} = \sum_{k=1}^{n} [B]_{r,k} [A]_{k,s} = \sum_{k=1}^{n} \tau(g_r g_k)\, \sigma(g_k^{-1} g_s)$$

$$= \sum_{k=1}^{n} \tau(g_r g_k)\, \sigma((g_r g_k)^{-1} g_r g_s) = \sum_{k=1}^{n} \tau(g_k)\, \sigma(g_k^{-1}(g_r g_s))$$

$$= v(g_r g_s), \quad \text{for } r, s = 1, 2, \ldots, n,$$

where v is the function from G to $\mathfrak{F}$ having $v(x) = \sum_{k=1}^{n} \tau(g_k)\, \sigma(g_k^{-1} x)$, for x in G.
Thus, BA is an auxiliary-pattern matrix.

We apply (17.5) and (17.6) to verify, for $r, s = 1, 2, \ldots, n$, that

$$(17.7) \qquad [NA]_{r,s} = \sum_{k=1}^{n} [N]_{r,k} [A]_{k,s} = \sum_{k=1}^{n} \phi(g_r g_k)\, \sigma(g_k^{-1} g_s) = \sigma(g_r g_s)$$

and

$$(17.8) \qquad [NB]_{r,s} = \sum_{k=1}^{n} [N]_{r,k} [B]_{k,s} = \sum_{k=1}^{n} \phi(g_r g_k)\, \tau(g_k g_s) = \tau(g_r^{-1} g_s).$$

Formula (17.7) shows that NA is an auxiliary-pattern matrix; and, formula (17.8)
shows that NB is a group-pattern matrix. Suppose that $\sigma = \tau$. Then, (17.7) and
(17.6) yield $NA = B$ while (17.8) and (17.6) give $NB = A$. This completes the
proof. $\qquad \square$

EXAMPLE 17.5. The dihedral group $\mathcal{D}_3$ of order 6 is nonabelian and its list

$$\mathscr{L}_{\mathcal{D}_3} = \left(g_1,\, g_2,\, g_3,\, g_4,\, g_5,\, g_6\right) = \left(e,\, \alpha,\, \alpha^2,\, \beta,\, \beta\alpha,\, \beta\alpha^2\right)$$

from (15.10) for $n = 3$, enables us to easily construct the multiplication tables

$\cdot$	g_1	g_2	g_3	g_4	g_5	g_6
g_1^{-1}	g_1	g_2	g_3	g_4	g_5	g_6
g_2^{-1}	g_3	g_1	g_2	g_5	g_6	g_4
g_3^{-1}	g_2	g_3	g_1	g_6	g_4	g_5
g_4^{-1}	g_4	g_5	g_6	g_1	g_2	g_3
g_5^{-1}	g_5	g_6	g_4	g_3	g_1	g_2
g_6^{-1}	g_6	g_4	g_5	g_2	g_3	g_1

and

$\cdot$	g_1	g_2	g_3	g_4	g_5	g_6
g_1	g_1	g_2	g_3	g_4	g_5	g_6
g_2	g_2	g_3	g_1	g_6	g_4	g_5
g_3	g_3	g_1	g_2	g_5	g_6	g_4
g_4	g_4	g_5	g_6	g_1	g_2	g_3
g_5	g_5	g_6	g_4	g_3	g_1	g_2
g_6	g_6	g_4	g_5	g_2	g_3	g_1

that correspond to (17.1) and (17.2). These patterns specify the matrices

$$A = \left[\begin{array}{ccc|ccc} 1 & 2 & 3 & 4 & 5 & 6 \\ 3 & 1 & 2 & 5 & 6 & 4 \\ 2 & 3 & 1 & 6 & 4 & 5 \\ \hline 4 & 5 & 6 & 1 & 2 & 3 \\ 5 & 6 & 4 & 3 & 1 & 2 \\ 6 & 4 & 5 & 2 & 3 & 1 \end{array}\right], \quad B = \left[\begin{array}{ccc|ccc} 1 & 2 & 3 & 4 & 5 & 6 \\ 2 & 3 & 1 & 6 & 4 & 5 \\ 3 & 1 & 2 & 5 & 6 & 4 \\ \hline 4 & 5 & 6 & 1 & 2 & 3 \\ 5 & 6 & 4 & 3 & 1 & 2 \\ 6 & 4 & 5 & 2 & 3 & 1 \end{array}\right], \quad N = \left[\begin{array}{cccccc} 1 & 0 & 0 & 0 & 0 & 0 \\ 0 & 0 & 1 & 0 & 0 & 0 \\ 0 & 1 & 0 & 0 & 0 & 0 \\ 0 & 0 & 0 & 1 & 0 & 0 \\ 0 & 0 & 0 & 0 & 1 & 0 \\ 0 & 0 & 0 & 0 & 0 & 1 \end{array}\right]$$

where A is the group-pattern matrix for $\mathcal{D}_3$ and $\mathscr{L}_{\mathcal{D}_3}$ with $\sigma(g_k) = k$, for $1 \le k \le 6$, B is the auxiliary-pattern matrix for $\mathcal{D}_3$ and $\mathscr{L}_{\mathcal{D}_3}$ having $\tau(g_k) = k$, for $1 \le k \le 6$, and N is given by (17.5). Here, we have $NA = B$, $NB = A$, and see that the matrix

$$BA = \left[\begin{array}{ccc|ccc} 90 & 87 & 84 & 63 & 60 & 57 \\ 87 & 84 & 90 & 57 & 63 & 60 \\ 84 & 90 & 87 & 60 & 57 & 63 \\ \hline 63 & 60 & 57 & 90 & 87 & 84 \\ 60 & 57 & 63 & 84 & 90 & 87 \\ 57 & 63 & 60 & 87 & 84 & 90 \end{array}\right]$$

has the pattern of the second multiplication table above. It is therefore an auxiliary-pattern matrix for $\mathcal{D}_3$ and $\mathscr{L}_{\mathcal{D}_3}$.

OBSERVATION 17.6. One can use computer algebra with the matrices A, B, and N of Example 17.5 to check that each of AN, BN, AB, B^2, B^T, and NAN is neither a group-pattern matrix nor an auxiliary-pattern matrix for $\mathcal{D}_3$ and $\mathscr{L}_{\mathcal{D}_3}$.

17.3. The situation where G is abelian

THEOREM 17.7. For a field $\mathfrak{F}$ and a list $\mathscr{L} = \left(g_1, g_2, \ldots, g_n\right)$ of the elements in an abelian group G of order n, let A, A_1, A_2 be group-pattern matrices for G with $\mathscr{L}$ over $\mathfrak{F}$; and, let B, B_1, B_2 be auxiliary-pattern matrices for G with $\mathscr{L}$ over $\mathfrak{F}$. Then, the matrices NB, BN, A_1A_2, B_1B_2, A^T, and $NAN = A^T$ are group pattern matrices for G with $\mathscr{L}$; the matrices NA, AN, BA, AB, and B^T are auxiliary-pattern matrices for G with $\mathscr{L}$; and, $B^T = B$.

PROOF. Let σ and τ be the functions from G to $\mathfrak{F}$ such that

(17.9) $\quad \left[A\right]_{r,s} = \sigma(g_r^{-1}g_s)$ and $\left[B\right]_{r,s} = \tau(g_r\, g_s)$, for $r,\, s = 1, 2, \ldots, n$.

We use (17.9) with N from (17.5) to deduce, for r, $s = 1, 2, \ldots, n$, that

$$\left[BN\right]_{r,s} = \sum_{k=1}^{n} \left[B\right]_{r,k} \left[N\right]_{k,s} = \sum_{k=1}^{n} \tau(g_r \, g_k) \, \phi(g_k \, g_s)$$

$$= \tau(g_r \, g_s^{-1}) = \tau(g_s^{-1} g_r) = \tau\left(\left(g_r^{-1} g_s\right)^{-1}\right) = \upsilon_1(g_r^{-1} g_s),$$

where υ_1 is the function from G to $\mathfrak{F}$ defined by $\upsilon_1(x) = \tau(x^{-1})$, for each x in G. Thus, BN is a group-pattern matrix for G and $\mathscr{L}$. For r, $s = 1, 2, \ldots, n$, we have

$$\left[AN\right]_{r,s} = \sum_{k=1}^{n} \left[A\right]_{r,k} \left[N\right]_{k,s} = \sum_{k=1}^{n} \sigma(g_r^{-1} g_k) \, \phi(g_k \, g_s)$$

$$= \sigma(g_r^{-1} g_s^{-1}) = \sigma(g_s^{-1} g_r^{-1}) = \sigma\left(\left(g_r \, g_s\right)^{-1}\right) = \upsilon_2(g_r \, g_s),$$

where υ_2 is the function from G to $\mathfrak{F}$ defined by $\upsilon_2(x) = \sigma(x^{-1})$, for each x in G. Thus, AN is an auxiliary-pattern matrix for G and $\mathscr{L}$.

For verifications of the assertions about NA, NB, BA, $A_1 A_2$, A^T, and NAN, see Proposition 17.4, Theorem 4.1, and Proposition 8.10.

We observe that A, BN, and $A(BN)$ are group-pattern matrices. Therefore, the identity $AB = \left(A(BN)\right)N$ shows that AB is an auxiliary-pattern matrix. Because $B_1 N$, NB_2, and $(B_1 N)(NB_2)$ are group-pattern matrices, we employ the identity $B_1 B_2 = B_1(NN)B_2$ to see that $B_1 B_2$ is a group-pattern matrix.

Since G is abelian, the table in (17.2) is symmetric with respect to its principal diagonal. Thus, we have $B^T = B$ and conclude that B^T is an auxiliary-pattern matrix. That completes the proof. $\qquad\square$

17.4. Injective Auxiliary-Pattern Matrices

THEOREM 17.8. *Suppose that $\mathcal{Z}$ is an injective auxiliary-pattern matrix. Then, the following six assertions about $\mathcal{Z}$ are valid.*

A: *$\mathcal{Z}$ is a matrix of size $n \times n$ for some positive integer n;*

B: *each row of $\mathcal{Z}$ and each column of $\mathcal{Z}$ has precisely n distinct elements;*

C: *the number of distinct elements that are components of $\mathcal{Z}$ is n;*

D: *there is an integer d such that $1 \leq d \leq n$ and the dth column of $\mathcal{Z}$ is equal to the transpose of the dth row of $\mathcal{Z}$;*

E: *for $k = 1, 2, \ldots, n$, an $n \times n$ permutation matrix R_k is obtained from $\mathcal{Z}$ by replacing each of the n components of $\mathcal{Z}$ equal to $\left[\mathcal{Z}\right]_{d,k}$ with 1 from the field $\mathbb{Q}$ of rational numbers and replacing each of the $n^2 - n$ components of $\mathcal{Z}$ that are unequal to $\left[\mathcal{Z}\right]_{d,k}$ with 0 from $\mathbb{Q}$;*

F: *for the n permutation matrices $P_1, P_2, \ldots, P_n$ defined by $P_k = R_d R_k$, when $1 \leq k \leq n$, the set $\mathfrak{P} = \left\{P_1, P_2, \ldots, P_n\right\}$ is closed under matrix multiplication.*

PROOF. Let $\mathcal{Z}$ be an injective auxiliary-pattern matrix for a group G of order n and a list $\mathscr{L} = \left(g_1, g_2, \ldots, g_n\right)$ of the elements in G. Then, we use Definition 17.2 to see that there is a one-to-one function τ from G onto a set S such that the component $\left[\mathcal{Z}\right]_{r,s}$ of $\mathcal{Z}$ in the rth row and sth column of $\mathcal{Z}$ is given by

(17.10) $\left[\mathcal{Z}\right]_{r,s} = \tau\left(g_r \, g_s\right)$,   for r, $s = 1, 2, \ldots, n$.

Let d denote the integer such that $1 \leq d \leq n$ and g_d is the identity element e of G.

(A). In view of (17.10), $\mathcal{Z}$ is an $n \times n$ matrix and $\mathbf{A}$ is satisfied.

(B) For each fixed integer r_0 satisfying $1 \leq r_0 \leq n$, we see that: as s ranges over the n integers from 1 through n, $g_{r_0} g_s$ ranges through the n elements of G and $\tau(g_{r_0} g_s)$ ranges through n distinct elements of S. Thus, each row of $\mathcal{Z}$ has n distinct elements. Also, for each s_0 satisfying $1 \leq s_0 \leq n$, we note that: as r ranges over the integers from 1 through n, $g_r g_{s_0}$ ranges through the n elements of G and $\tau(g_r g_{s_0})$ ranges through n distinct elements of S. Thus, each column of $\mathcal{Z}$ has n distinct elements. Since each row and column has n distinct elements, $\mathbf{B}$ is satisfied

(C) Due to $\mathbf{B}$, $\mathcal{Z}$ possesses at least n distinct components that are elements of $\tau(G)$. Thus, there are precisely n distinct components of $\mathcal{Z}$ and $\mathbf{C}$ is valid.

(D) We have $\left[\mathcal{Z}\right]_{d,k} = \tau(g_d g_k) = \tau(g_k) = \tau(g_k g_d) = \left[\mathcal{Z}\right]_{k,d}$, for $1 \leq k \leq n$. This shows that $\mathbf{D}$ is satisfied.

(E) For $k = 1, 2, \ldots, n$, we introduce $z_k = \left[\mathcal{Z}\right]_{d,k}$ as the component of $\mathcal{Z}$ in its dth row and kth column. The results labeled $\mathbf{A}$, $\mathbf{B}$, and $\mathbf{C}$ show that each component of $\mathcal{Z}$ is an element of $\{z_1, z_2, \ldots, z_n\}$. For $1 \leq k \leq n$, they also show that: the element z_k appears precisely one time as a component in any row of $\mathcal{Z}$ and z_k appears precisely one time as a component in any column of $\mathcal{Z}$. Thus, for 0, 1 in $\mathbb{Q}$, an $n \times n$ matrix R_k is defined, for $1 \leq k, r, s \leq n$, by

$$\left[R_k\right]_{r,s} = \begin{cases} 1, & \text{if } [Z]_{r,s} = z_k, \\ 0, & \text{if } [Z]_{r,s} \neq z_k, \end{cases} = \begin{cases} 1, & \text{if } \tau(g_r g_s) = \tau(g_k), \\ 0, & \text{if } \tau(g_r g_s) \neq \tau(g_k), \end{cases} = \begin{cases} 1, & \text{if } g_r g_s = g_k, \\ 0, & \text{if } g_r g_s \neq g_k. \end{cases}$$

Thus, R_k is an $n \times n$ permutation matrix and $\mathbf{E}$ is therefore valid.

(F) For $k = 1, 2, \ldots, n$, let ϕ_k be the function from G to $\mathbb{Q}$ having

$$(17.11) \qquad \phi_k(x) = \begin{cases} 1, & \text{if } x = g_k, \\ 0, & \text{if } x \neq g_k, \end{cases}, \quad \text{for each } x \text{ in } G.$$

In view of the two preceding formulas, we see that

$$(17.12) \qquad \left[R_k\right]_{r,s} = \phi_k(g_r g_s), \quad \text{for } k, r, s = 1, 2, \ldots, n.$$

For $k, r, s = 1, 2, \ldots, n$ and $P_k = R_d R_k$, we employ (17.11) and (17.12) to obtain

$$(17.13) \qquad \left[P_k\right]_{r,s} = \sum_{\mu=1}^{n} \left[R_d\right]_{r,\mu} \left[R_k\right]_{\mu,s} = \sum_{\mu=1}^{n} \phi_d(g_r g_\mu) \, \phi_k(g_\mu g_s)$$

Since we have $\phi_d(g_r g_\mu) \neq 0$ only when $g_r g_\mu = g_d = e$, $g_\mu = g_r^{-1}$, and $\phi_d(g_r g_\mu) = 1$, we apply (17.13) to deduce

$$(17.14) \qquad \left[P_k\right]_{r,s} = \phi_k(g_r^{-1} g_s), \quad \text{for } r, s = 1, 2, \ldots, n.$$

In particular, for $k = d$ and $1 \leq r, s \leq n$, we find that (17.14) and (17.11) yield

$$\left[P_d\right]_{r,s} = \begin{cases} 1, & \text{if } g_r^{-1} g_s = g_d = e, \\ 0, & \text{if } g_r^{-1} g_s \neq g_d = e, \end{cases} = \begin{cases} 1, & \text{if } r = s, \\ 0, & \text{if } r \neq s. \end{cases}$$

Thus, the matrix $P_d = (R_d)^2$ is the $n \times n$ identity matrix.

With $\mathfrak{F} = \mathbb{Q}$ and $\widehat{g}_k = \phi_k$, for $1 \leq k \leq n$, we apply Theorem 2.16 on page 22 to (17.14). Thus, $\mathfrak{P}$ forms a group that is isomorphic to G. In particular, $\mathfrak{P}$ is closed under matrix multiplication and $\mathbf{F}$ is satisfied. This completes the proof. $\qquad \square$

THEOREM 17.9. *Suppose a matrix $\mathcal{Z}$ satisfies the conditions* **A, B, C, D, E, F** *of* Theorem 17.8. *Then, the set* $\mathfrak{P}$ *of permutation matrices for* **F** *is a group under matrix multiplication with* $P_d = I_n$*; and,* $\mathcal{Z}$ *is an injective auxiliary-pattern matrix for the group* $\mathfrak{P}$ *with respect to its list* $\mathscr{L}_{\mathfrak{P}} = (P_1, P_2, \ldots, P_n)$.

PROOF. Condition **F** shows $\mathfrak{P}$ is closed under matrix multiplication and P_d is the $n \times n$ identity matrix. For $1 \leq k \leq n$, the element P_k of $\mathfrak{P}$ specifies two lists

$$(P_1 P_k, P_2 P_k, \ldots, P_n P_k) \quad \text{and} \quad (P_k P_1, P_k P_2, \ldots, P_k P_n)$$

of the n elements for $\mathfrak{P}$ in some order. Thus, there are elements U_k and V_k in $\mathfrak{P}$ that yield $U_k P_k = P_d = P_k V_k$ and $U_k = U_k P_d = U_k P_k V_k = P_d V_k = V_k$. Thus, each element P_k in $\mathfrak{P}$ has an inverse in $\mathfrak{P}$. Consequently, $\mathfrak{P}$ is a group.

The components of $\mathcal{Z}$ may not be elements of the ring $\mathfrak{R} = \mathbb{Q}[X_1, X_2, \ldots, X_n]$ of polynomials in the variables $X_1, X_2, \ldots, X_n$ over the field $\mathbb{Q}$ of rational numbers. To remedy that inconvenience, we introduce in the next paragraph a matrix $\mathcal{A}$ having the pattern of $\mathcal{Z}$ and possessing components in $\mathfrak{R}$.

Let $z_1, z_2, \ldots, z_n$ denote the n distinct components in the dth row of $\mathcal{Z}$. We have $z_k = [\mathcal{Z}]_{d,k}$, for $1 \leq k \leq n$. We introduce $S_1 = \{z_1, z_2, \ldots, z_n\}$ and the subset $S_2 = \{X_1, X_2, \ldots, X_n\}$ of $\mathfrak{R}$. Let υ be the function from S_1 onto S_2 having $\upsilon(z_k) = X_k$, for $1 \leq k \leq n$. Let A be the $n \times n$ matrix over $\mathfrak{R}$ given by

$$(17.15) \qquad [A]_{r,s} = \upsilon\big([\mathcal{Z}]_{r,s}\big), \quad \text{for } r, s = 1, 2, \ldots, n,$$

and let **$\mathcal{A}$, $\mathcal{B}$, $\mathcal{C}$, $\mathcal{D}$, $\mathcal{E}$, $\mathcal{F}$** denote the assertions about A obtained by replacing $\mathcal{Z}$ with A throughout each of **A, B, C, D, E, F**. We apply (17.15) to see that each of **$\mathcal{A}$, $\mathcal{B}$, $\mathcal{C}$, $\mathcal{D}$, $\mathcal{E}$, $\mathcal{F}$** is a valid assertion about A.

The components of A, $R_1, R_2, \ldots, R_n$ belong to $\mathfrak{R}$. Thus, (17.15) and **$\mathcal{E}$** yield

$$(17.16) \qquad A = X_1 R_1 + X_2 R_2 + \cdots + X_n R_n.$$

The multiplication table of the form (17.2) on page 165 for the group $\mathfrak{P}$ is given by

$$(17.17)$$

$\cdot$	P_1	P_2	P_3	$\ldots$	P_s	$\ldots$	P_n
P_1	$P_1 P_1$	$P_1 P_2$	$P_1 P_3$	$\ldots$	$P_1 P_s$	$\ldots$	$P_1 P_n$
P_2	$P_2 P_1$	$P_2 P_2$	$P_2 P_3$	$\ldots$	$P_2 P_s$	$\ldots$	$P_2 P_n$
P_3	$P_3 P_1$	$P_3 P_2$	$P_3 P_3$	$\ldots$	$P_3 P_s$	$\ldots$	$P_3 P_n$
$\vdots$	$\vdots$	$\vdots$	$\vdots$	$\vdots$	$\vdots$	$\vdots$	$\vdots$
P_r	$P_r P_1$	$P_r P_2$	$P_r P_3$	$\ldots$	$P_r P_s$	$\ldots$	$P_r P_n$
$\vdots$	$\vdots$	$\vdots$	$\vdots$	$\vdots$	$\vdots$	$\vdots$	$\vdots$
P_n	$P_n P_1$	$P_n P_2$	$P_n P_3$	$\ldots$	$P_n P_s$	$\ldots$	$P_n P_n$

with respect to the list $\mathscr{L}_{\mathfrak{P}} = (P_1, P_2, \ldots, P_n)$ of its elements. For $k = 1, 2, \ldots, n$, let f_k be the function from $\mathfrak{P}$ to $\mathfrak{R}$ defined by

$$(17.18) \qquad f_k(P) = \begin{cases} 1, & \text{if } P = P_k, \\ 0, & \text{if } P \neq P_k, \end{cases} \quad \text{for each } P \text{ in } \mathfrak{P},$$

and, for $k = 1, 2, \ldots, n$, let Q_k be the $n \times n$ matrix with components in $\mathfrak{R}$ having

$$(17.19) \qquad [Q_k]_{r,s} = f_k(P_r P_s), \quad \text{for } r, s = 1, 2, \ldots, n.$$

We use (17.19), (17.18), and properties of $\mathfrak{P}$ as a group to see that Q_k is an $n \times n$ permutation matrix, for $1 \leq k \leq n$.

Suppose k, r, s are integers such that $1 \leq k$, r, $s \leq n$ and $\left[Q_k\right]_{r,s} = 1$. Then, we have $P_k = P_r P_s$, $R_d R_k = P_k = P_r P_s = (R_d R_r)(R_d R_s)$, $R_k = R_r R_d R_s$, and

$$\left[R_k\right]_{r,s} = \sum_{\mu=1}^{n} \sum_{\nu=1}^{n} \left[R_r\right]_{r,\mu} \left[R_d\right]_{\mu,\nu} \left[R_s\right]_{\nu,s}.$$

Due to $\left[R_s\right]_{\nu,s} \neq 0$ only when $\nu = d$ and $\left[R_s\right]_{d,s} = 1$, we find that

$$\left[R_k\right]_{r,s} = \sum_{\mu=1}^{n} \left[R_r\right]_{r,\mu} \left[R_d\right]_{\mu,d}.$$

Due to $\left[R_d\right]_{\mu,d} \neq 0$ only when $\mu = d$ and $\left[R_d\right]_{d,d} = 1$, this gives $\left[R_k\right]_{r,s} = \left[R_r\right]_{r,d}$. Since R_r can be obtained from A by replacing each of X_1, X_2, ..., X_n in A with corresponding element from $\mathbb{Q}$, Condition $\mathcal{D}$ shows that the transpose of the dth column of R_r is equal to the dth row of R_r. Thus, we have

$$\left[R_k\right]_{r,s} = \left[R_r\right]_{r,d} = \left[R_r\right]_{d,r} = 1.$$

The preceding argument establishes that: whenever a component of the $n \times n$ permutation matrix Q_k is 1, the corresponding component of the $n \times n$ permutation matrix R_k is 1. Since an $n \times n$ permutation matrix has precisely n components equal to 1, each of the other $n^2 - n$ components of R_k is equal 0. This yields

(17.20) $$R_k = Q_k, \quad \text{for } k = 1, 2, \ldots, n.$$

Thus, each of the matrices R_1, R_2, ..., R_n is an auxiliary-pattern matrix with respect to the list $\mathscr{L}_{\mathfrak{P}}$ for $\mathfrak{P}$. In particular, (17.20) and (17.19) yield

(17.21) $$\left[R_k\right]_{r,s} = f_k(P_r P_s), \quad \text{for } r, s = 1, 2, \ldots, n.$$

We employ $A = X_1 R_1 + X_2 R_2 + \cdots + X_n R_n$ from (17.16) with (17.21) to deduce

(17.22) $$[A]_{r,s} = \left[\sum_{\mu=1}^{n} X_\mu R_\mu\right]_{r,s}$$

$$= \sum_{\mu=1}^{n} X_\mu [R_\mu]_{r,s} = \sum_{\mu=1}^{n} X_\mu f_\mu(P_r P_s)$$

$$= \tau(P_r P_s), \quad \text{for } r, s = 1, 2, \ldots, n,$$

where τ is the function from $\mathfrak{P}$ to $\mathfrak{R}$ defined by $\tau = X_1 f_1 + X_2 f_2 + \cdots + X_n f_n$ in terms of (17.18) and the components in the dth row of A. We observe that

(17.23) $$\tau(P_k) = \sum_{\mu=1}^{n} X_\mu f_\mu(P_k) = X_k f_k(P_k) = X_k, \quad \text{for } k = 1, 2, \ldots, n.$$

For $1 \leq i < j \leq n$, we use (17.23) to obtain $\tau(P_i) = X_i \neq X_j = \tau(P_j)$ and see that τ is a one-to-one function from $\mathfrak{P}$ onto S_2. In terms of (17.15), (17.22) and the one-to-one function ψ from S_2 onto S_1 with $\psi(X_k) = z_k$, for $1 \leq k \leq n$, we have

$$[\mathcal{Z}]_{r,s} = \psi\Big([A]_{r,s}\Big) = \psi\Big(\tau(P_r P_s)\Big) = (\psi \circ \tau)(P_r P_s), \quad \text{for } 1 \leq r, s \leq n,$$

where $\psi \circ \tau$ is a one-to-one function from $\mathfrak{P}$ onto S_1. Thus, Definition 17.2 shows that $\mathcal{Z}$ is an injective auxiliary-pattern matrix. This completes the proof. $\qquad \square$

EXAMPLE 17.10. The transpose of the second column in the matrix

$$(17.24) \qquad \mathcal{Z} = \left[\begin{array}{ccc|ccc} b_1 & b_2 & b_3 & b_4 & b_5 & b_6 \\ b_2 & b_3 & b_1 & b_6 & b_4 & b_5 \\ b_3 & b_1 & b_2 & b_5 & b_6 & b_4 \\ \hline b_5 & b_6 & b_4 & b_3 & b_1 & b_2 \\ b_6 & b_4 & b_5 & b_2 & b_3 & b_1 \\ b_4 & b_5 & b_6 & b_1 & b_2 & b_3 \end{array}\right]$$

equals its second row. Thus, $\mathcal{Z}$ satisfies **A**, **B**, **C**, and **D** with $d = 2$. We have

$$R_1 = \begin{bmatrix} 0&1&0&0&0&0 \\ 1&0&0&0&0&0 \\ 0&0&1&0&0&0 \\ 0&0&0&0&0&1 \\ 0&0&0&1&0&0 \\ 0&0&0&0&1&0 \end{bmatrix}, \quad R_2 = \begin{bmatrix} 0&0&1&0&0&0 \\ 0&1&0&0&0&0 \\ 1&0&0&0&0&0 \\ 0&0&0&1&0&0 \\ 0&0&0&0&1&0 \\ 0&0&0&0&0&1 \end{bmatrix}, \quad R_3 = \begin{bmatrix} 1&0&0&0&0&0 \\ 0&0&1&0&0&0 \\ 0&1&0&0&0&0 \\ 0&0&0&0&1&0 \\ 0&0&0&0&0&1 \\ 0&0&0&1&0&0 \end{bmatrix},$$

$$R_4 = \begin{bmatrix} 0&0&0&0&0&1 \\ 0&0&0&1&0&0 \\ 0&0&0&0&1&0 \\ 0&1&0&0&0&0 \\ 1&0&0&0&0&0 \\ 0&0&1&0&0&0 \end{bmatrix}, \quad R_5 = \begin{bmatrix} 0&0&0&1&0&0 \\ 0&0&0&0&1&0 \\ 0&0&0&0&0&1 \\ 0&0&1&0&0&0 \\ 0&1&0&0&0&0 \\ 1&0&0&0&0&0 \end{bmatrix}, \quad R_6 = \begin{bmatrix} 0&0&0&0&1&0 \\ 0&0&0&0&0&1 \\ 0&0&0&1&0&0 \\ 1&0&0&0&0&0 \\ 0&0&1&0&0&0 \\ 0&1&0&0&0&0 \end{bmatrix}$$

with $d = 2$ for **E**. We compute $P_k = R_2 R_k$, for $1 \le k \le 6$, and obtain

$$P_1 = \begin{bmatrix} 0&0&1&0&0&0 \\ 1&0&0&0&0&0 \\ 0&1&0&0&0&0 \\ 0&0&0&0&0&1 \\ 0&0&0&1&0&0 \\ 0&0&0&0&1&0 \end{bmatrix}, \quad P_2 = \begin{bmatrix} 1&0&0&0&0&0 \\ 0&1&0&0&0&0 \\ 0&0&1&0&0&0 \\ 0&0&0&1&0&0 \\ 0&0&0&0&1&0 \\ 0&0&0&0&0&1 \end{bmatrix}, \quad P_3 = \begin{bmatrix} 0&1&0&0&0&0 \\ 0&0&1&0&0&0 \\ 1&0&0&0&0&0 \\ 0&0&0&0&1&0 \\ 0&0&0&0&0&1 \\ 0&0&0&1&0&0 \end{bmatrix},$$

$$P_4 = \begin{bmatrix} 0&0&0&0&1&0 \\ 0&0&0&1&0&0 \\ 0&0&0&0&0&1 \\ 0&1&0&0&0&0 \\ 1&0&0&0&0&0 \\ 0&0&1&0&0&0 \end{bmatrix}, \quad P_5 = \begin{bmatrix} 0&0&0&0&0&1 \\ 0&0&0&0&1&0 \\ 0&0&0&1&0&0 \\ 0&0&1&0&0&0 \\ 0&1&0&0&0&0 \\ 1&0&0&0&0&0 \end{bmatrix}, \quad P_6 = \begin{bmatrix} 0&0&0&1&0&0 \\ 0&0&0&0&0&1 \\ 0&0&0&0&1&0 \\ 1&0&0&0&0&0 \\ 0&0&1&0&0&0 \\ 0&1&0&0&0&0 \end{bmatrix}.$$

With respect to $\mathscr{L}_{\mathfrak{P}} = (P_1, P_2, P_3, P_4, P_5, P_6)$, the multiplication table

(17.25)

$\cdot$	P_1	P_2	P_3	P_4	P_5	P_6
P_1	P_3	P_1	P_2	P_5	P_6	P_4
P_2	P_1	P_2	P_3	P_4	P_5	P_6
P_3	P_2	P_3	P_1	P_6	P_4	P_5
P_4	P_6	P_4	P_5	P_2	P_3	P_1
P_5	P_4	P_5	P_6	P_1	P_2	P_3
P_6	P_5	P_6	P_4	P_3	P_1	P_2

establishes closure of $\mathfrak{P} = \{P_1, P_2, P_3, P_4, P_5, P_6\}$ for **F**. Consequently, $\mathcal{Z}$ satisfies **A**, **B**, **C**, **D**, **E**, **F**. Therefore, with $d = 2$, Theorem 17.9 shows that $\mathcal{Z}$ is an injective auxiliary-pattern matrix with respect to the group $\mathfrak{P}$ and its list $\mathscr{L}_{\mathfrak{P}}$.

17.5. Computer program that implements Theorems 17.8 and 17.9

The following program checks whether a given matrix Z satisfies the conditions **A**, **B**, **C**, **D**, **E**, **F** of page 168 for Theorems 17.8–17.9 and is therefore an injective auxiliary-pattern matrix.. We use a version of *Mathematica* such as [48] to enter in a *Mathematica* notebook the following input statement

```
injectiveAuxiliaryPatternQ[Z_] :=
  Module[{n, r, c, listOne, listTwo, F, rule,
         R, Per, S, t, d, r1, r2, table},
  If[ MatrixQ[Z], {}, (Print["It is not a matrix."]; Abort[])];
  If[Dimensions[Z][[1]] == Dimensions[Z][[2]], {},
       (Print["It is not a square matrix."]; Abort[])];
  n = Dimensions[Z][[1]];
  Do[r[k] = Length[Union[Z[[k]]]], {k, 1, n}];
  Do[c[k] = Length[Union[Transpose[Z][[k]]]], {k, 1, n}];
  listOne = Join[Table[r[k]==n, {k,1,n}], Table[c[k]==n, {k,1,n}]];
  If[Apply[And, listOne], {}, (Print["Some row or column has
          fewer than ", n, " distinct elements."]; Abort[])];
  If[Length[Union[Flatten[Z]]] == n, {}, (Print["It does not have
          precisely ", n, " distinct elements."]; Abort[])];
  listTwo = Table[TrueQ[Z[[k]] == Transpose[Z][[k]]], {k,1,n}];
  If[Apply[Or,listTwo], {},
          (Print["Condition D is not satisfied."]; Abort[])];
  F[d_] := ( Do[rule[k] = Table[If[i == k, Z[[d, i]] -> 1,
     Z[[d, i]] -> 0], {i,1,n}], {k,1,n}];
     Do[R[k] = Z /. rule[k], {k,1,n}];
     Do[Per[k] = R[d].R[k], {k,1,n}]; S = Table[Per[k], {k,1,n}];
     Do[t[i, j] = MemberQ[S, Per[i].Per[j]], {i,1,n}, {j,1,n}];
     Apply[And, Flatten[Table[t[i, j], {i, 1, n}, {j, 1, n}]]] );
  d = Catch[ Do[ If [TrueQ[F[i]], Throw[i], {}], {i, 1, n} ]];
  If[IntegerQ[d],
  Print["It is an injective auxiliary-pattern matrix with d = ",d],
     (Print["Closure for multiplication is violated."]; Abort[])];
  Do[If[TrueQ[Per[i].Per[j] == Per[k]],
     f[i,j] = P[k], {}], {i,1,n}, {j,1,n}, {k,1,n}];
  Do[r1[i] = Table[f[i, j], {j, 1, n}], {i, 1, n}]; Clear[P];
  Do[r2[i] = Prepend[r1[i], (P[i])], {i, 1, n}];
  r2[0] = Prepend[Table[P[j], {j, 1, n}], "\[CenterDot]"];
  table = Table[r2[i], {i, 0, n}];
  Print["Its multiplication table for (P[1], P[2], ..., P[", n,
     "] is:"];  Print[""];
  Print[Grid[table, Frame -> All]]; Print[""];  Print["where"];
  Do[(Print[""]; Print["P[",k,"] = ",
     Per[k] //MatrixForm]),{k,1,n}]    ]
```

and evaluate it. Then, after the evaluation of a *Mathematica* representation Z for Z, the evaluation of `injectiveAuxiliaryPatternQ[Z]` either establishes that Z is an injective auxiliary-pattern matrix or it gives a reason why not.

EXAMPLE 17.11. When Z is a representation of $\mathcal{Z}$ in (17.24), the evaluations of the program on page 173 as well as Z and `injectiveAuxiliaryPatternQ[Z]` yield output where the assertion "It is an injective auxiliary-pattern matrix with d = 2" is followed by the multiplication table (17.25) and the explicit expressions for the matrices P_1, P_2, P_3, P_4, P_5, P_6 given on page 172.

OBSERVATION 17.12. A matrix Z may satisfy Conditions **A**-**F** of Theorem 17.8 for several values of the integer d. Then, the least such value of d is used and reported by the program of page 173 when it is applied to a representation Z for Z.

As an illustration, let Z be an injective auxiliary-pattern matrix for some abelian group G of order n and list $\mathscr{L}$ of its elements. Then, with $Z^T = Z$, Conditions **A**-**F** are satisfied by each d having $1 \leq d \leq n$. When the program of page 173 is applied, it is $d = 1$ that is used and reported.

COROLLARY 17.13. *A matrix $\mathcal{Z}$ is an injective auxiliary-pattern matrix if and only if it satisfies* **A, B, C, D, E, F** *of Theorem* 17.8.

PROOF. Theorems 17.9 and 17.8 immediately yield this result. □

17.6. Some lists are not replaceable by ones having $g_1 = e$

To have a contrast, we first include a consequence of Proposition 2.7.

PROPOSITION 17.14. *Suppose that A is a group-pattern matrix for a group G of order n and a list $\mathscr{L}$ of its elements. Then, A is a group-pattern matrix for G and a list $\mathscr{L}_1$ in which the first element of $\mathscr{L}_1$ is the identity element e of G.*

PROOF. Let A be given in terms of $\mathscr{L} = (g_1, g_2, \ldots, g_n)$ by

$$(17.26) \qquad [A]_{r,s} = \sigma(g_r^{-1} g_s), \quad \text{for } r, s = 1, 2, \ldots, n,$$

where σ is a function from G to a set S. Then, a list $\mathscr{L}_1$ for G is defined by

$$\mathscr{L}_1 = (h_1, h_2, \ldots, h_n), \quad \text{where } h_k = g_1^{-1} g_k, \text{ when } k = 1, 2, \ldots, n.$$

It yields $h_1 = e$ as well as $\sigma(h_k) = \sigma(g_1^{-1} g_k)$, for $1 \leq k \leq n$, and

$$[A]_{r,s} = \sigma(g_r^{-1} g_s) = \sigma\left(\left(g_1^{-1} g_r \right)^{-1} \left(g_1^{-1} g_s \right) \right) = \sigma(h_r^{-1} h_s), \quad \text{for } r, s = 1, 2, \ldots, n.$$

Thus, A is a group-pattern matrix for G and $\mathscr{L}_1$. This completes the proof. □

PROPOSITION 17.15. *Suppose B is an auxiliary-pattern matrix for a group G of order n and a list $\mathscr{L} = (g_1, g_2, \ldots, g_n)$ of its elements in which $g_1 = e$. Then, the first row R of B is equal to the transpose of the first column C of B.*

PROOF. Let τ be a function from G to a set S such that

$$[B]_{r,s} = \tau(g_r g_s), \quad \text{for } 1 \leq r, s \leq n.$$

Then, for $1 \leq k \leq n$, we have

$$[R]_{1,k} = [B]_{1,k} = \tau(g_1 g_k) = \tau(g_k) = \tau(g_k g_1) = [B]_{k,1} = [C]_{k,1} = [C^T]_{1,k}$$

and $R = C^T$. This completes the proof. □

OBSERVATION 17.16. Example 17.10 on page 172 establishes that the matrix $\mathcal{Z}$ in (17.24) is an injective auxiliary-pattern matrix for the group $\mathfrak{P}$ and the list $\mathscr{L}_{\mathfrak{P}} = \left(P_1, P_2, P_3, P_4, P_5, P_6\right)$ of the permutation matrices in $\mathfrak{P}$. There, the identity element of $\mathfrak{P}$ is the matrix $P_2 = I_6$. Since the first row of the matrix $\mathcal{Z}$ is not equal to the transpose of its first column, Proposition 17.15 shows that: for any group G of order n and any list $\mathscr{L}_1$ of the elements in G such that the identity element of G is the first element of $\mathscr{L}_1$, the matrix $\mathcal{Z}$ in (17.24) is not an auxiliary-pattern matrix for G and $\mathscr{L}_1$. Thus, *a general result analogous to* Proposition 17.14 *does not exist for auxiliary-pattern matrices*.

EXAMPLE 17.17. Let G denote any finite nonabelian group. It is easy to specify an injective auxiliary-pattern matrix B for G such that its first row R is not equal to the transpose of its first column C. Namely, let α, β be elements in G such that $\beta\alpha \neq \alpha\beta$, let $\mathscr{L} = \left(g_1, g_2, \ldots, g_n\right)$ be a list for G having $g_1 = \alpha$ as well as $g_2 = \beta$, and let τ be a one-to-one function from G to a field $\mathfrak{F}$. Then, we have

$$[R]_{1,2} = [B]_{1,2} = \tau(g_1 g_2) = \tau(\alpha\beta) \neq \tau(\beta\alpha) = \tau(g_2 g_1) = [B]_{2,1} = [C]_{2,1} = [C^T]_{1,2}$$

and $R \neq C^T$.

EXERCISES

1. Use the program of Section 17.5 to verify that the matrix

$$B_1 = \left[\begin{array}{cc|cc|cc} b_1 & b_2 & b_3 & b_4 & b_5 & b_6 \\ b_2 & b_1 & b_4 & b_3 & b_6 & b_5 \\ \hline b_3 & b_4 & b_5 & b_6 & b_1 & b_2 \\ b_4 & b_3 & b_6 & b_5 & b_2 & b_1 \\ \hline b_5 & b_6 & b_1 & b_2 & b_3 & b_4 \\ b_6 & b_5 & b_2 & b_1 & b_4 & b_3 \end{array}\right]$$

is an injective auxiliary-pattern matrix for the cyclic group $\mathcal{C}_6$ of order 6.

2. Use the program of Section 17.5 to verify that the matrix

$$B_2 = \left[\begin{array}{ccc|ccc} b_1 & b_2 & b_3 & b_4 & b_5 & b_6 \\ b_2 & b_3 & b_1 & b_5 & b_6 & b_4 \\ b_3 & b_1 & b_2 & b_6 & b_4 & b_5 \\ \hline b_4 & b_5 & b_6 & b_1 & b_2 & b_3 \\ b_5 & b_6 & b_4 & b_2 & b_3 & b_1 \\ b_6 & b_4 & b_5 & b_3 & b_1 & b_2 \end{array}\right]$$

is an injective auxiliary-pattern matrix for the cyclic group $\mathcal{C}_6$ of order 6.

3. Use the program of Section 17.5 to verify that the matrix

$$B_3 = \left[\begin{array}{ccc|ccc} b_1 & b_2 & b_3 & b_4 & b_5 & b_6 \\ b_2 & b_3 & b_1 & b_6 & b_4 & b_5 \\ b_3 & b_1 & b_2 & b_5 & b_6 & b_4 \\ \hline b_4 & b_5 & b_6 & b_1 & b_2 & b_3 \\ b_5 & b_6 & b_4 & b_3 & b_1 & b_2 \\ b_6 & b_4 & b_5 & b_2 & b_3 & b_1 \end{array}\right]$$

is an injective auxiliary-pattern matrix for the dihedral group $\mathcal{D}_3$ of order 6.

CHAPTER 18

Standard Auxiliary-Pattern Matrices

The automorphisms for a group G of order n are efficiently provided by the program of page 34 that is based on an injective group-pattern matrix for G. An equivalent program based on a standard auxiliary-pattern matrix for G is presented in the next chapter.

18.1. The standard auxiliary-pattern matrix for G and $\mathscr{L}$

Let G be any group of order n and let $X_1, X_2, \ldots, X_n$ be n distinct variables.

DEFINITION 18.1. For any list $\mathscr{L} = (g_1, g_2, \ldots, g_n)$ of the n elements in G with $g_1 = e$, there is a unique $n \times n$ matrix $Stan_{\mathfrak{A}}(G, \mathscr{L})$ defined by

$$(18.1) \qquad \big[Stan_{\mathfrak{A}}(G, \mathscr{L})\big]_{r,s} = f_{\mathscr{L}}(g_r\, g_s), \quad \text{for } r, s = 1, 2, \ldots, n,$$

where $f_{\mathscr{L}}$ is the function from G to $\{X_1, X_2, \ldots, X_n\}$ having

$$(18.2) \qquad f_{\mathscr{L}}(g_k) = X_k, \quad \text{for } k = 1, 2, \ldots, n.$$

It is the ***standard auxiliary-pattern matrix*** for G and $\mathscr{L}$.

In view of $g_1 = e$, (18.1), and (18.2), the first row of $Stan_{\mathfrak{A}}(G, \mathscr{L})$ consists of the components $\big[Stan_{\mathfrak{A}}(G, \mathscr{L})\big]_{1,k} = f_{\mathscr{L}}(g_k) = X_k$, for $k = 1, 2, \ldots, n$.

18.2. Examples of standard auxiliary-pattern matrices

EXAMPLE 18.2. For $G = \mathcal{C}_1 = \{e\}$, the only list is $\mathscr{L} = (e)$ and we have

$$Stan_{\mathfrak{A}}(\mathcal{C}_1, \mathscr{L}) = [X_1].$$

Thus, there is just one standard auxiliary-pattern matrix for $\mathcal{C}_1$.

For $G = \mathcal{C}_2 = \{e, \alpha\}$ with $\alpha^2 = e$, the only list is $\mathscr{L} = (e, \alpha)$ and we have

$$Stan_{\mathfrak{A}}(\mathcal{C}_2, \mathscr{L}) = \begin{bmatrix} X_1 & X_2 \\ X_2 & X_1 \end{bmatrix}.$$

Thus, there is just one standard auxiliary-pattern matrix for $\mathcal{C}_2$.

For $\mathcal{C}_3 = \{1, \omega, \omega^2\}$, there are two lists: $\mathscr{L}_1 = (1, \omega, \omega^2)$ and $\mathscr{L}_2 = (1, \omega^2, \omega)$. We use each of the corresponding multiplication tables

$\cdot$	1	ω	ω^2
1	1	ω	ω^2
ω	ω	ω^2	1
ω^2	ω^2	1	ω

$\cdot$	1	ω^2	ω
1	1	ω^2	ω
ω^2	ω^2	ω	1
ω	ω	1	ω^2

to conclude that

$$Stan_{\mathfrak{A}}(G, \mathscr{L}_1) = Stan_{\mathfrak{A}}(G, \mathscr{L}_2) = \begin{bmatrix} X_1 \ X_2 \ X_3 \\ X_2 \ X_3 \ X_1 \\ X_3 \ X_1 \ X_2 \end{bmatrix}.$$

Thus, there is just one standard auxiliary-pattern matrix for $\mathcal{C}_3$.

EXAMPLE 18.3. For $\mathcal{C}_4 = \{1, i, -1, -i\}$ with $i^2 = -1$, there are six lists

$$\mathscr{L}_1 = (1, i, -1, -i), \qquad \mathscr{L}_4 = (1, -i, -1, i),$$
$$\mathscr{L}_2 = (1, i, -i, -1), \qquad \mathscr{L}_5 = (1, -i, i, -1),$$
$$\mathscr{L}_3 = (1, -1, i, -i), \qquad \mathscr{L}_6 = (1, -1, -i, i).$$

The lists $\mathscr{L}_1$ and $\mathscr{L}_4$ have the respective multiplication tables

·	1	i	-1	$-i$
1	1	i	-1	$-i$
i	i	-1	$-i$	1
-1	-1	$-i$	1	i
$-i$	$-i$	1	i	-1

·	1	$-i$	-1	i
1	1	$-i$	-1	i
$-i$	$-i$	-1	i	1
-1	-1	i	1	$-i$
i	i	1	$-i$	-1

and they specify the same standard auxiliary-pattern matrix

$$(18.3) \qquad Stan_{\mathfrak{A}}(\mathcal{C}_4, \mathscr{L}_1) = Stan_{\mathfrak{A}}(\mathcal{C}_4, \mathscr{L}_4) = \begin{bmatrix} X_1 \ X_2 \ X_3 \ X_4 \\ X_2 \ X_3 \ X_4 \ X_1 \\ X_3 \ X_4 \ X_1 \ X_2 \\ X_4 \ X_1 \ X_2 \ X_3 \end{bmatrix}.$$

The lists $\mathscr{L}_2$ and $\mathscr{L}_5$ have the respective multiplication tables

·	1	i	$-i$	-1
1	1	i	$-i$	-1
i	i	-1	1	$-i$
$-i$	$-i$	1	-1	i
-1	-1	$-i$	i	1

·	1	$-i$	i	-1
1	1	$-i$	i	-1
$-i$	$-i$	-1	1	i
i	i	1	-1	$-i$
-1	-1	i	$-i$	1

and they specify the same standard auxiliary-pattern matrix

$$(18.4) \qquad Stan_{\mathfrak{A}}(\mathcal{C}_4, \mathscr{L}_2) = Stan_{\mathfrak{A}}(\mathcal{C}_4, \mathscr{L}_5) = \begin{bmatrix} X_1 \ X_2 \ X_3 \ X_4 \\ X_2 \ X_4 \ X_1 \ X_3 \\ X_3 \ X_1 \ X_4 \ X_2 \\ X_4 \ X_3 \ X_2 \ X_1 \end{bmatrix}.$$

The lists $\mathscr{L}_3$ and $\mathscr{L}_6$ have the respective multiplication tables

·	1	-1	i	$-i$
1	1	-1	i	$-i$
-1	-1	1	$-i$	i
i	i	$-i$	-1	1
$-i$	$-i$	i	1	-1

·	1	-1	$-i$	i
1	1	-1	$-i$	i
-1	-1	1	i	$-i$
$-i$	$-i$	i	-1	1
i	i	$-i$	1	-1

and they specify the same standard auxiliary-pattern matrix

$$(18.5) \qquad Stan_{\mathfrak{A}}(\mathcal{C}_4, \mathscr{L}_3) = Stan_{\mathfrak{A}}(\mathcal{C}_4, \mathscr{L}_6) = \begin{bmatrix} X_1 \ X_2 \ X_3 \ X_4 \\ X_2 \ X_1 \ X_4 \ X_3 \\ X_3 \ X_4 \ X_2 \ X_1 \\ X_4 \ X_3 \ X_1 \ X_2 \end{bmatrix}.$$

Consequently, there are three standard auxiliary-pattern matrices for $\mathcal{C}_4$.

EXAMPLE 18.4. The $3! = 6$ lists for the elements of $\mathcal{C}_2 \times \mathcal{C}_2$ are

$$\mathscr{L}_1 = (e, \alpha, \beta, \gamma), \quad \mathscr{L}_2 = (e, \beta, \alpha,, \gamma), \quad \mathscr{L}_3 = (e, \gamma, \alpha, \beta),$$

$$\mathscr{L}_4 = (e, \alpha, \gamma, \beta), \quad \mathscr{L}_5 = (e, \beta, \gamma, \alpha), \quad \mathscr{L}_6 = (e, \gamma, \beta, \alpha).$$

These lists specify the corresponding multiplication tables

$\cdot$	e	α	β	γ
e	e	α	β	γ
α	α	e	γ	β
β	β	γ	e	α
γ	γ	β	α	e

$\cdot$	e	β	α	γ
e	e	β	α	γ
β	β	e	γ	α
α	α	γ	e	β
γ	γ	α	β	e

$\cdot$	e	γ	α	β
e	e	γ	α	β
γ	γ	e	β	α
α	α	β	e	γ
β	β	α	γ	e

$\cdot$	e	α	γ	β
e	e	α	γ	β
α	α	e	β	γ
γ	γ	β	e	α
β	β	γ	α	e

$\cdot$	e	β	γ	α
e	e	β	γ	α
β	β	e	α	γ
γ	γ	α	e	β
α	α	γ	β	e

$\cdot$	e	γ	β	α
e	e	γ	β	α
γ	γ	e	α	β
β	β	α	e	γ
α	α	β	γ	e

and each of these six tables yields

$$Stan_{\mathfrak{A}}(\mathcal{C}_2 \times \mathcal{C}_2, \mathscr{L}_k) = \begin{bmatrix} X_1 \ X_2 \ X_3 \ X_4 \\ X_2 \ X_1 \ X_4 \ X_3 \\ X_3 \ X_4 \ X_1 \ X_2 \\ X_4 \ X_3 \ X_2 \ X_1 \end{bmatrix}, \quad \text{for } k = 1, 2, \ldots, 6.$$

Consequently, there is just one standard auxiliary-pattern matrix for $\mathcal{C}_2 \times \mathcal{C}_2$.

18.3. Number of standard auxiliary-pattern matrices for G

For a group G of order n, let $\mathcal{S}_{\mathfrak{G}}$ be the set of standard group-pattern matrices for G as defined by Definition 3.1 on page 27 and let $\mathcal{S}_{\mathfrak{A}}$ denote the set of standard auxiliary-pattern matrices for G. We established in Theorem 3.7 on page 30 that the number $\mathfrak{N}_{\mathfrak{G}}(G)$ of elements in $\mathcal{S}_{\mathfrak{G}}$ is equal to $(n-1)!/m$, where m is the number of automorphisms of G. Slight modifications for the proof of that theorem enable us to establish independently of the concept of a group-pattern matrix that the number $\mathfrak{N}_{\mathfrak{A}}(G)$ of elements of $\mathcal{S}_{\mathfrak{A}}$ is also given by $(n-1)!/m$. Instead, we shall deduce $\mathfrak{N}_{\mathfrak{A}}(G) = \mathfrak{N}_{\mathfrak{G}}(G)$ by defining a bijection of $\mathcal{S}_{\mathfrak{G}}$ onto $\mathcal{S}_{\mathfrak{A}}$.

CONTEXT 18.5. Let $\mathscr{L}_1 = (g_1, g_2, \ldots, g_n)$ and $\mathscr{L}_2 = (h_1, h_2, \ldots, h_n)$ be lists for G that have $g_1 = e$ and $h_1 = e$. Let the elements of $S = \{X_1, X_2, \ldots X_n\}$ be n distinct variables. Let σ_1 and σ_2 be the one-to-one functions from G onto S defined by $\sigma_1(g_k) = X_k$ and $\sigma_2(h_k) = X_k$, for $1 \le k \le n$. Let ψ be the one-to-one function from S onto G defined by $\psi(X_k) = h_k$, for $1 \le k \le n$. Let ϕ be the one-to-one function from G onto G having $\phi(g_k) = h_k$, for $1 \le k \le n$.

PROPOSITION 18.6. *In terms of* Context 18.5, *the condition*

$$(18.6) \qquad \sigma_1\big(g_r^{-1}g_s\big) = \sigma_2\big(h_r^{-1}h_s\big), \quad for\ r,\ s = 1,\ 2,\ \ldots,\ n,$$

is satisfied if and only if ϕ *is an automorphism of* G. *Moreover, the condition*

$$(18.7) \qquad \sigma_1\big(g_r\,g_s\big) = \sigma_2\big(h_r\,h_s\big), \quad for\ r,\ s = 1,\ 2,\ \ldots,\ n,$$

is satisfied if and only if ϕ *is an automorphism of* G.

PROOF. We observes that

$$(18.8) \qquad \phi = \psi \circ \sigma_1, \ \ \psi \circ \sigma_2 = id_G, \ \text{ and } \ \sigma_2 \circ \psi = id_s.$$

(i) Suppose (18.6) is satisfied. Then, (18.6) and (18.8) give

$$\phi\big(g_r^{-1}g_s\big) = (\psi \circ \sigma_1)\big(g_r^{-1}g_s\big) = (\psi \circ \sigma_2)\big(h_r^{-1}h_s\big) = h_r^{-1}h_s, \quad \text{for } 1 \le r,\ s \le n.$$

We use $g_1 = h_1 = e$ with $s = 1$ to obtain $\phi\big(g_r^{-1}\big) = h_r^{-1}$ and

$$\phi\big(g_r^{-1}g_s\big) = \phi\big(g_r^{-1}\big)\,\phi(g_s), \quad \text{for } r,\ s = 1,\ 2,\ \ldots,\ n.$$

This yields $\phi(x\,y) = \phi(x)\,\phi(y)$, for x, y in G. Thus, ϕ is an automorphism of G.

(ii) Suppose that ϕ is an automorphism of G. We use (18.8) with that to obtain

$$(\psi \circ \sigma_1)\big(g_r^{-1}g_s\big) = \phi\big(g_r^{-1}g_s\big) = \big(\phi(g_r)\big)^{-1}\phi(g_s) = h_r^{-1}h_s,$$

and $\sigma_2\big(h_r^{-1}h_s\big) = \big(\sigma_2 \circ (\psi \circ \sigma_1)\big)\big(g_r^{-1}g_s\big) = \sigma_1\big(g_r^{-1}g_s\big)$. for $1 \le r,\ s \le n$. Thus, (18.6) is satisfied.

(iii) Suppose (18.7) is satisfied. Then, for $1 \le r,\ s \le n$, (18.8) and (18.7) yield

$$\phi\big(g_r\,g_s\big) = (\psi \circ \sigma_1)\big(g_r\,g_s\big) = (\psi \circ \sigma_2)\big(h_r\,h_s\big) = h_r\,h_s = \phi(g_r)\,\phi(g_s).$$

This shows that $\phi(x\,y) = \phi(x)\,\phi(y)$, for x, y in G, and ϕ is an automorphism of G.

(iv) Suppose that ϕ is an automorphism of G. With (18.8), that gives

$$(\psi \circ \sigma_1)\big(g_r\,g_s\big) = \phi\big(g_r\,g_s\big) = \phi(g_r)\,\phi(g_s) = h_r\,h_s$$

and $\sigma_2\big(h_r\,h_s\big) = \big(\sigma_2 \circ (\psi \circ \sigma_1)\big)\big(g_r\,g_s\big) = \sigma_1\big(g_r\,g_s\big)$, for $1 \le r,\ s \le n$. Thus, (18.7) is satisfied. This completes the proof. $\qquad \square$

PROPOSITION 18.7. *There is a one-to-one correspondence between* $\mathcal{S}_{\mathfrak{G}}$ *and* $\mathcal{S}_{\mathfrak{A}}$.

PROOF. Proposition 18.6 establishes that (18.7) is satisfied if and only if (18.6) is satisfied. Therefore, a one-to-one function Φ from $\mathcal{S}_{\mathfrak{G}}$ onto $\mathcal{S}_{\mathfrak{A}}$ is well-defined by

$$\Phi\Big(Stan_{\mathfrak{G}}(G, \mathscr{L})\Big) = Stan_{\mathfrak{A}}(G, \mathscr{L}),$$

where $\mathscr{L}$ denotes any list for the n elements of G in which the first component is e. This completes the proof. $\qquad \square$

COROLLARY 18.8. *For a finite group* G *of order* n *that has* m *automorphisms, the number of standard auxiliary-pattern matrices is given by* $\mathfrak{N}_{\mathfrak{A}}(G) = (n-1)!/m$.

PROOF. Proposition 18.7 and Theorem 3.7 give $\mathfrak{N}_{\mathfrak{A}}(G) = \mathfrak{N}_{\mathfrak{G}}(G) = (n-1)!/m$. This completes the proof. $\qquad \square$

18.4. Perspective

To specify the automorphisms for a group G of order n, we have two techniques. Both use a list $\mathscr{L}_0 = (g_1, g_2, \ldots, g_n)$ for the elements in G of the special type where g_1 is the identity element e of G. Then, the automorphisms for G are given by particular permutations of $g_1, g_2, \ldots, g_n$ that leave g_1 fixed. The first step is to use either the technique of Section 3.4 or that of Chapter 19. They are equivalent.

(1) If the technique of Section 3.4 is selected, then the next step is to use $\mathscr{L}_0$ with G to construct an injective group-pattern matrix A for G and $\mathscr{L}_0$. To obtain the automorphisms for G, evaluate a *Mathematica* representation A for A, evaluate the program on page 34 for automorphisms[A_], and evaluate automorphisms[A].

(2) If the technique of Chapter 19 is selected, then the next step is to use $\mathscr{L}_0$ with G to construct an injective auxiliary-pattern matrix B for G and $\mathscr{L}_0$. Then, to obtain the automorphisms for G, evaluate a *Mathematica* representation B for B, evaluate the program on page 185 for automorphisms[A_], and then evaluate automorphisms[B].

OBSERVATION 18.9. Theorem 19.3 on page 184 shows that the program on page 38 can be used without alteration to exhibit all of the standard auxiliary-pattern matrices for G. Simply evaluate any *Mathematica* representation B for a standard auxiliary-pattern matrix B for G, evaluate the program on page 38 for standard[A_], and then evaluate standard[B].

EXAMPLE 18.10. The 4×4 leftward-circulant matrix B with (X_1, X_2, X_3, X_4) as its first row is a standard auxiliary-pattern matrix for the cyclic group C_4. To find all of the standard auxiliary-pattern matrices for C_4, evaluate a *Mathematica* representation B for B, evaluate the program on page 38 for standard[A_], and then evaluate standard[B]. The output provides the three standard auxiliary-pattern matrices of (18.3), (18.4), and (18.5) for C_4.

EXERCISE

1. The 5×5 leftward circulant matrix

$$B = \begin{bmatrix} X_1 & X_2 & X_3 & X_4 & X_5 \\ X_2 & X_3 & X_4 & X_5 & X_1 \\ X_3 & X_4 & X_5 & X_1 & X_2 \\ X_4 & X_5 & X_1 & X_2 & X_3 \\ X_5 & X_1 & X_2 & X_3 & X_4 \end{bmatrix}$$

is a standard auxiliary-pattern matrix for the cyclic group C_5. Find all of the standard auxiliary-pattern matrices for C_5

CHAPTER 19

Deduction of Automorphisms
via Standard Auxiliary-Pattern Matrices

Let $\mathscr{L}_0 = (g_1, g_2, \ldots, g_n)$ with $g_1 = e$ be a list of the elements in a group G having order n and set $\mathcal{Z}_0 = Stan_{\mathfrak{A}}(G, \mathscr{L}_0)$. Then, we have

(19.1) $$[\mathcal{Z}_0]_{r,s} = f_{\mathscr{L}_0}(g_r g_s), \quad \text{for } r, s = 1, 2, \ldots, n,$$

where $f_{\mathscr{L}_0}$ is the function from G to the ring $\mathfrak{R} = \mathbb{Q}[X_1, X_2, \ldots, X_n]$ of polynomials in the variables $X_1, X_2, \ldots, X_n$ over $\mathbb{Q}$ such that

(19.2) $$f_{\mathscr{L}_0}(g_k) = X_k, \quad \text{for } k = 1, 2, \ldots, n.$$

For each $n \times n$ permutation matrix P having $[P]_{1,1} = 1$, there is a corresponding permutation π of $\{1, 2, \ldots, n\}$, with $\pi(1) = 1$, given by

(19.3) $$[\pi(1), \pi(2), \ldots, \pi(n)] = [1, 2, \ldots, n]P;$$

there is a one-to-one function ϕ_π from G onto G, with $\phi_\pi(e) = e$, defined by

$$\phi_\pi(g_k) = g_{\pi(k)}, \quad \text{for } k = 1, 2, \ldots, n,$$

there is a corresponding list

$$\mathscr{L}_\pi = (g_{\pi(1)}, g_{\pi(2)}, \ldots, g_{\pi(n)}) = (\phi_\pi(g_1), \phi_\pi(g_2), \ldots, \phi_\pi(g_n)),$$

and there is a corresponding standard auxiliary-pattern matrix

$$S_\pi = Stan_{\mathfrak{A}}(G, \mathscr{L}_\pi),$$

where

(19.4) $$[S_\pi]_{r,s} = f_{\mathscr{L}_\pi}(g_{\pi(r)} g_{\pi(s)}) = f_{\mathscr{L}_\pi}(\phi_\pi(g_r) \phi_\pi(g_s)), \quad \text{for } 1 \le r, s \le n,$$
and

(19.5) $$f_{\mathscr{L}_\pi}(\phi_\pi(g_k)) = X_k, \quad \text{for } k = 1, 2, \ldots, n.$$

Moreover, as P ranges over the $n \times n$ permutation matrices P having $[P]_{1,1} = 1$, we see that π, ϕ_π, $\mathscr{L}_\pi$, and S_π range over all of the objects of their type.

THEOREM 19.1. *In order for the function ϕ_π to be an automorphism of G it is necessary and sufficient that $S_\pi = \mathcal{Z}_0$.*

PROOF. We observe that (19.5) and (19.2) yield

(19.6) $$f_{\mathscr{L}_\pi} \circ \phi_\pi = f_{\mathscr{L}_0}.$$

(i) Suppose $S_\pi = \mathcal{Z}_0$. We use (19.6), (19.1), $\mathcal{Z}_0 = S_\pi$, and (19.4) to obtain

$$f_{\mathscr{L}_\pi}(\phi_\pi(g_r g_s)) = f_{\mathscr{L}_0}(g_r g_s) = [\mathcal{Z}_0]_{r,s} = [S_\pi]_{r,s} = f_{\mathscr{L}_\pi}(\phi_\pi(g_r) \phi_\pi(g_s)),$$

for $1 \le r, s \le n$. Since $f_{\mathscr{L}_\pi}$ is one-to-one, this yields $\phi_\pi(g_r g_s) = \phi_\pi(g_r) \phi_\pi(g_s)$, for $r, s = 1, 2, \ldots, n$. Thus, ϕ_π is an automorphism of G.

(ii) Suppose that ϕ_π is an automorphism of G. Then, we find that (19.4), the automorphism property of ϕ_π, (19.6), and (19.1) establish, for $1 \leq r,\, s \leq n$, that

$$\left[\mathcal{S}_\pi\right]_{r,s} = f_{\mathscr{L}_\pi}\!\left(\phi_\pi(g_r)\,\phi_\pi(g_s)\right) = f_{\mathscr{L}_\pi}\!\left(\phi_\pi(g_r\, g_s)\right) = f_{\mathscr{L}_0}\!\left(g_r\, g_s\right) = \left[\mathcal{Z}_0\right]_{r,s}.$$

Thus, we have $\mathcal{S}_\pi = \mathcal{Z}_0$. This completes the proof. $\qquad\square$

PROPOSITION 19.2. *If an $n \times n$ permutation matrix P and a permutation π of $\{1,\, 2,\, \ldots,\, n\}$ are related by $\left[\pi(1),\, \pi(2),\, \ldots,\, \pi(n)\right] = \left[1,\, 2,\, \ldots,\, n\right]P$, then they are also related by*

(19.7) $$\left[P\right]_{r,s} = \left\{\begin{array}{ll} 1, & \text{if } r = \pi(s), \\ 0, & \text{if } r \neq \pi(s), \end{array}\right\} \cdot \quad \text{for } r,\, s = 1,\, 2,\, \ldots,\, n.$$

PROOF. Let Q be the $n \times n$ permutation matrix defined by

(19.8) $$\left[Q\right]_{r,s} = \left\{\begin{array}{ll} 1, & \text{if } r = \pi(s), \\ 0, & \text{if } r \neq \pi(s), \end{array}\right\}, \quad \text{for } r,\, s = 1,\, 2\, \ldots,\, n.$$

For $k = 1,\, 2,\, \ldots,\, n$, we use (19.8) to obtain

$$\left[\left[1,\, 2,\, \ldots,\, n\right]Q\right]_{1,k} = \sum_{\mu=1}^{n} \left[\left[1,\, 2,\, \ldots,\, n\right]\right]_{1,\mu}\left[Q\right]_{\mu,k} = \left[\left[1,\, 2,\, \ldots,\, n\right]\right]_{1,\pi(k)} = \pi(k).$$

With (19.3), this yields

(19.9) $$\left[1,\, 2,\, \ldots,\, n\right]Q = \left[\pi(1),\, \pi(2),\, \ldots,\, \pi(n)\right] = \left[1,\, 2,\, \ldots,\, n\right]P.$$

We write

(19.10) $$Q = \left[\,q_1 \,|\, q_2 \,|\, \cdots \,|\, q_n\,\right] \quad \text{and} \quad P = \left[\,p_1 \,|\, p_2 \,|\, \cdots \,|\, p_n\,\right],$$

where, for $k = 1,\, 2,\, \ldots,\, n$, q_k and p_k are $n \times 1$ matrices having a single component equal to 1 and $n - 1$ components equal to 0. Since (19.9) and (19.10) require

$$\left[1,\, 2,\, \ldots,\, n\right]q_k = \left[1,\, 2,\, \ldots,\, n\right]p_k, \quad \text{for } k = 1,\, 2,\, \ldots,\, n,$$

we must have $q_k = p_k$, for $1 \leq k \leq n$, and $Q = P$. This completes the proof. $\qquad\square$

THEOREM 19.3. *For the context of (19.1)–(19.5), the standard auxiliary-pattern matrix $\mathcal{S}_\pi$ is obtained from the matrix $P^T \mathcal{Z}_0 P$ and its first row $(b_1,\, b_2,\, \ldots,\, b_n)$ by replacing each b_k in $P^T \mathcal{Z}_0 P$ with X_k, for $k = 1,\, 2,\, \ldots,\, n$.*

PROOF. We use (19.7), and (19.1) to obtain

(19.11) $$\left[P^T \mathcal{Z}_0 P\right]_{r,s} = \sum_{\mu=1}^{n}\sum_{\nu=1}^{n}\left[P^T\right]_{r,\mu}\left[\mathcal{Z}_0\right]_{\mu,\nu}\left[P\right]_{\nu,s} = \sum_{\mu=1}^{n}\sum_{\nu=1}^{n}\left[P\right]_{\mu,r}\left[\mathcal{Z}_0\right]_{\mu,\nu}\left[P\right]_{\nu,s}$$

$$= \left[\mathcal{Z}_0\right]_{\pi(r),\pi(s)} = f_{\mathscr{L}_0}\!\left(g_{\pi(r)}\, g_{\pi(s)}\right), \quad \text{for } r,\, s = 1,\, 2,\, \ldots,\, n.$$

With $\pi(1) = 1$, $g_1 = e$, and (19.2), we see that (19.11) yields

(19.12) $$\text{(the first row of } P^T \mathcal{Z}_0 P) = \left(X_{\pi(1)},\, X_{\pi(2)},\, \ldots,\, X_{\pi(n)}\right).$$

We compare

$$\left[\mathcal{S}_\pi\right]_{r,s} = f_{\mathscr{L}_\pi}\!\left(g_{\pi(r)}\, g_{\pi(s)}\right), \quad \text{for } 1 \leq r,\, s \leq n,$$

and

$$\text{(the first row of } \mathcal{S}_\pi) = \left(X_1,\, X_2,\, \ldots,\, X_n\right)$$

from (19.4) and (19.5) with (19.11) and (19.12) to see that $\mathcal{S}_\pi$ is obtained from

$P^T \mathcal{Z}_0 P$ by replacing each $X_{\pi(k)}$ in $P^T \mathcal{Z}_0 P$ with X_k, for $k = 1, 2, \ldots, n$. Thus, when computer algebra picks out $b_1, b_2, \ldots, b_n$ as the n distinct respective components in the first row of $P^T \mathcal{Z}_0 P$, the standard auxiliary-pattern matrix S_π is obtained from $P^T \mathcal{Z}_0 P$ by replacing each b_k in $P^T \mathcal{Z}_0 P$ with X_k, for $k = 1, 2, \ldots, n$. This completes the proof.   □

19.1. Computer program for automorphisms

As P ranges through the $(n-1)!$ permutation matrices of size $n \times n$ having $[P]_{1,1} = 1$, a computer can check whether the matrix $\mathcal{Z}_0$ is equal to the matrix obtained from $P^T \mathcal{Z}_0 P$ by replacing each b_k in $P^T \mathcal{Z}_0 P$ with X_k, where

$$b_k = \left[P^T \mathcal{Z}_0 P \right]_{1,k}, \quad \text{for } 1 \leq k \leq n.$$

In each situation where $\mathcal{Z}_0$ is equal to the standard auxiliary-pattern matrix S_π specified by P, the computer can evaluate $[1, 2, \ldots, n] P$ and print that out in order to have $[\pi(1), \pi(2), \ldots, \pi(n)]$ as a representation for the automorphism of G that P specifies as $g_k \mapsto g_{\pi(k)}$, for $k = 1, 2, \ldots, n$.

If A is an injective auxiliary-pattern matrix for G and a list $\mathcal{L}_0$ for G where it is essential that $g_1 = e$, then a suitable Z_0 can be obtained from A and the first row $(a_1, a_2, \ldots, a_n)$ of A by replacing each a_k in A with X_k, for $1 \leq k \leq n$.

For a version of *Mathematica* such as [48], we enter

```
automorphisms[A_] :=
    Module[ {n, id, list, P, counter, a, rule1,
             Z0, L0, B, rule2, b},
    n = Dimensions[A][[1]];
    id = IdentityMatrix[n];
    list = Permutations[id];
    Do[P[i] = list[[i]], {i, 1, (n - 1)!}];
    Clear[counter];  counter =  0;
    Do[a[k] = A[[1, k]], {k, 1, n}];
    rule1 = Table[a[k] -> X[k], {k, 1, n}];
    Z0 = A /. rule1;
    L0 = Table[k, {k, 1, n}];
    Do[( B[j] = Transpose[P[j]].Z0.P[j];
       Do[b[k] = B[j][[1, k]], {k, 1, n}];
       rule2 = Table[b[k] -> X[k], {k, 1, n}];
       B[j] = B[j] /. rule2;
       If[TrueQ[B[j] == Z0], (lastCount = counter + 1;
          counter = counter + 1;
          Print["L[", counter, "] = " , L0.P[j]] ), {} ];
       Clear[B]      ), {j, 1, (n - 1)!}];
    Print["Number m of automorphisms is ", lastCount];
    Print["Number of standard matrices is ",
       (n - 1)!/lastCount ] ]
```

as input in a *Mathematica* notebook and evaluate it. Thus, if A is an injective auxiliary-pattern matrix for G and $\mathcal{L}_0 = (g_1, g_2, \ldots, g_n)$ with $g_1 = e$, then the automorphisms of G are obtained with respect to $\mathcal{L}_0$ by evaluating a representation A of A and then evaluating automorphisms[A].

OBSERVATION 19.4. The preceding program is the same as the one on page 34. In fact, the output for `automorphisms[A1]` and the output for `automorphisms[A2]` are the same when A1 is a *Mathematica* representations for an injective group-pattern matrix A_1 for G with $\mathscr{L}_0$ and A2 is a *Mathematica* representation for an injective auxiliary-pattern matrix A_2 for G with $\mathscr{L}_0$. (For that, $\mathscr{L}_0$ has $g_1 = e$.)

EXAMPLE 19.5. In terms of $\mathcal{C}_2 = \{e, \alpha\}$ with $\alpha^2 = e$. we introduce

$$g_1 = (e, e, e), \qquad g_2 = (e, e, \alpha), \qquad g_3 = (e, \alpha, e), \qquad g_4 = (e, \alpha, \alpha),$$
$$g_5 = (\alpha, e, e), \qquad g_6 = (\alpha, e, \alpha), \qquad g_7 = (\alpha, \alpha, e), \qquad g_8 = (\alpha, \alpha, \alpha)$$

as the elements of the group $G = \mathcal{C}_2 \times \mathcal{C}_2 \times \mathcal{C}_2$ with the multiplication table

(19.13)

$\cdot$	g_1	g_2	g_3	g_4	g_5	g_6	g_7	g_8
g_1	g_1	g_2	g_3	g_4	g_5	g_6	g_7	g_8
g_2	g_2	g_1	g_4	g_3	g_6	g_5	g_8	g_7
g_3	g_3	g_4	g_1	g_2	g_7	g_8	g_5	g_6
g_4	g_4	g_3	g_2	g_1	g_8	g_7	g_6	g_5
g_5	g_5	g_6	g_7	g_8	g_1	g_2	g_3	g_4
g_6	g_6	g_5	g_8	g_7	g_2	g_1	g_4	g_3
g_7	g_7	g_8	g_5	g_6	g_3	g_4	g_1	g_2
g_8	g_8	g_7	g_6	g_5	g_4	g_3	g_2	g_1

that is isomorphic to the multiplication table of (1.12) on page 7. In terms of the list $\mathscr{L}_0 = (g_1, g_2, g_3, g_4, g_5, g_6, g_7, g_8)$ for G, we use (19.13) to see that the matrix

(19.14)
$$A = \left[\begin{array}{cccc|cccc} a_1 & a_2 & a_3 & a_4 & a_5 & a_6 & a_7 & a_8 \\ a_2 & a_1 & a_4 & a_3 & a_6 & a_5 & a_8 & a_7 \\ a_3 & a_4 & a_1 & a_2 & a_7 & a_8 & a_5 & a_6 \\ a_4 & a_3 & a_2 & a_1 & a_8 & a_7 & a_6 & a_5 \\ \hline a_5 & a_6 & a_7 & a_8 & a_1 & a_2 & a_3 & a_4 \\ a_6 & a_5 & a_8 & a_7 & a_2 & a_1 & a_4 & a_3 \\ a_7 & a_8 & a_5 & a_6 & a_3 & a_4 & a_1 & a_2 \\ a_8 & a_7 & a_6 & a_5 & a_4 & a_3 & a_2 & a_1 \end{array}\right]$$

is an injective auxiliary-pattern matrix and an injective group-pattern matrix for G and $\mathscr{L}_0$. We evaluate a *Mathematica* representation A of A and `automorphisms[A_]` as well as `automorphisms[A]` to obtain 168 automorphisms specified with respect to $\mathscr{L}_0$ as output. The first six are

```
L[1] = [1,2,3,4,5,6,7,8],
L[2] = [1,2,3,4,6,5,8,7],
L[3] = [1,2,3,4,7,8,5,6],
L[4] = [1,2,3,4,8,7,6,5],
L[5] = [1,2,4,3,5,6,8,7],
L[6] = [1,2,4,3,6,5,7,8]
```

and all one hundred and sixty-eight are listed in the notebook P-186.nb that can be downloaded according to the directions of page 235. The notebook P-186.nb shows that this computation was performed in less than two seconds.

19.2. Auxiliary-pattern viewpoint for the quaternion group Q_8

For Q_8, we use the list $\mathscr{L}_0 = (g_1, g_2, g_3, g_4, g_5, g_6, g_7, g_8)$ of elements where $g_1 = 1$, $g_2 = i$, $g_3 = -1$, $g_4 = -i$, $g_5 = j$, $g_6 = k$, $g_7 = -j$, $g_8 = -k$.. This is consistent with Example 3.9 on page 35. Now, the equivalent multiplication tables

·	1	i	-1	$-i$	j	k	$-j$	$-k$
1	1	i	-1	$-i$	j	k	$-j$	$-k$
i	i	-1	$-i$	1	k	$-j$	$-k$	j
-1	-1	$-i$	1	i	$-j$	$-k$	j	k
$-i$	$-i$	1	i	-1	$-k$	j	k	$-j$
j	j	$-k$	$-j$	k	-1	i	1	$-i$
k	k	j	$-k$	$-j$	$-i$	-1	i	1
$-j$	$-j$	k	j	$-k$	1	$-i$	-1	i
$-k$	$-k$	$-j$	k	j	i	1	$-i$	-1

·	g_1	g_2	g_3	g_4	g_5	g_6	g_7	g_8
g_1	g_1	g_2	g_3	g_4	g_5	g_6	g_7	g_8
g_2	g_2	g_3	g_4	g_1	g_6	g_7	g_8	g_5
g_3	g_3	g_4	g_1	g_2	g_7	g_8	g_5	g_6
g_4	g_4	g_1	g_2	g_3	g_8	g_5	g_6	g_7
g_5	g_5	g_8	g_7	g_6	g_3	g_2	g_1	g_4
g_6	g_6	g_5	g_8	g_7	g_4	g_3	g_2	g_1
g_7	g_7	g_6	g_5	g_8	g_1	g_4	g_3	g_2
g_8	g_8	g_7	g_6	g_5	g_2	g_1	g_4	g_3

for Q_8 and $\mathscr{L}_0$ specify the standard auxiliary-pattern matrix

$$(19.15) \qquad A = \begin{bmatrix} X_1 & X_2 & X_3 & X_4 & X_5 & X_6 & X_7 & X_8 \\ X_2 & X_3 & X_4 & X_1 & X_6 & X_7 & X_8 & X_5 \\ X_3 & X_4 & X_1 & X_2 & X_7 & X_8 & X_5 & X_6 \\ X_4 & X_1 & X_2 & X_3 & X_8 & X_5 & X_6 & X_7 \\ X_5 & X_8 & X_7 & X_6 & X_3 & X_2 & X_1 & X_4 \\ X_6 & X_5 & X_8 & X_7 & X_4 & X_3 & X_2 & X_1 \\ X_7 & X_6 & X_5 & X_8 & X_1 & X_4 & X_3 & X_2 \\ X_8 & X_7 & X_6 & X_5 & X_2 & X_1 & X_4 & X_3 \end{bmatrix}$$

for Q_8 and $\mathscr{L}_0$. When the program of Section 17.5 on page 173 is applied to A as a guard against misprints, the output indicates that A is an injective auxiliary-pattern matrix with respect to a group $\mathfrak{P}$ of $n \times n$ permutation matrices that have

$$(19.16)$$

·	P_1	P_2	P_3	P_4	P_5	P_6	P_7	P_8
P_1	P_1	P_2	P_3	P_4	P_5	P_6	P_7	P_8
P_2	P_2	P_3	P_4	P_1	P_6	P_7	P_8	P_5
P_3	P_3	P_4	P_1	P_2	P_7	P_8	P_5	P_6
P_4	P_4	P_1	P_2	P_3	P_8	P_5	P_6	P_7
P_5	P_5	P_8	P_7	P_6	P_3	P_2	P_1	P_4
P_6	P_6	P_5	P_8	P_7	P_4	P_3	P_2	P_1
P_7	P_7	P_6	P_5	P_8	P_1	P_4	P_3	P_2
P_8	P_8	P_7	P_6	P_5	P_2	P_1	P_4	P_3

as their multiplication table. Independent of previous knowledge, (19.16) shows that the two multiplication tables above provide a group structure for Q_8.

To specify the automorphisms of Q_8 based on the technique of Section 19.1 on page 185, we first obtain a *Mathematica* representation A of A and then successively evaluate A as well as the program on page 185 and automorphisms[A]. The output is the same as that in Example 3.9 on page 35. Of course, to represent

$$(1, i, -1, -i, j, k, -j, -k) = (g_1, g_2, g_3, g_4, g_5, g_6, g_7, g_8)$$
$$\mapsto (g_{\pi(1)}, g_{\pi(2)}, g_{\pi(3)}, g_{\pi(4)}, g_{\pi(5)}, g_{\pi(6)}, g_{\pi(7)}, g_{\pi(8)}),$$

it suffices to use $(\pi(1), \pi(2), \pi(3), \pi(4), \pi(5), \pi(6), \pi(7), \pi(8))$.

CHAPTER 20

Unequal Injective Matrices for Dissimilar Groups

We show that: if a matrix is an injective group-pattern matrix or an injective auxiliary-pattern matrix for groups G and H, then G and H are isomorphic.

20.1. Injective group-pattern matrix for each of two groups

THEOREM 20.1. *Suppose that an $n \times n$ matrix A is an injective group-pattern matrix for both a group G and a group H. Then, G and H are isomorphic.*

PROOF. Let $a_1, a_2, \ldots, a_n$ denote the n distinct components in the first row of A and set $S = \{a_1, a_2, \ldots, a_n\}$. We use Proposition 2.7 on page 13 to obtain a list $\mathscr{L}_1 = (g_1, g_2, \ldots, g_n)$ for G and a list $\mathscr{L}_2 = (h_1, h_2, \ldots, h_n)$ for H such that: g_1 is the identity element e_G of G, h_1 is the identity element e_H of H, and there are one-to-one functions σ_1 from G onto S and σ_2 from H onto S for which

(20.1) $\qquad [A]_{r,s} = \sigma_1\big(g_r^{-1}g_s\big) = \sigma_2\big(h_r^{-1}h_s\big), \quad \text{for } r,\, s = 1,\, 2,\, \ldots,\, n,$

and $a_k = [A]_{1,k} = \sigma_1(g_k) = \sigma_2(h_k)$, for $k = 1, 2, \ldots, n$.

Let ψ be the function from S onto H having $\psi(a_k) = h_k$, for $k = 1, 2, \ldots, n$. It yields $\psi \circ \sigma_2 = id_H$. Let ϕ be the function $\phi = \psi \circ \sigma_1$ from G onto H. We have

$$\phi(g_k) = (\psi \circ \sigma_1)(g_k) = \psi(a_k) = h_k, \quad \text{for } k = 1, 2, \ldots, n.$$

Thus, ϕ is a one-to-one function from G onto H and, with (20.1), we obtain

(20.2) $\quad \phi\big(g_r^{-1}g_s\big) = \psi\Big(\sigma_1\big(g_r^{-1}g_s\big)\Big) = \psi\Big(\sigma_2\big(h_r^{-1}h_s\big)\Big) = h_r^{-1}h_s = \big(\phi(g_r)\big)^{-1}\phi(g_s),$
$$\text{for } r,\, s = 1,\, 2,\, \ldots,\, n.$$

By setting $s = 1$ in (20.2), we use $g_1 = e_G$ and $\phi(g_1) = h_1 = e_H$ to establish that $\phi\big(g_r^{-1}\big) = \big(\phi(g_r)\big)^{-1}$, for $r = 1, 2, \ldots, n$. Thus, (20.2) gives

$$\phi\big(g_r^{-1}g_s\big) = \phi\big(g_r^{-1}\big)\,\phi(g_s), \quad \text{for } 1 \le r,\, s \le n.$$

Since this yields $\phi(xy) = \phi(x)\,\phi(y)$, for each x, y in G, ϕ is an isomorphism of G onto H. This completes the proof. $\qquad \square$

20.2. Injective group-pattern matrices that are auxiliary ones

THEOREM 20.2. *For an integer $n \ge 2$, suppose that an $n \times n$ matrix A is an injective group-pattern matrix for a group G and suppose that A is an injective auxiliary-pattern matrix for a group H. Then, there is a positive integer k such that $n = 2^k$ and the groups G, H are isomorphic to the direct product*

$$\overbrace{C_2 \times C_2 \times \cdots \times C_2}^{k-factors}$$

(20.3)

of k cyclic groups of order 2.

PROOF. Let S be the set of the n distinct components for A. Definition 2.2 on page 12 shows that there is a list $\mathscr{L}_1 = (g_1, g_2, \ldots, g_n)$ for G and there is a one-to-one function σ from G onto S such that

$$(20.4) \qquad [A]_{r,s} = \sigma(g_r^{-1} g_s), \quad \text{for } r, s = 1, 2, \ldots, n.$$

Definition 18.1 on page 177 shows there is a list $\mathscr{L}_2 = (h_1, h_2, \ldots, h_n)$ for H and there is a one-to-one function τ from H onto S such that

$$(20.5) \qquad [A]_{r,s} = \tau(h_r h_s), \quad \text{for } r, s = 1, 2, \ldots, n.$$

We use (20.5) and (20.4) to deduce

$$(20.6) \qquad \tau(h_r h_s) = \sigma(g_r^{-1} g_s), \quad \text{for } r, s = 1, 2, \ldots, n.$$

By setting $r = s = k$ in (20.6), we obtain

$$\tau\big((h_k)^2\big) = \sigma(g_k^{-1} g_k) = \sigma(e_G), \quad \text{for } k = 1, 2, \ldots, n,$$

and therefore

$$(20.7) \qquad \tau\big((h_1)^2\big) = \tau\big((h_2)^2\big) = \cdots = \tau\big((h_n)^2\big).$$

Since τ is one-to-one and e_H is one of $h_1, h_2, \ldots, h_n$, we see that (20.7) yields

$$(20.8) \qquad h_1^2 = h_2^2 = \cdots = h_n^2 = e_H.$$

For x, y in H, (20.8) yields

$$yx = yx\,e_H = yx(xy)^2 = yxxyxy = yyxy = xy.$$

Consequently, H is abelian. The basis theorem for finite abelian groups shows that there are cyclic subgroups $C_{n_1}, C_{n_2}, \ldots, C_{n_k}$ of H whose respective orders $n_1, n_2, \ldots, n_k$ satisfy $n_1 \geq 2, n_2 \geq 2, \ldots, n_k \geq 2$ such that H is expressible as the internal direct product

$$H = C_{n_1} \times C_{n_2} \times \cdots \times C_{n_k}.$$

Due to $x^2 = e_H$, for each x in H, we note that $n_1 = n_2 = \cdots = n_k = 2$ and $n = 2^k$. Thus, H has the structure of (20.3). We rewrite (20.5) as

$$(20.9) \qquad [A]_{r,s} = \tau(h_r h_s) = \tau(h_r^{-1} h_s), \quad \text{for } r, s = 1, 2, \ldots, n,$$

to see that A is an injective group-pattern matrix for both G and H. Therefore, Theorem 20.1 shows that G is isomorphic to H. This completes the proof. $\square$

OBSERVATION 20.3. We note that each of the matrices

$$\begin{bmatrix} X_1 \end{bmatrix}, \quad \begin{bmatrix} X_1 & X_2 \\ X_2 & X_1 \end{bmatrix}, \quad \left[\begin{array}{cc|cc} X_1 & X_2 & X_3 & X_4 \\ X_2 & X_1 & X_4 & X_3 \\ \hline X_3 & X_4 & X_1 & X_2 \\ X_4 & X_3 & X_2 & X_1 \end{array}\right], \quad \left[\begin{array}{cccc|cccc} X_1 & X_2 & X_3 & X_4 & X_5 & X_6 & X_7 & X_8 \\ X_2 & X_1 & X_4 & X_3 & X_6 & X_5 & X_8 & X_7 \\ X_3 & X_4 & X_1 & X_2 & X_7 & X_8 & X_5 & X_6 \\ X_4 & X_3 & X_2 & X_1 & X_8 & X_7 & X_6 & X_5 \\ \hline X_5 & X_6 & X_7 & X_8 & X_1 & X_2 & X_3 & X_4 \\ X_6 & X_5 & X_8 & X_7 & X_2 & X_1 & X_4 & X_3 \\ X_7 & X_8 & X_5 & X_6 & X_3 & X_4 & X_1 & X_2 \\ X_8 & X_7 & X_6 & X_5 & X_4 & X_3 & X_2 & X_1 \end{array}\right], \quad \cdots$$

is an injective group-pattern matrix and an injective auxiliary-pattern matrix.

20.3. Injective auxiliary-pattern matrix for each of two groups

THEOREM 20.4. *Suppose that an $n \times n$ matrix B is an injective auxiliary-pattern matrix for both a group G and a group H. Then, G and H are isomorphic.*

PROOF. Let S denote the set consisting of the n distinct components of B. Then, there is a list $\mathscr{L}_1 = (g_1, g_2, \ldots, g_n)$ for G and there is a one-to-one function τ_1 from G onto S such that

$$(20.10) \qquad [B]_{r,s} = \tau_1(g_r g_s), \quad \text{for } r, s = 1, 2, \ldots, n.$$

Moreover, let d denote the unique integer subject to $1 \leq d \leq n$ such that g_d is the identity element e_G for G.

There is a list $\mathscr{L}_2 = (h_1, h_2, \ldots, h_n)$ for H and there is a one-to-one function τ_2 from H onto S such that

$$(20.11) \qquad [B]_{r,s} = \tau_2(h_r h_s), \quad \text{for } r, s = 1, 2, \ldots, n.$$

Let ψ be the inverse function of τ_2. It is defined from S onto H and yields

$$(20.12) \qquad \psi \circ \tau_2 = id_H.$$

Thus, $\psi \circ \tau_1$ is a one-to-one function from G onto H. We use (20.10), (20.11), and (20.12) to deduce, for $r, s = 1, 2, \ldots, n$, that

$$(20.13) \qquad (\psi \circ \tau_1)(g_r g_s) = \psi(\tau_1(g_r g_s)) = \psi(\tau_2(h_r h_s)) = h_r h_s.$$

Let υ be the one-to-one function from H onto H defined by

$$\upsilon(h) = h_d^{-1} h h_d^{-1}, \quad \text{for each } h \text{ in } H,$$

and let ϕ be the one-to-one function from G onto H defined by $\phi = \upsilon \circ (\psi \circ \tau_1)$. We use this with (20.13) to verify, for $r, s = 1, 2, \ldots, n$, that

$$(20.14) \qquad \phi(g_r g_s) = \upsilon\big((\psi \circ \tau_1)(g_r g_s)\big) = \upsilon(h_r h_s) = \big(h_d^{-1} h_r\big)\big(h_s h_d^{-1}\big).$$

We set $s = d$ in (20.14) and employ $g_d = e_G$ to see that

$$(20.15) \qquad \phi(g_r) = h_d^{-1} h_r, \quad \text{for } r = 1, 2, \ldots, n.$$

We set $r = d$ in (20.14) and apply $g_d = e_G$ to show that

$$(20.16) \qquad \phi(g_s) = h_s h_d^{-1}, \quad \text{for } s = 1, 2, \ldots, n.$$

By combining (20.14), (20.15), and (20.16), we obtain

$$\phi(g_r g_s) = \phi(g_r) \phi(g_s), \quad \text{for } r, s = 1, 2, \ldots, n,$$

Thus, ϕ is an isomorphism of G onto H. This completes the proof. $\square$

CHAPTER 21

Particular Advancements of Richard Dedekind

A series of letters written by Richard Dedekind and sent to Georg Frobenius during the years 1882–1896 are included in [20, pages 414–442]. Key portions of two principal letters are presented here. Currently, the web page

https://gdz.sub.uni-goettingen.de/id/PPN235693928?tify

enables the entire publication [20] to be downloaded.

21.0.1. Portion of letter in [20, pages 420–421] dated March 25, 1896.

"25. März 1896.

... Da ich einmal von Gruppen spreche, so möchte ich noch eine andere Betrachtung erwähnen, auf die ich im Februar 1886 gekommen bin. Zu jeder Gruppe n^{ten} Grades G bilde ich eine Form n^{ten} Grades H mit n Variabelen, die ich die Determinante von G nenne: sind 1, 2, ..., n die in irgend einer Ordnung aufgeschriebenen Elemente von G, so lasse ich jedem Elemente r der Gruppe G eine Variabele x_r entprechen, und bilde die Determinante

$$H = \begin{vmatrix} x_{11'} & x_{21'} & \cdots & x_{n1'} \\ x_{12'} & x_{22'} & \cdots & x_{n2'} \\ \vdots & \vdots & \ddots & \vdots \\ x_{1n'} & x_{2n'} & \cdots & x_{nn'} \end{vmatrix},$$

wo r' das zu r reciproke Element von G bedeutet. Ist G eine Abel'sche Gruppe, und sind ψ', ψ'', ..., $\psi^{(n)}$ die ihr entsprechenden Charaktere (Einheitswurzeln), so ist die Determinante H eine zerlegbare Form, nämlich das Product der n linearen Factoren

$$\sum^r \psi^{(s)}(r)\, x_r = \psi^{(s)}(1)\, x_1 + \cdots + \psi^{(s)}(n)\, x_n,$$

die den n Werthen von s entsprechen (ein Satz welcher in dieser Allgemeinheit, wie ich glaube, noch nicht ausgesprochen ist). Wenn aber G keine Abel'sche Gruppe ist, so besitzt ihre Determinante H, soweit ich es untersucht habe, ausser linearen Factoren (wie z. B. immer $z_1 + x_2 + \cdots + x_n$) auch Factoren höheren Grades, die im gewöhnlichen Sinne unzerlegbar sind; ..."

(Richard Dedekind)

21.0.2. Portion of letter in [20, pages 423–425] dated April 6, 1896.

"6. April 1896.

... Für den Fall, dass Sie sich noch näher mit den Gruppen-Determinanten beschäftigen wollen, erlaube ich mir hiermit, Ihnen wenigstens zwei von den Beispielen zu senden, die ich im

Februar 1886 ausgerechnet habe; doch übergehe ich die über-
complexe Zerlegung der nicht linearen Factoren.

Beispiel 1.

Gruppe V_3 der sechs Versetzungen von drei Buchstaben a, b, c.
Bezeichnung und Composition der Substitutionen:

	a	b	c		1^{-1}	2^{-1}	3^{-1}	4^{-1}	5^{-1}	6^{-1}
1	a	b	c	1	1	3	2	4	5	6
2	b	c	a	2	2	1	3	5	6	4
3	c	a	b	3	3	2	1	6	4	5
4	a	c	b	4	4	5	6	1	3	2
5	c	b	a	5	5	6	4	2	1	3
6	b	a	c	6	6	4	5	3	2	1

Setz man
$$1 + \rho + \rho^2 = 0$$

und
$$u = x_1 + x_2 + x_3, \qquad v = x_4 + x_5 + x_6,$$
$$u_1 = x_1 + \rho x_2 + \rho^2 x_3, \qquad v_1 = x_4 + \rho x_5 + \rho^2 x_6,$$
$$u_2 = x_1 + \rho^2 x_2 + \rho x_3, \qquad v_2 = x_4 + \rho^2 x_5 + \rho x_6,$$

so wird die Gruppen-Determinante

$$
\begin{vmatrix}
x_1 & x_3 & x_2 & x_4 & x_5 & x_6 \\
x_2 & x_1 & x_3 & x_5 & x_6 & x_4 \\
x_3 & x_2 & x_1 & x_6 & x_4 & x_5 \\
x_4 & x_5 & x_6 & x_1 & x_3 & x_2 \\
x_5 & x_6 & x_4 & x_2 & x_1 & x_3 \\
x_6 & x_4 & x_5 & x_3 & x_2 & x_1
\end{vmatrix}
= (u+v)(u-v)(u_1 u_2 - v_1 v_2)^2
$$

am kürzesten wohl durch Multiplication mit der Determinante

$$
\begin{vmatrix}
1 & 1 & 1 & 1 & 1 & 1 \\
1 & \rho & \rho^2 & 1 & \rho & \rho^2 \\
1 & \rho^2 & \rho & 1 & \rho^2 & \rho \\
1 & 1 & 1 & -1 & -1 & -1 \\
1 & \rho & \rho^2 & -1 & -\rho & -\rho^2 \\
1 & \rho^2 & \rho & -1 & -\rho^2 & -\rho
\end{vmatrix}
= 6^3 = 216.
$$

Beispiel 2.

Bezeichnet man die Elemente 1, ϵ, α, α^{-1}, β, β^{-1}, γ, γ^{-1} der
Quaternion-Gruppe Q mit $1, 2, 3, 4, 5, 6, 7, 8$, so ist die ent-
sprechende Gruppen-Determinante

$$
\begin{vmatrix}
x_1 & x_2 & x_4 & x_3 & x_6 & x_5 & x_8 & x_7 \\
x_2 & x_1 & x_3 & x_4 & x_5 & x_6 & x_7 & x_8 \\
x_3 & x_4 & x_1 & x_2 & x_8 & x_7 & x_5 & x_6 \\
x_4 & x_3 & x_2 & x_1 & x_7 & x_8 & x_6 & x_5 \\
x_5 & x_6 & x_7 & x_8 & x_1 & x_2 & x_4 & x_3 \\
x_6 & x_5 & x_8 & x_7 & x_2 & x_1 & x_3 & x_4 \\
x_7 & x_8 & x_6 & x_5 & x_3 & x_4 & x_1 & x_2 \\
x_8 & x_7 & x_5 & x_6 & x_4 & x_3 & x_2 & x_1
\end{vmatrix}
=
$$

$$= \begin{vmatrix} u_1 & u_2 & u_3 & u_4 \\ u_2 & u_1 & u_4 & u_3 \\ u_3 & u_4 & u_1 & u_2 \\ u_4 & u_3 & u_2 & u_1 \end{vmatrix} \times \begin{vmatrix} v_1 & -v_2 & -v_3 & -v_4 \\ v_2 & v_1 & -v_4 & v_3 \\ v_3 & v_4 & v_1 & -v_2 \\ v_4 & -v_3 & v_2 & v_1 \end{vmatrix}$$

$$= \begin{cases} \left(u_1 + u_2 + u_3 + u_4\right)\left(u_1 + u_2 - u_3 - u_4\right) \\ \times \left(u_1 - u_2 + u_3 - u_4\right)\left(u_1 - u_2 - u_3 + u_4\right) \\ \times \left(v_1^2 + v_2^2 + v_3^2 + v_4^2\right)^2, \end{cases}$$

wo

$$\begin{Bmatrix} u_1 \\ v_1 \end{Bmatrix} = x_1 \pm x_2, \quad \begin{Bmatrix} u_2 \\ v_2 \end{Bmatrix} = x_3 \pm x_4,$$

$$\begin{Bmatrix} u_3 \\ v_3 \end{Bmatrix} = x_5 \pm x_6, \quad \begin{Bmatrix} u_4 \\ v_4 \end{Bmatrix} = x_7 \pm x_8."$$

(Richard Dedekind)

21.0.3. Observations. The preceding letters indicate that:

(1) Richard Dedekind's concept of a group-determinant in February of 1886 provided a unifying structure to explain numerous earlier isolated results about factorizations for the determinants of various matrices;

(2) for any finite abelian group G of order n, there are n group characters (i.e., homomorphisms of G into $\mathbb{C}^*$) and each character is explicitly expressible in terms of a primitive nth root of unity in $\mathbb{C}$;

(3) for an abelian group G of order n, the n group characters enable the group-determinant for G to be expressed as a product of n linear combinations of its components over $\mathbb{C}$;

(4) if G is a nonabelian group, Dedekind suspected that the factorization for its group-determinant would necessarily have some irreducible factors over $\mathbb{C}$ of degree greater than 1.

Note the subscript notation for the components of the determinant H on page 193 where two group elements are to be multiplied together to obtain the true subscript. Notation like that was also employed by Georg Frobenius in [**24, 25**]. Throughout the mathematical literature after 1896, the term *group-determinant* or *Gruppendeterminante* has been used for that concept of Richard Dedekind made known by the publications [**24, 25**] of Georg Frobenius.

In the precedng letter, Richard Dedekind presents two unusually instructive examples of factorizations for the determinants of group-matrices where the groups are nonabelian. Not all of their factors are linear combinations of the components. However the quadratic factor $v_1^2 + v_2^2 + v_3^2 + v_4^2$ in the second example resembles the product of a quaternion with its conjugate and suggested to Dedekind that nonlinear factors may still be expressible as products of linear combinations of the components if the coefficients are permitted to be elements of some noncommutative ring. That idea has been abandoned.

Numerous persons would certainly have found it interesting to learn about the insights that Dedekind had in February of 1886. It seems likely that Dedekind avoided their publication because he hoped to also acquire complete results when the finite group is not abelian. The footnote of [**20**, page 414] indicates that letters from Frobenius to Dedekind were unavailable. However, Frobenius responded in each of [**23, 24, 25**] to Dedekind's contributions mentioned above.

CHAPTER 22

Particular Advancements of Georg Frobenius

While letters from Frobenius to Richard Dedekind were unavailable for [20], the influence of Dedekind's research is reflected in each of the three papers [23, 24, 25] or [26] by Georg Frobenius during the years 1896–1897. They can be downloaded from the internet in the manner for [20]. Here, we point out: (i) the notation that Frobenius used for group-matrices in [24]; and, (ii) the modification by Frobenius in [25, pages 1007–1008] of Dedekind's Beispiel 1.

22.0.1. First page of [24] for notation of Frobenius.

"'Die Theorie der Charaktere einer Gruppe, deren Grundlagen ich in meiner letzten Arbeit entwickelt habe, erfordert zu ihrer weiteren Ausgestaltung die Untersuchung einer Determinante, deren Grad der Ordnung der Gruppe gleich ist. Nach dem Vorgange von Dedekind, der zuerst ihre Bedeutung für die Theorie der Gruppen erkannt und meine Aufmerksamkeit auf sie gelenkt hat, nenne ich sie die der Gruppe entsprechende *Gruppendeterminante*.

Die h Elemente A, B, C, ... der Gruppe $\mathfrak{H}$ benutze ich als Indices für h unabhängige Variabele x_A, x_B, x_C, ... Indem ich diese Bezeichnung wähle, treffe ich die Festsetzung, dass, wenn $L = MN$ ist, auch $x_L = x_{MN}$ sein soll. Aus diesen h Grössen, die durch einen Index von einander unterschieden sind, bilde ich h^2 Grössen, die mit zwei Indices versehen sind, indem ich $x_{P,Q} = x_{PQ^{-1}}$ setze. Sind G_1, G_2, ..., G_h die h Elemente von $\mathfrak{H}$ in irgend einer bestimmten Reihenfolge, so betrachte ich die Matrix $(x_{P,Q}) = (x_{PQ^{-1}})$, deren h Zeilen man erhält, indem man für P der Reihe nach die h Elemente G_1, G_2, ..., G_h setzt, und deren h Spalten man erhält, indem man für Q dieselben h Elemente in derselben Reihenfolge setzt. Diese Matrix besitzt gewisse, durch die Constitution der Gruppe $\mathfrak{H}$ bedingte Symmetrieeigenschaften. In jeder Zeile finden sich die h Variabelen sämmtlich und ebenso in jeder Spalte. Die verschiedenen Zeilen (Spalten) unterscheiden sich von einander nur durch die Anordnung der Variabelen. Die Gruppendeterminante, die der Gruppe $\mathfrak{H}$ entspricht, ist die Determinante dieser Matrix

$$\Theta - \left| x_{P,Q} \right| - \left| x_{PQ^{-1}} \right|.$$

Addirt man zu den Elementen der ersten Zeile die aller anderen Zeilen, so werden jene Elemente alle gleich $\sum x_R = \xi$. ...'"

(Georg Frobenius)

197

22.0.2. Paragraph of [25, pages 1007–1008] where Frobenius provided wider perspective about Beispiel 1 on page 194.

"Zur Erläuterung dieser Transformation der Gruppenmatrix wähle ich das Beispiel, das Dedekind im Jahre 1886 gefunden und mir im April 1896 mitgetheilt hat. Seien

1. abc   2. bca   3. cab   4. acb   5. cba   6. bac

die 6 Permutationen von 3 Symbolen. Die Substitutionen, die als in diese 6 Permutationen überführen, mögen statt mit $A, B, C, \ldots$ mit den Ziffern 1, 2, $\ldots$, 6 bezeichnet werden. Sei ρ eine primitiv dritte Wurzel der Einheit und

$$u = x_1 + x_2 + x_3, \qquad\qquad v = x_4 + x_5 + x_6,$$
$$u_1 = x_1 + \rho x_2 + \rho^2 x_3, \qquad\qquad v_1 = x_4 + \rho x_5 + \rho^2 x_6,$$
$$u_2 = x_1 + \rho^2 x_2 + \rho x_3, \qquad\qquad v_2 = x_4 + \rho^2 x_5 + \rho x_6.$$

Ferner seien X, L und U die drei Matrizen

$$X = \begin{bmatrix} x_1 & x_3 & x_2 & x_4 & x_5 & x_6 \\ x_2 & x_1 & x_3 & x_5 & x_6 & x_4 \\ x_3 & x_2 & x_1 & x_6 & x_4 & x_5 \\ x_4 & x_5 & x_6 & x_1 & x_3 & x_2 \\ x_5 & x_6 & x_4 & x_2 & x_1 & x_3 \\ x_6 & x_4 & x_5 & x_3 & x_2 & x_1 \end{bmatrix}, \quad L = \begin{bmatrix} 1 & -1 & 1 & 0 & 0 & 1 \\ 1 & -1 & \rho^2 & 0 & 0 & \rho \\ 1 & -1 & \rho & 0 & 0 & \rho^2 \\ 1 & 1 & 0 & 1 & 1 & 0 \\ 1 & 1 & 0 & \rho & \rho^2 & 0 \\ 1 & 1 & 0 & \rho^2 & \rho & 0 \end{bmatrix},$$

$$U = \begin{bmatrix} u+v & 0 & 0 & 0 & 0 & 0 \\ 0 & u-v & 0 & 0 & 0 & 0 \\ 0 & 0 & u_1 & v_1 & 0 & 0 \\ 0 & 0 & v_2 & u_2 & 0 & 0 \\ 0 & 0 & 0 & 0 & u_1 & v_1 \\ 0 & 0 & 0 & 0 & v_2 & u_2 \end{bmatrix}.$$

Dann ist X die Gruppenmatrix, und es ist

$$XL = LU, \quad L^{-1}XL = U,$$

und indem man $L^T L$ bildet, erkennt man, dass die Determinante von L nicht verschwindet."   (Georg Frobenius)

22.0.3. Observations. For more detail about the Gruppendeterminante Θ in Subsection 22.0.1, see Section 10.1.

Of course, the relation $XL = LU$ with $\det(L) \neq 0$ gives the factorization

$$\det(X) = \det\left(LUL^{-1}\right) = \det(U) = (u+v)(u-v)\left(u_1 u_2 - v_1 v_2\right)^2$$

in agreement with Beispiel 1 of Subsection 21.0.2. However, emphasis should be placed on the block-diagonalization $U = L^{-1}XL$ of X as a remarkable discovery. Namely, in terms of group-pattern matrices for the nonabelian group of order 6, it corresponds to the diagonalizations for abelian-group-pattern matrices.

CHAPTER 23

Group-Pattern Matrices for Q_8

23.1. Application of Theorem 13.3
for second example of Dedekind

Since the determinants of a square matrix and its transpose are equal, we introduce the matrix

(23.1)
$$A = \begin{bmatrix}
x_1 & x_2 & x_3 & x_4 & x_5 & x_6 & x_7 & x_8 \\
x_2 & x_1 & x_4 & x_3 & x_6 & x_5 & x_8 & x_7 \\
x_4 & x_3 & x_1 & x_2 & x_7 & x_8 & x_6 & x_5 \\
x_3 & x_4 & x_2 & x_1 & x_8 & x_7 & x_5 & x_6 \\
x_6 & x_5 & x_8 & x_7 & x_1 & x_2 & x_3 & x_4 \\
x_5 & x_6 & x_7 & x_8 & x_2 & x_1 & x_4 & x_3 \\
x_8 & x_7 & x_5 & x_6 & x_4 & x_3 & x_1 & x_2 \\
x_7 & x_8 & x_6 & x_5 & x_3 & x_4 & x_2 & x_1
\end{bmatrix}$$

as the transpose of the matrix whose determinant Richard Dedekind factored in his second example of [**20**, pages 423–425]; namely, see Beispiel 2 of Subsection 21.0.2. For the quaternion group Q_8 introduced through (2.21) on page 21, we set

(23.2) $g_1 = 1,\ g_2 = -1,\ g_3 = -i,\ g_4 = i,\ g_5 = -j,\ g_6 = j,\ g_7 = -k,\ g_8 = k$

and observe that the multiplication table

(23.3)

$\cdot$	g_1	g_2	g_3	g_4	g_5	g_6	g_7	g_8
$g_1 = g_1^{-1}$	g_1	g_2	g_3	g_4	g_5	g_6	g_7	g_8
$g_2 = g_2^{-1}$	g_2	g_1	g_4	g_3	g_6	g_5	g_8	g_7
$g_4 = g_3^{-1}$	g_4	g_3	g_1	g_2	g_7	g_8	g_6	g_5
$g_3 = g_4^{-1}$	g_3	g_4	g_2	g_1	g_8	g_7	g_5	g_6
$g_6 = g_5^{-1}$	g_6	g_5	g_8	g_7	g_1	g_2	g_3	g_4
$g_5 = g_6^{-1}$	g_5	g_6	g_7	g_8	g_2	g_1	g_4	g_3
$g_8 = g_7^{-1}$	g_8	g_7	g_5	g_6	g_4	g_3	g_1	g_2
$g_7 = g_8^{-1}$	g_7	g_8	g_6	g_5	g_3	g_4	g_2	g_1

for Q_8 corresponds to the list $\mathscr{L} = (g_1, g_2, g_3, g_4, g_5, g_6, g_7, g_8)$ of its elements. Thus, A in (23.1) is the group-pattern matrix for Q_8 and $\mathscr{L}$ that has

$$[A]_{r,s} = \sigma\big(g_r^{-1}g_s\big), \quad \text{for } r,\, s = 1, 2, \ldots, 8,$$

where σ is the function from Q_8 to $\mathbb{F}$ defined by $\sigma(g_k) = x_k$, for $1 \le k \le 8$.

23.1.1. An irreducible matrix representation of degree 2. We observe that, with i in $\mathbb{C}$ having $i^2 = -1$, the nonsingular matrices

$$(23.4) \quad \begin{cases} w_1 = \begin{bmatrix} 1 & 0 \\ 0 & 1 \end{bmatrix}, & w_2 = \begin{bmatrix} -1 & 0 \\ 0 & -1 \end{bmatrix}, & w_3 = \begin{bmatrix} -i & 0 \\ 0 & i \end{bmatrix}, & w_4 = \begin{bmatrix} i & 0 \\ 0 & -i \end{bmatrix}, \\[2ex] w_5 = \begin{bmatrix} 0 & -1 \\ 1 & 0 \end{bmatrix}, & w_6 = \begin{bmatrix} 0 & 1 \\ -1 & 0 \end{bmatrix}, & w_7 = \begin{bmatrix} 0 & -i \\ -i & 0 \end{bmatrix}, & w_8 = \begin{bmatrix} 0 & i \\ i & 0 \end{bmatrix} \end{cases}$$

yield the multiplication table

$$(23.5)$$

$\cdot$	w_1	w_2	w_3	w_4	w_5	w_6	w_7	w_8
$w_1 = w_1^{-1}$	w_1	w_2	w_3	w_4	w_5	w_6	w_7	w_8
$w_2 = w_2^{-1}$	w_2	w_1	w_4	w_3	w_6	w_5	w_8	w_7
$w_4 = w_3^{-1}$	w_4	w_3	w_1	w_2	w_7	w_8	w_6	w_5
$w_3 = w_4^{-1}$	w_3	w_4	w_2	w_1	w_8	w_7	w_5	w_6
$w_6 = w_5^{-1}$	w_6	w_5	w_8	w_7	w_1	w_2	w_3	w_4
$w_5 = w_6^{-1}$	w_5	w_6	w_7	w_8	w_2	w_1	w_4	w_3
$w_8 = w_7^{-1}$	w_8	w_7	w_5	w_6	w_4	w_3	w_1	w_2
$w_7 = w_8^{-1}$	w_7	w_8	w_6	w_5	w_3	w_4	w_2	w_1

that repeats exactly the pattern of the table (23.3). Thus, the set $\mathcal{M}$ of the eight matrices $w_1, w_2, \ldots, w_8$ forms a group that is isomorphic to Q_8. In particular, a one-to-one matrix representation for Q_8 of degree 2 is provided by $g_k \mapsto w_k$, for $1 \leq k \leq 8$. Since Q_8 is nonabelian, this representation is irreducible.

23.1.2. Four inequivalent matrix representations of degree 1. We use Theorems 10.14 and 10.16 on page 99 to see that: with the preceding irreducible matrix representation of degree 2 and the matrix representation of degree 1 given by $g_k \mapsto [1]$, for $1 \leq k \leq 8$, formula (10.34) on page 99 requires $8 = 2^2 + 1^2 + S$, where S is a sum of squares of positive integers. Thus, we have $S = 1^2 + 1^2 + 1^2$ and conclude Q_8 has four pairwise-inequivalent matrix representations of degree 1.

Suppose that $g_k \mapsto \big[f(g_k)\big]$, for $1 \leq k \leq 8$, is a matrix representation of degree 1 for Q_8. Then, f is a group-homomorphism of Q_8 into the multiplicative group $\mathbb{C}^*$ of nonzero elements in $\mathbb{C}$. For $\alpha = g_4 = i$ and $\beta = g_6 = j$, (23.2) and (23.3) yield

$$g_1 = e, \ g_2 = \alpha^2, \ g_3 = \alpha^3, \ g_4 = \alpha, \ g_5 = \alpha^2\beta, \ g_6 = \beta, \ g_7 = \alpha^3\beta, \ g_8 = \alpha\beta$$

and, in regard to any homomorphism f for Q_8, we therefore have

$$f(g_1) = 1, \qquad f(g_3) = \big(f(\alpha)\big)^3, \quad f(g_5) = \big(f(\alpha)\big)^2 f(\beta), \quad f(g_7) = \big(f(\alpha)\big)^3 f(\beta),$$
$$f(g_2) = \big(f(\alpha)\big)^2, \quad f(g_4) = f(\alpha), \qquad f(g_6) = f(\beta), \qquad \qquad f(g_8) = f(\alpha)f(\beta).$$

With $\alpha^4 = e$ and $\beta^4 = e$, we obtain $\big(f(\alpha)\big)^4 = f(e) = 1$ and $\big(f(\beta)\big)^4 = f(e) = 1$. Thus, there are sixteen possibilities for $(u, v) = \big(f(\alpha), f(\beta)\big)$ as u and v range independently over the four complex numbers $1, -1, i, -i$. The table

$f_{u,v}(g_k)$	g_1	g_2	g_3	g_4	g_5	g_6	g_7	g_8
$f_{1,1}$	1	1	1	1	1	1	1	1
$f_{1,-1}$	1	1	1	1	-1	-1	-1	-1
$f_{-1,1}$	1	1	-1	-1	1	1	-1	-1
$f_{-1,-1}$	1	1	-1	-1	-1	-1	1	1

(23.6)

includes those four of the sixteen that yield the desired matrix representations. Namely, for $(u, v) = (1, 1), (1, -1), (-1, 1), (-1, -1)$, machine computations give

$$f_{u,v}(g_i)\, f_{u,v}(g_j) - f_{u,v}(g_k) = 0, \quad \text{whenever } 1 \le i, j,\, k \le 8 \text{ and } g_i g_j = g_k.$$

To describe the computations for a version of *Mathematica* such as [48], we first enter and evaluate representations labeled w[1], w[2], ..., w[8] for the eight 2×2 matrices in (23.4). Then, we enter each of the eight input statements

```
F[1,k_]  :=  {{1}}   /;   1 <= k <= 8

F[2,k_]  :=  {{1}}   /;   1 <= k <= 4

F[2,k_]  :=  {{-1}}  /;   5 <= k <= 8

F[3,k_]  :=  {{1}}   /;  (k==1) || (k==2) || (k==5) || (k==6)

F[3,k_]  :=  {{-1}}  /:  (k==3) || (k==4) || (k==7) || (k==8)

F[4,k_]  :=  {{1}}   /;  (k==1) || (k==2) || (k==7) || (k==8)

F[4,k_]  :=  {{-1}}  /;  (k==3) || (k==4) || (k==5) || (k==6)

repQ[r_]  :=  (
     Do[ If[w[i].w[j]-w[k]=={{0,0,0},{0,0,0},{0,0,0}},
         Print[ "F[", r, ",",  i, "].F[", r, ",", j,
             "] - F[", r, ",",  k, "] = ",
             Expand[ F[r,i]*F[r,j]-F[r,k] ]], {}],
     {i,1,12}, {j,1,12}, {k,1,12}] )
```

and evaluate each one. Finally, we enter repQ[1], repQ[2], repQ[3], repQ[4] and find that each of their evaluations consists of zeros. That completes the verification.

23.1.3. Frobenius block-diagonalization of the matrix A in (23.1).
We use the preceding subsections to see that Q_8 has the matrix representations

(23.7)
$$\begin{cases}
F_1(g_k) = [1], \text{ for } 1 \le k \le 8, \\
F_2(g_k) = [1], \text{ for } k - 1, 2, 3, 4 \quad \text{and} \quad F_2(g_k) = [-1], \text{ for } k = 5, 6, 7, 8, \\
F_3(g_k) = [1], \text{ for } k = 1, 2, 5, 6 \quad \text{and} \quad F_3(g_k) = [-1], \text{ for } k = 3, 4, 7, 8, \\
F_4(g_k) = [1], \text{ for } k = 1, 2, 7, 8 \quad \text{and} \quad F_4(g_k) = [-1], \text{ for } k = 3, 4, 5, 6, \\
F_5(g_k) = w_k, \text{ for } 1 \le k \le 8.
\end{cases}$$

With $n_1 = n_2 = n_3 = n_4 = 1$, $n_5 = 2$, and $n_1^2 + n_2^2 + n_3^2 + n_4^2 + n_5^2 = 8 = n$, we conclude that $\{F_1, F_2, F_3, F_4, F_5\}$ is a complete set of pairwise-inequivalent irreducible matrix representations for Q_8 and the abbreviation

$$\kappa_s = \big(\lambda(s), \mu(s), \nu(s)\big), \quad 1 \le s \le 8,$$

yields

(23.8) $\quad \begin{cases} \kappa_1 = (1, 1, 1), & \kappa_2 = (2, 1, 1), & \kappa_3 = (3, 1, 1), & \kappa_4 = (4, 1, 1), \\ \kappa_5 = (5, 1, 1), & \kappa_6 = (5, 1, 2), & \kappa_7 = (5, 2, 1), & \kappa_8 = (5, 2, 2) \end{cases}$

for $\lambda(s)$, $\mu(s)$, $\nu(s)$ of Section 11.2. To Frobenius block-diagonalize A in (23.1) with M for Theorem 13.3 on page 125, we observe that (13.6) yields M through

$$[M]_{r,s} = \Big[F_{\lambda(s)}(g_r)\Big]_{\mu(s),\nu(s)}, \quad \text{for } r, s = 1, 2, \ldots, 8,$$

and (23.7)–(23.8) as

(23.9)
$$M = \begin{bmatrix} 1 & 1 & 1 & 1 & 1 & 0 & 0 & 1 \\ 1 & 1 & 1 & 1 & -1 & 0 & 0 & -1 \\ 1 & 1 & -1 & -1 & -i & 0 & 0 & i \\ 1 & 1 & -1 & -1 & i & 0 & 0 & -i \\ 1 & -1 & 1 & -1 & 0 & -1 & 1 & 0 \\ 1 & -1 & 1 & -1 & 0 & 1 & -1 & 0 \\ 1 & -1 & -1 & 1 & 0 & -i & -i & 0 \\ 1 & -1 & -1 & 1 & 0 & i & i & 0 \end{bmatrix},$$

where i in $\mathbb{C}$ has $i^2 = -1$. For A in (23.1) and $B = M^{-1}AM$, we find that (23.9) gives

(23.10)
$$B = \left[\begin{array}{c|c|c|c|cc|cc} d_1 & 0 & 0 & 0 & 0 & 0 & 0 & 0 \\ \hline 0 & d_2 & 0 & 0 & 0 & 0 & 0 & 0 \\ \hline 0 & 0 & d_3 & 0 & 0 & 0 & 0 & 0 \\ \hline 0 & 0 & 0 & d_4 & 0 & 0 & 0 & 0 \\ \hline 0 & 0 & 0 & 0 & d_5 & d_6 & 0 & 0 \\ 0 & 0 & 0 & 0 & d_7 & d_8 & 0 & 0 \\ \hline 0 & 0 & 0 & 0 & 0 & 0 & d_5 & d_6 \\ 0 & 0 & 0 & 0 & 0 & 0 & d_7 & d_8 \end{array} \right]$$

as the Frobenius block-diagonalization for A where

(23.11)
$$\begin{cases} d_1 = x_1 + x_2 + x_3 + x_4 + x_5 + x_6 + x_7 + x_8, \\ d_2 = x_1 + x_2 + x_3 + x_4 - x_5 - x_6 - x_7 - x_8, \\ d_3 = x_1 + x_2 - x_3 - x_4 + x_5 + x_6 - x_7 - x_8, \\ d_4 = x_1 + x_2 - x_3 - x_4 - x_5 - x_6 + x_7 + x_8, \\ d_5 = x_1 - x_2 - i\,x_3 + i\,x_4, \\ d_6 = -x_5 + x_6 - i\,x_7 + i\,x_8, \\ d_7 = x_5 - x_6 - i\,x_7 + i\,x_8, \\ d_8 = x_1 - x_2 + i\,x_3 - i\,x_4. \end{cases}$$

In particular, we use $B = M^{-1}AM$, (23.10) and (23.11) to obtain

$$(23.12) \quad \det(A) = \det(B) = d_1 d_2 d_3 d_4 \big(d_5 d_8 - d_6 d_7\big)^2$$

$$= d_1 d_2 d_3 d_4 \Big(\big(x_1 - x_2\big)^2 + \big(x_3 - x_4\big)^2 + \big(x_5 - x_6\big)^2 + \big(x_7 - x_8\big)^2\Big)^2 .$$

While the factorization (23.12) for $\det(A)$ is unique, we note that the Frobenius block-diagonalization B for A in (23.10) depends upon the particular complete set $\{F_1, F_2, F_3, F_4, F_5\}$ of pairwise-inequivalent irreducible matrix representations for Q_8 as well as our preferred selection of (13.6) for M in Theorem 13.3.

With $\det\big(A^T\big) = \det(A)$, we see that (23.12) is identical with the factorization that Dedekind presented at the lower portion of [20, page 424] as shown on page 195 where he wrote the quadratic factor as $v_1^2 + v_2^2 + v_3^2 + v_4^2$.

23.2. A Frobenius block-diagonalization based on Section 13.5

As an example of a group-pattern matrix for Q_8, we selected

$$(23.13) \qquad A_* = \left[\begin{array}{cccc|cccc} y_1 & y_2 & y_3 & y_4 & y_5 & y_6 & y_7 & y_8 \\ y_4 & y_1 & y_2 & y_3 & y_8 & y_5 & y_6 & y_7 \\ y_3 & y_4 & y_1 & y_2 & y_7 & y_8 & y_5 & y_6 \\ y_2 & y_3 & y_4 & y_1 & y_6 & y_7 & y_8 & y_5 \\ \hline y_7 & y_6 & y_5 & y_8 & y_1 & y_4 & y_3 & y_2 \\ y_8 & y_7 & y_6 & y_5 & y_2 & y_1 & y_4 & y_3 \\ y_5 & y_8 & y_7 & y_6 & y_3 & y_2 & y_1 & y_4 \\ y_6 & y_5 & y_8 & y_7 & y_4 & y_3 & y_2 & y_1 \end{array} \right] ,$$

in (2.22) on page 21. Here, with

$$(23.14) \quad h_1 = 1, \ h_2 = i, \ h_3 = -1, \ h_4 = -i, \ h_5 = j, \ h_6 = k, \ h_7 = -j, \ h_8 = -k,$$

we see that the multiplication table (2.21) on page 21 for A_* can be written as

$$(23.15)$$

$\cdot$	h_1	h_2	h_3	h_4	h_5	h_6	h_7	h_8
$h_1 = h_1^{-1}$	h_1	h_2	h_3	h_4	h_5	h_6	h_7	h_8
$h_4 = h_2^{-1}$	h_4	h_1	h_2	h_3	h_8	h_5	h_6	h_7
$h_3 = h_3^{-1}$	h_3	h_4	h_1	h_2	h_7	h_8	h_5	h_6
$h_2 = h_4^{-1}$	h_2	h_3	h_4	h_1	h_6	h_7	h_8	h_5
$h_7 = h_5^{-1}$	h_7	h_6	h_5	h_8	h_1	h_4	h_3	h_2
$h_8 = h_6^{-1}$	h_8	h_7	h_6	h_5	h_2	h_1	h_4	h_3
$h_5 = h_7^{-1}$	h_5	h_8	h_7	h_6	h_3	h_2	h_1	h_4
$h_6 = h_8^{-1}$	h_6	h_5	h_8	h_7	h_4	h_3	h_2	h_1

with respect to the list $\mathscr{L}_* = (h_1, h_2, h_3, h_4, h_5, h_6, h_7, h_8)$.

We observe that the permutation π of $\{1, 2, \ldots, 8\}$ that relates (23.2) for $\mathscr{L}$ and (23.14) for $\mathscr{L}_*$ according to $h_k = g_{\pi(k)}$, for $k = 1, 2, \ldots, 8$, is given by

$$(23.16) \qquad \begin{cases} \pi(1) = 1, & \pi(2) = 4, & \pi(3) = 2, & \pi(4) = 3, \\ \pi(5) = 6, & \pi(6) = 8, & \pi(7) = 5, & \pi(8) = 7. \end{cases}$$

Since A in (23.1) and A_* above are group-pattern matrices for the group Q_8, Proposition 13.10 on page 132 shows how the Frobenius block-diagonalization for A given by $B = M^{-1}AM$ in (23.10) can be modified to obtain a Frobenius block-diagonalization for A_*. To apply that procedure here, we use (13.33) and π in (23.16) to obtain the 8×8 permutation matrix

$$P_* = \begin{bmatrix} 1 & 0 & 0 & 0 & 0 & 0 & 0 & 0 \\ 0 & 0 & 0 & 1 & 0 & 0 & 0 & 0 \\ 0 & 1 & 0 & 0 & 0 & 0 & 0 & 0 \\ 0 & 0 & 1 & 0 & 0 & 0 & 0 & 0 \\ 0 & 0 & 0 & 0 & 0 & 1 & 0 & 0 \\ 0 & 0 & 0 & 0 & 0 & 0 & 0 & 1 \\ 0 & 0 & 0 & 0 & 1 & 0 & 0 & 0 \\ 0 & 0 & 0 & 0 & 0 & 0 & 1 & G \end{bmatrix}.$$

Proposition 13.10 shows that a Frobenius block-diagonalization for A_* is given by the matrix $M_* = P_*M$. In particular, P_* permutes the rows of M in (23.9) to yield

$$(23.17) \qquad M_* = P_*M = \begin{bmatrix} 1 & 1 & 1 & 1 & 1 & 0 & 0 & 1 \\ 1 & 1 & -1 & -1 & i & 0 & 0 & -i \\ 1 & 1 & 1 & 1 & -1 & 0 & 0 & -1 \\ 1 & 1 & -1 & -1 & -i & 0 & 0 & i \\ 1 & -1 & 1 & -1 & 0 & 1 & -1 & 0 \\ 1 & -1 & -1 & 1 & 0 & i & i & 0 \\ 1 & -1 & 1 & -1 & 0 & -1 & 1 & 0 \\ 1 & -1 & -1 & 1 & 0 & -i & -i & 0 \end{bmatrix}$$

and M_* transforms A_* into the Frobenius block-diagonal form $B_* = M_*^{-1}A_*M_*$ given by

$$(23.18) \qquad B_* = \left[\begin{array}{c|c|c|c|cc|cc} \delta_1 & 0 & 0 & 0 & 0 & 0 & 0 & 0 \\ \hline 0 & \delta_2 & 0 & 0 & 0 & 0 & 0 & 0 \\ \hline 0 & 0 & \delta_3 & 0 & 0 & 0 & 0 & 0 \\ \hline 0 & 0 & 0 & \delta_4 & 0 & 0 & 0 & 0 \\ \hline 0 & 0 & 0 & 0 & \delta_5 & \delta_6 & 0 & 0 \\ 0 & 0 & 0 & 0 & \delta_7 & \delta_8 & 0 & 0 \\ \hline 0 & 0 & 0 & 0 & 0 & 0 & \delta_5 & \delta_6 \\ 0 & 0 & 0 & 0 & 0 & 0 & \delta_7 & \delta_8 \end{array} \right],$$

where

$$(23.19) \qquad \begin{cases} \delta_1 = y_1 + y_2 + y_3 + y_4 + y_5 + y_6 + y_7 + y_8, \\ \delta_2 = y_1 + y_2 + y_3 + y_4 - y_5 - y_6 - y_7 - y_8, \\ \delta_3 = y_1 - y_2 + y_3 - y_4 + y_5 - y_6 + y_7 - y_8, \\ \delta_4 = y_1 - y_2 + y_3 - y_4 - y_5 + y_6 - y_7 + y_8, \\ \delta_5 = y_1 + i\,y_2 - y_3 - i\,y_4, \\ \delta_6 = y_5 + i\,y_6 - y_7 - i\,y_8, \\ \delta_7 = -y_5 + i\,y_6 + y_7 - i\,y_8, \\ \delta_8 = y_1 - i\,y_2 - y_3 + i\,y_4. \end{cases}$$

Proposition 13.11 shows that: the right-hand members of the formulas in (23.19) can be obtained from those in (23.11) by replacing each x_k with $y_{\pi^{-1}(k)}$, for $1 \le k \le 8$. We have

$$\pi = \begin{pmatrix} 1 & 2 & 3 & 4 & 5 & 6 & 7 & 8 \\ 1 & 4 & 2 & 3 & 6 & 8 & 5 & 7 \end{pmatrix}, \quad \text{and} \quad \pi^{-1} = \begin{pmatrix} 1 & 2 & 3 & 4 & 5 & 6 & 7 & 8 \\ 1 & 3 & 4 & 2 & 7 & 5 & 8 & 6 \end{pmatrix}.$$

The formulas $B_* = M_*^{-1} A_* M_*$, (23.18), and (23.19) yield

$$(23.20) \quad \det(A_*) = \det(B_*) = \delta_1 \delta_2 \delta_3 \delta_4 \big(\delta_5 \delta_8 - \delta_6 \delta_7\big)^2$$
$$= \delta_1 \delta_2 \delta_3 \delta_4 \Big(\big(y_1 - y_3\big)^2 + \big(y_2 - y_4\big)^2 + \big(y_5 - y_7\big)^2 + \big(y_6 - y_8\big)^2\Big)^2.$$

Nothing has been assumed to relate y_1, y_2, ..., y_8 with the variables x_1, x_2, ..., x_8.

23.3. The Frobenius *Gruppendeterminante* for Q_8

The notation that Frobenius used for his definition of the *Gruppendeterminante* of a finite group is indicated on page 197 and interpreted on pages 91–92. It provides an efficient assignment of a unique polynomial to each finite group. Our present context for Q_8 illustrates it advantageously.

The set $S_1 = \{1, i, j, k, -1, -i, -j, -k\}$ specifies the elements of Q_8 without there being any implied ordering for them. They uniquely yield corresponding variables of the set $S_2 = \{X_1, X_i, X_j, X_k, X_{-1}, X_{-i}, X_{-j}, X_{-k}\}$. There are 8! ways the eight elements of S_1 can be placed in the eight circled positions

(23.21)

·	①	②	③	④	⑤	⑥	⑦	⑧
1	1							
2		1						
3			1					
4				1				
5					1			
6						1		
7							1	
8								1

of (23.21). Each such selection specifies a unique multiplication table for Q_8 having the identity element 1 of Q_8 in each diagonal position. Namely, for $k = 1, 2, \ldots, 8$, the element of Q_8 placed in $\boxed{k}$ must be the inverse of the element placed in ⓚ and this yields unique entries for the remainder of the table.

The 8×8 interior $\mathcal{P}$ of any such multiplication table yields a corresponding 8×8 matrix $A_{\mathcal{P}}$ where the (r, s)-component of $A_{\mathcal{P}}$ is X_g if and only if the (r, s)-position of $\mathcal{P}$ is occupied by g. Theorem 10.1 on page 92 shows that the determinants of all those various matrices are equal. The Frobenius *Gruppendeterminante* for Q_8 is the polynomial given by any one of those determinants.

EXPLANATION 23.1. While there are 8! multiplication tables given by (23.21), Proposition 2.7 on page 13 shows that all of the 8×8 interior patterns are provided when the occupants of positions ① and $\boxed{1}$ are restricted to the identity element 1. This yields 7! patterns and they specify 7! group-matrices for Q_8 in agreement with Proposition 11.1 on age 103. Each group-matrix specifies a corresponding standard group-pattern matrix. Example 3.9 on page 35 shows that Q_8 has 210 standard group-pattern matrices. Therefore, each standard group-pattern matrix for Q_8 is specified by some 24 of the group-matrices for Q_8 and those 24 group-matrices are related by the 24 automorphisms for Q_8.

EXAMPLE 23.2. Let the places ①, ②, ③, ④, ⑤, ⑥, ⑦, ⑧ of (23.21) be respectively occupied by $1, -1, -i, i, -j, j, -k, k$. The multiplication table (23.21) is then identifiable with (23.3) on page 199 and the *Gruppendeterminante* polynomial $\mathfrak{P}(Q_8)$ for Q_8 is given by the determinant of (23.1) with $x_1 = X_1$, $x_2 = X_{-1}$, $x_3 = X_{-i}$, $x_4 = X_i$, $x_5 = X_{-j}$, $x_6 = X_j$, $x_7 = X_{-k}$, $x_8 = X_k$

EXAMPLE 23.3. Let the places ①, ②, ③, ④, ⑤, ⑥, ⑦, ⑧ of (23.21) be respectively occupied by $1, i, -1, -i, j, k, -j, -k$. The multiplication table (23.21) is then identifiable with (23.15) and the *Gruppendeterminante* polynomial $\mathfrak{P}(Q_8)$ for Q_8 is given by the determinant of (23.13) with $y_1 = X_1$, $y_2 = X_i$, $y_3 = X_{-1}$, $y_4 = X_{-i}$, $y_5 = X_j$, $y_6 = X_k$, $y_7 = X_{-j}$, $y_8 = X_{-k}$.

OBSERVATION 23.4. A *Gruppenmatrix* for Q_8 is given by any one of the 7! matrices indicated above whose determinant yields the unique polynomial termed the *Gruppendeterminante* of Q_8. Similar observations apply for any finite group G. The term *group-matrix* is a direct translation of *Gruppenmatrix* and we believe its use should continue to be restricted to that context as we have consistently done since Item (4) on page xiv.

EXERCISES

1. Use computer algebra with (23.11)–(23.12) and (23.19)–(23.20) to check that the same polynomial $\mathfrak{P}(Q_8)$ is given by Example 23.2 and Example 23.3. (See page 235.)

2. The lists $\mathscr{L} = (g_1, g_2, \ldots, g_8)$ and $\mathscr{L}_* = (h_1, h_2, \ldots, h_8)$ for Q_8 defined by (23.2) and (23.14) specify a function ϕ from Q_8 onto Q_8 having $\phi(g_k) = h_k$, for $1 \leq k \leq 8$. Check the standard group-pattern matrices for $\mathscr{L}$ and $\mathscr{L}_*$ derivable from (23.3) and (23.15) to verify that ϕ is not an automorphism of Q_8.

APPENDIX A

Use of Determinants to Factor Polynomials

In 1955, when Professor Emeritus Chao-Hui Yang at SUNY, Oneonta, was a graduate student at the University of Cincinnati, he mentioned that: to easily derive Cardano's formulas for the roots of a cubic equation, one merely needed to recall the identity

$$(A.1) \quad X^3 + (-3ab)X + (a^3 + b^3) \equiv \big(X + a + b\big)\big(X + a\omega + b\omega^2\big)\big(X + a\omega^2 + b\omega\big),$$

where a, b, ω are complex numbers and $\omega^2 + \omega + 1 = 0$. At that time, an old Chinese mathematical encyclopedia of his displayed the determinant

$$(A.2) \qquad \begin{vmatrix} X & a & b \\ b & X & a \\ a & b & X \end{vmatrix}$$

amid numerous Chinese characters. Chao-Hui explained that an ordinary expansion of (A.2) yields the left member of (A.1) while addition to the first row for (A.2) of the sums of the products of the second and third rows by 1 and 1 or by ω and ω^2 or by ω^2 and ω shows that the left member of (A.1) has the factors displayed in its right member.

The publications [21, pages 138–140] of 1877, [45] of 1880, [5] of 1882, [32] of 1883, [34] of 1883, and [33] of 1884 relate the solving of polynomial equations to determinants. In modern times as well as in [7] of 1975 or [27] of 2001, the main connection with these earlier publications is through the treatise [36] by Thomas Muir. However, Muir omits any mention in [36] of Dostor's contribution in [21, pages 138–140].

The research [21, pages 138–140] by Georges Dostor in 1877 is the earliest we have found where a determinant is applied to solve a polynomial equation. There, Dostor employs the determinant of a 3×3 leftward-circulant matrix to recall (A.1) and derive Cardano's formulas. The web page

    https://babel.hathitrust.org/cgi/pt?id=nyp.33433069082893&view=1up&seq=176

may be visited for those three pages or for the entire content of [21].

The paper [32] of 1883 by Alphonse Legoux provides an unusually clear and concise general plan how determinants of circulant matrices can be employed for the solutions of polynomial equations. The web address

    https://babel.hathitrust.org/cgi/pt?id=msu.31293002358780&view=1up&seq=51

enables it to be downloaded.

207

Change of Viewpoint

**B.0.1. The 1975 paper [7] titled *Circulant matrices and algebraic equations*
begins with the following five sentences.**

1. Introduction. For each monic polynomial

$$(1) \qquad f(X) = X^n + c_1 X^{n-1} + \cdots + c_n$$

of degree $n \geq 1$ over the field $\mathbb{C}$ of complex numbers, there are elements $a_1, a_2, \ldots, a_n$ in $\mathbb{C}$ such that the $n \times n$ circulant matrix

$$(2) \qquad A = \begin{bmatrix} a_1 & a_2 & a_3 & \cdots & a_n \\ a_n & a_1 & a_2 & \cdots & a_{n-1} \\ a_{n-1} & a_n & a_1 & \cdots & a_{n-2} \\ \vdots & \vdots & \vdots & \cdots & \vdots \\ a_2 & a_3 & a_4 & \cdots & a_1 \end{bmatrix}$$

has $f(X)$ as its characteristic polynomial in the sense that

$$(3) \qquad f(X) = \det(X I_n - A).$$

In Section 2, we prove the preceding statement and the identity

$$(4) \qquad \det(X I_n - A) = \prod_{s=1}^{n} \left(X - \sum_{k=1}^{n} a_k \zeta_n^{(k-1)(s-1)} \right),$$

where ζ_n denotes a primitive nth root of unity. Thus, the eigenvalues

$$(5) \qquad \xi_s = \sum_{k=1}^{n} a_k \zeta_n^{(k-1)(s-1)}, \quad \text{for } s = 1, 2, \ldots, n,$$

of A are the roots of $f(X)$ in $\mathbb{C}$. *The problem to solve $f(X) = 0$ in $\mathbb{C}$
can therefore be replaced by the problem to find an $n \times n$ circulant
matrix over $\mathbb{C}$ that has $f(X)$ as its characteristic polynomial. For $n =
1, 2, 3, 4$, all such $n \times n$ circulant matrices are presented in Section 4.*

Attention was directed to circulant matrices as interesting objects for study independent of
their long involvement with determinants. Moreover, the verifications required the results
presented in both Theorems 6.4 and 6.6 on pages 65 and 66. When [7] was written, the
results of Theorem 6.4 were well known. However, to prove the existence of a circulant
matrix A for (2) that satisfies (3) for $f(X)$ in (1), the results of Theorem 6.6 were needed
and developed in [7]. Explicit mention of a group was not required.

Before selecting the name *circulant matrix* for use in [7], we recognized that others
have used the term *cyclic matrix* or, in German, *zyklische Matrix* as in [44, page 501].
The influence of Thomas Muir in [36] was decisive.

B.0.2. The principal results in the paper [9] of 1976. When [7] was written, we recognized that the visual definition for an $n \times n$ circulant matrix, as in (2) of the previous page, can be improved for algebraic computations. Thus, we arrived at the definitions involving (0.7) and (0.8) on page xvi of the Preface. However, the multiplication table (0.7) is that of a cyclic group having order n in which the elements are ordered by $\mathscr{L} = \left(1, \alpha, \alpha^2, \ldots, \alpha^{n-1}\right)$. This led naturally to the concept of a group-pattern matrix for a finite group G of order n and a list $\mathscr{L} = \left(g_1, g_2, \ldots g_n\right)$ of its elements as formulated in Definition 2.2 on page 12. Without naming such matrices, we introduced that defining concept in [9] of 1976 and then limited the groups to ones that were abelian.

The principal result in [9] is that the diagonalization of circulant matrices can be extended directly to the diagonalization of abelian-group-pattern matrices in the precise sense that Theorem 8.3 on page 81 includes Theorems 6.4 and 6.6 as special cases.

B.0.3. Connections to the 1979 treatise [19] by Philip J. Davis. Shortly after [7] was published, Philip J. Davis applied for an NSF grant with Emilie V. Haynsworth to develop non-algebraic applications for circulant matrices. Then, in 1979, he published his monograph [19] on *Circulant Matrices*. He described clearly his interests in the preface. The following paragraph, of particular significance, appears in [19, page viii of Preface].

> "It would have been possible to develop the theory of circulants and their generalizations from the point of view of finite abelian groups and group matrices. However, my interest in the subject has a strong numerical and geometric base, which pointed me in the direction taken. The interested reader will find references to these algebraic matters." (Philip J. Davis)

There are numerous references in [19, pages 235–246]. However, the only one that fits the description *"finite abelian groups and group matrices"* is the one to our paper [9] of 1976 titled *Matrices derived from finite abelian groups*. In that paper [9], we should have introduced the terminology *group-pattern matrix*. Then, in reference to [9], Philip Davis could have written *"finite abelian groups and group-pattern matrices."* Then, that terminology would have been continued in [10].

Some algebraic aspects of [19] are considered in Appendices C and D.

B.0.4. The paper [10] of 1981. The key result in [10] is a version of Theorem 13.3 on page 125. In fact, Ω_0 of [10, page 132] is equal to M in (13.6) on page 124. Also, Theorem 3.7 on page 30 was discovered for [10, page 124, Theorem 1] in 1981. The main results of [10] were not mentioned in *Mathematical Reviews*. Thus, they have remained for a thorough reexamination.

APPENDIX C

Circulant Matrices without Mention of a Group

C.1. A suitable context

(I) Let $\mathcal{F}$ denote a field that contains a primitive nth root ρ of unity having period n. Thus, the elements of $S = \{1, \rho, \rho^2, \ldots, \rho^{n-1}\}$ are n distinct roots for the equation $X^n - 1$. For the definition of a circulant matrix motivated by page xv of the Preface, we can select ρ in place of α and see that an $n \times n$ matrix A having components in $\mathcal{F}$ is a circulant matrix if and only if there is a function f from S to $\mathcal{F}$ such that the component $[A]_{r,s}$ of A in its rth row and sth column is given by

$$(C.1) \qquad [A]_{r,s} = f\left(g_r^{-1} g_s\right), \quad \text{for } r, s = 1, 2, \ldots, n,$$

where $g_k = \rho^{k-1}$, for $1 \leq k \leq n$.

(II) An $n \times n$ matrix having components in $\mathcal{F}$ is defined by

$$(C.2) \qquad [M]_{r,s} = \rho^{(r-1)(s-1)}, \quad \text{for } r, s = 1, 2, \ldots, n.$$

(III) Since $\phi(X) \equiv X^n - 1$ factors as $\phi(X) \equiv (X - \rho)\psi(X)$ with $\psi(\rho) \neq 0$, we have $nX^{n-1} \equiv \phi^{(1)}(X) \equiv \psi(X) + (X - \rho)\psi^{(1)}(X)$ as well as $n\rho^{n-1} = \psi(\rho) \neq 0$ and

$$n \cdot 1 = \overbrace{1 + 1 + \cdots + 1}^{n} \neq 0, \quad \text{with respect to 1 in } \mathcal{F}.$$

Thus, the element $n \cdot 1$ in $\mathcal{F}$ (i.e., n in $\mathcal{F}$) has the inverse $1/(n \cdot 1)$ in $\mathcal{F}$ (i.e., $1/n$ in $\mathcal{F}$) and an $n \times n$ matrix L is defined by

$$(C.3) \qquad [L]_{r,s} = \frac{1}{n}\rho^{-(r-1)(s-1)}, \quad \text{for } r, s = 1, 2, \ldots, n.$$

PROPOSITION C.1. *The matrix M is nonsingular and L is its inverse.*

PROOF. We first establish the identity

$$(C.4) \qquad \frac{1}{n}\sum_{\mu=1}^{n} \rho^{(\mu-1)(s-r)} = \begin{cases} 1, & \text{if } r = s, \\ 0, & \text{if } r \neq s, \end{cases} \quad \text{for } 1 \leq r, s \leq n.$$

To do that, we set $\gamma = \rho^{s-r}$. For $1 \leq r, s \leq n$ and $r \neq s$, we have $\gamma \neq 1$, $\gamma^n = 1$, and

$$\frac{1}{n}\sum_{\mu=1}^{n} \rho^{(\mu-1)(s-r)} = \frac{1}{n}\left(1 + \gamma + \gamma^2 + \cdots + \gamma^{n-1}\right) = \frac{\gamma^n - 1}{n(\gamma - 1)} = 0.$$

If $r = s$, then $\gamma = 1$ and the left member of (C.4) equals 1. Consequently, (C.4) is valid. We use (C.3), (C.2), and (C.4) to obtain

$$[LM]_{r,s} = \sum_{\mu=1}^{n} [L]_{r,\mu}\, [M]_{\mu,s} = \frac{1}{n}\sum_{\mu=1}^{n} \rho^{(\mu-1)(s-r)}, \quad \text{for } r, s = 1, 2, \ldots, n,$$

and see that LM is the $n \times n$ identity matrix. Thus, M is nonsingular and $M^{-1} = L$. This competes the proof. $\qquad \square$

C.2. The principal result

THEOREM C.2. *If A is an $n \times n$ circulant matrix over $\mathcal{F}$, then the matrix $M^{-1}AM$ is an $n \times n$ diagonal matrix with components $d_r = \left[M^{-1}AM\right]_{r,r}$, for $1 \leq r \leq n$, given by*

$$\text{(C.5)} \qquad [d_1, d_2, \ldots, d_n] = [a_1, a_2, \ldots, a_n]M,$$

where $a_1, a_2, \ldots, a_n$ are the components in the first row of A. Moreover, if D is an $n \times n$ diagonal matrix over $\mathcal{F}$, then the matrix MDM^{-1} is an $n \times n$ circulant matrix A and the components in the first row of A are given in terms of $d_k = \left[D\right]_{k,k}$, for $1 \leq k \leq n$, by

$$\text{(C.6)} \qquad [a_1, a_2, \ldots, a_n] = [d_1, d_2, \ldots, d_n]M^{-1}.$$

PROOF. (i) Suppose that A is an $n \times n$ circulant matrix over $\mathcal{F}$ and let A be given by (C.1). After setting $D = M^{-1}AM$, we observe, for $r, s = 1, 2, \ldots, n$, that

$$\text{(C.7)} \qquad \left[D\right]_{r,s} = \left[M^{-1}AM\right]_{r,s} = \sum_{\mu=1}^{n} \left[M^{-1}\right]_{r,\mu} \sum_{\nu=1}^{n} \left[A\right]_{\mu,\nu} \left[M\right]_{\nu,s}.$$

For $1 \leq \mu, \nu \leq n$, we have

$$g_\mu^{-1} g_\nu = \left(\rho^{\mu-1}\right)^{-1} \rho^{\nu-1} = \rho^{\nu-\mu} = \begin{cases} g_{\nu-\mu+1}, & \text{if } \nu \geq \mu, \\ g_{n+\nu-\mu+1}, & \text{if } \nu < \mu. \end{cases}$$

This and (C.1) yield

$$\text{(C.8)} \qquad \left[A\right]_{\mu,\nu} = \begin{cases} f(\rho^{\nu-\mu}), & \text{if } \nu \geq \mu, \\ f(\rho^{n+\nu-\mu}), & \text{if } \nu < \mu. \end{cases}$$

For fixed μ and s subject to $1 \leq \mu, s \leq n$, we use (C.8), (C.2), and $\rho^n = 1$ to deduce

$$
\begin{aligned}
\text{(C.9)} \quad \sum_{\nu=1}^{n} \left[A\right]_{\mu,\nu} \left[M\right]_{\nu,s} &= \sum_{\nu=1}^{\mu-1} \left[A\right]_{\mu,\nu} \left[M\right]_{\nu,s} + \sum_{\nu=\mu}^{n} \left[A\right]_{\mu,\nu} \left[M\right]_{\nu,s} \\
&= \sum_{\nu=1}^{\mu-1} f(\rho^{n+\nu-\mu}) \, \rho^{(\nu-1)(s-1)} + \sum_{\nu=\mu}^{n} f(\rho^{\nu-\mu}) \, \rho^{(\nu-1)(s-1)} \\
&= \sum_{\nu=n+1}^{n+\mu-1} f(\rho^{\nu-\mu}) \, \rho^{(\nu-1-n)(s-1)} + \sum_{\nu=\mu}^{n} f(\rho^{\nu-\mu}) \, \rho^{(\nu-1)(s-1)} \\
&= \sum_{\nu=\mu}^{n+\mu-1} f(\rho^{\nu-\mu}) \, \rho^{(\nu-1)(s-1)} \\
&= \sum_{\nu=1}^{n} f(\rho^{\nu-1}) \, \rho^{(\mu-1+\nu-1)(s-1)}. \\
&= \rho^{(\mu-1)(s-1)} \sum_{\nu=1}^{n} f(\rho^{\nu-1}) \, \rho^{(\nu-1)(s-1)}.
\end{aligned}
$$

For $r, s = 1, 2, \ldots, n$, we combine (C.7), (C.3), and (C.9) to obtain

$$\text{(C.10)} \qquad \left[D\right]_{r,s} = \left(\frac{1}{n} \sum_{\mu=1}^{n} \rho^{(\mu-1)(s-r)}\right) \left(\sum_{\nu=1}^{n} f(\rho^{\nu-1}) \, \rho^{(\nu-1)(s-1)}\right).$$

Since (C.10) and (C.4) give $\left[D\right]_{r,s} = 0$ when $r \neq s$, we conclude that D is a diagonal matrix. In view of (C.1), we have $a_\nu = \left[A\right]_{1,\nu} = f(g_1^{-1}g_\nu) = f(\rho^{\nu-1})$, for $1 \leq \nu \leq n$. Consequently, formulas (C.10), (C.4), and (C.2) give

$$d_s = \left[D\right]_{s,s} = \sum_{\nu=1}^{n} a_\nu \, \rho^{(\nu-1)(s-1)} = \sum_{k=1}^{n} a_\nu M_{\nu,s}, \quad \text{for } s = 1, 2, \ldots, n.$$

This shows that (C.5) is valid.

(ii) Suppose that D is an $n \times n$ diagonal matrix over $\mathcal{F}$ and let its diagonal components be given by $d_k = \left[D \right]_{k,k}$, for $1 \leq k \leq n$. With respect to $M^{-1} = L$ in (C.3), we define elements $a_1, a_2, \ldots, a_n$ in $\mathcal{F}$ through

(C.11) $$\left[a_1, a_2, \ldots, a_n \right] = \left[d_1, d_2, \ldots, d_n \right] M^{-1}$$

and we introduce A as the $n \times n$ circulant matrix over $\mathfrak{F}$ having $a_1, a_2, \ldots, a_n$ as the components in its first row. Part (i) shows that the matrix $\widehat{D} = M^{-1}AM$ is an $n \times n$ diagonal matrix and its diagonal components $\widehat{d_k} = \left[\widehat{D} \right]_{k,k}$, for $1 \leq k \leq n$, are given by

(C.12) $$\left[\widehat{d_1}, \widehat{d_2}, \ldots, \widehat{d_n} \right] = \left[a_1, a_2, \ldots, a_n \right] M.$$

Since (C.12) and (C.11) yield $\left[\widehat{d_1}, \widehat{d_2}, \ldots, \widehat{d_n} \right] = \left[a_1, a_2, \ldots, a_n \right] M = \left[d_1, d_2, \ldots, d_n \right]$, we have $\widehat{D} = D$. With $D = \widehat{D} = M^{-1}AM$, we see that MDM^{-1} is the circulant matrix A. In view of (C.11), (C.6) is valid. This completes the proof. $\qquad\square$

C.3. Two consequences from [**7**]

COROLLARY C.3. *For each monic polynomial*

$$f(X) \equiv X^n + c_1 X^{n-1} + \cdots + c_{n-1}X + c_n$$

of degree $n \geq 1$ over the field $\mathbb{C}$ of complex numbers, there is an $n \times n$ circulant matrix A over $\mathbb{C}$ that has $f(X)$ as its characteristic polynomial in the sense that

(C.13) $$f(X) \equiv \det(XI_n - A),$$

where I_n denotes the $n \times n$ identity matrix.

PROOF. With $\mathcal{F} = \mathbb{C}$ and $\rho = \cos(2\pi/n) + i\sin(2\pi/n)$, let $r_1, r_2, \ldots, r_n$ denote the n roots of $f(X)$ in $\mathbb{C}$. Let D be the $n \times n$ diagonal matrix having $\left[D \right]_{k,k} = r_k$, for $1 \leq k \leq n$. Theorem C.2 shows that the $n \times n$ matrix $A = MDM^{-1}$ is a circulant one that yields

$$\det(XI_n - A) = \det\left(XI_n - MDM^{-1}\right) = \det\left(M(XI_n - D)M^{-1}\right)$$
$$= \det(M)\det(XI_n - D)\det(M^{-1}) = \det(XI_n - D) = \prod_{k=1}^{n}\left(X - r_k\right) = f(X).$$

Thus, (C.13) is valid. This completes the proof. $\qquad\square$

For $1 \leq k \leq n$, let V_k be the kth column vector of M in (C.2). Thus, in terms of (C.2), we have $\left[V_k \right]_{r,1} = \left[M \right]_{r,k} = \rho^{(r-1)(k-1)}$, for $1 \leq k, r \leq n$.

COROLLARY C.4. *An $n \times n$ matrix A over $\mathcal{F}$ is a circulant matrix if and only if each of $V_1, V_2, \ldots, V_n$ is an eigenvector of A.*

PROOF. (i) To show that $V_1, V_2, \ldots, V_n$ are eigenvectors of A when A is circulant, we set $D = M^{-1}AM$ and use Theorem C.2 to see that D is a diagonal matrix over $\mathcal{F}$. Let the diagonal elements of D be denoted by $\lambda_k = \left[D \right]_{k,k}$, for $1 \leq k \leq n$. Then, we have

$$\left[AV_1 \middle| AV_2 \middle| \cdots \middle| AV_n \right] = AM = MD = \left[\lambda_1 V_1 \middle| \lambda_2 V_2 \middle| \cdots \middle| \lambda_n V_n \right] \quad \text{and}$$

(C.14) $$AV_k = \lambda_k V_k, \quad \text{for } k = 1, 2, \ldots, n.$$

In view of (C.14), we see that $V_1, V_2, \ldots, V_n$ are eigenvectors for A.

(ii) To show that A is a circulant matrix when $V_1, V_2, \ldots, V_n$ are eigenvectors of A, we observe that there are elements $\lambda_1, \lambda_2, \ldots, \lambda_n$ such that (C.14) is satisfied. Since the components of $V_1, V_2, \ldots, V_n$ and A are elements of $\mathcal{F}$, we find that $\lambda_1, \lambda_2, \ldots, \lambda_n$ are elements of $\mathcal{F}$. Let D be the $n \times n$ diagonal matrix having $\left[D \right]_{k,k} = \lambda_k$, for $1 \leq k \leq n$. Then, we employ (C.14) to deduce

$$AM = \left[AV_1 \middle| AV_2 \middle| \cdots \middle| AV_n \right] = \left[\lambda_1 V_1 \middle| \lambda_2 V_2 \middle| \cdots \middle| \lambda_n V_n \right] = MD$$

and obtain $A = MDM^{-1}$. Using Theorem C.2, we conclude that A is a circulant matrix. This completes the proof. $\qquad\square$

C.4. An unusual permissible selection for the field $\mathcal{F}$ when $n = 3$

The set $\mathbb{Z}$ of integers forms a ring with respect to their addition and multiplication. For a, b, and 7 in $\mathbb{Z}$, the notation

$$a \equiv b \mod 7 \qquad \text{(is read ``}a \text{ is congruent to } b \text{ modulo 7'')}$$

means that there exists an integer q such that $a - b = q \cdot 7$. It is easy to check that this congruency is an equivalence relation defined on $\mathbb{Z}$ and thereby partitions $\mathbb{Z}$ into equivalence classes. For any integer x, there are unique integers q and r such that

$$x = q \cdot 7 + r, \quad \text{where } 0 \le r \le 6, \text{ and we therefore have} \quad x \equiv r \mod 7$$

For $0 \le r \le 6$, let $\widehat{r}$ denote the set of integers x in $\mathbb{Z}$ such that $x \equiv r \mod 7$. Let $\mathcal{F}$ be the set consisting of the seven equivalence classes $\widehat{0}, \widehat{1}, \widehat{2}, \widehat{3}, \widehat{4}, \widehat{5}, \widehat{6}$ for $\mathbb{Z}$.

For any a, b, x, y in $\mathbb{Z}$, suppose that $a \equiv x \mod 7$ and $b \equiv y \mod 7$. Then, it is easy to verify that $a + b \equiv x + y \mod 7$ and $a \cdot b \equiv x \cdot y \mod 7$. Thus, for any $\widehat{a}, \widehat{b}$ in $\mathcal{F}$, there is a unique element $\widehat{a} \oplus \widehat{b}$ in $\mathcal{F}$ such that, whenever x is in $\widehat{a}$ and y is in $\widehat{b}$, the integer $x + y$ is in $\widehat{a} \oplus \widehat{b}$. Also, for any $\widehat{a}, \widehat{b}$ in $\mathcal{F}$, there is a unique element $\widehat{a} \odot \widehat{b}$ in $\mathcal{F}$ such that, whenever x is in $\widehat{a}$ and y is in $\widehat{b}$, the integer $x \cdot y$ is in $\widehat{a} \odot \widehat{b}$. The corresponding tables

$\oplus$	$\widehat{0}$	$\widehat{1}$	$\widehat{2}$	$\widehat{3}$	$\widehat{4}$	$\widehat{5}$	$\widehat{6}$
$\widehat{0}$	$\widehat{0}$	$\widehat{1}$	$\widehat{2}$	$\widehat{3}$	$\widehat{4}$	$\widehat{5}$	$\widehat{6}$
$\widehat{1}$	$\widehat{1}$	$\widehat{2}$	$\widehat{3}$	$\widehat{4}$	$\widehat{5}$	$\widehat{6}$	$\widehat{0}$
$\widehat{2}$	$\widehat{2}$	$\widehat{3}$	$\widehat{4}$	$\widehat{5}$	$\widehat{6}$	$\widehat{0}$	$\widehat{1}$
$\widehat{3}$	$\widehat{3}$	$\widehat{4}$	$\widehat{5}$	$\widehat{6}$	$\widehat{0}$	$\widehat{1}$	$\widehat{2}$
$\widehat{4}$	$\widehat{4}$	$\widehat{5}$	$\widehat{6}$	$\widehat{0}$	$\widehat{1}$	$\widehat{2}$	$\widehat{3}$
$\widehat{5}$	$\widehat{5}$	$\widehat{6}$	$\widehat{0}$	$\widehat{1}$	$\widehat{2}$	$\widehat{3}$	$\widehat{4}$
$\widehat{6}$	$\widehat{6}$	$\widehat{0}$	$\widehat{1}$	$\widehat{2}$	$\widehat{3}$	$\widehat{4}$	$\widehat{5}$

and

$\odot$	$\widehat{0}$	$\widehat{1}$	$\widehat{2}$	$\widehat{3}$	$\widehat{4}$	$\widehat{5}$	$\widehat{6}$
$\widehat{0}$	$\widehat{0}$	$\widehat{0}$	$\widehat{0}$	$\widehat{0}$	$\widehat{0}$	$\widehat{0}$	$\widehat{0}$
$\widehat{1}$	$\widehat{0}$	$\widehat{1}$	$\widehat{2}$	$\widehat{3}$	$\widehat{4}$	$\widehat{5}$	$\widehat{6}$
$\widehat{2}$	$\widehat{0}$	$\widehat{2}$	$\widehat{4}$	$\widehat{6}$	$\widehat{1}$	$\widehat{3}$	$\widehat{5}$
$\widehat{3}$	$\widehat{0}$	$\widehat{3}$	$\widehat{6}$	$\widehat{2}$	$\widehat{5}$	$\widehat{1}$	$\widehat{4}$
$\widehat{4}$	$\widehat{0}$	$\widehat{4}$	$\widehat{1}$	$\widehat{5}$	$\widehat{2}$	$\widehat{6}$	$\widehat{3}$
$\widehat{5}$	$\widehat{0}$	$\widehat{5}$	$\widehat{3}$	$\widehat{1}$	$\widehat{6}$	$\widehat{4}$	$\widehat{2}$
$\widehat{6}$	$\widehat{0}$	$\widehat{6}$	$\widehat{5}$	$\widehat{4}$	$\widehat{3}$	$\widehat{2}$	$\widehat{1}$

display addition and multiplication for $\mathcal{F}$. Since the structure of $\mathbb{Z}$ as a ring is transferred to $\mathcal{F}$, the preceding multiplication table for $\mathcal{F}$ shows that $\mathcal{F}$ is a field.

We note that $\left(\widehat{1}\right)^3 = \widehat{1}$, $\left(\widehat{2}\right)^3 = \widehat{1}$, and $\left(\widehat{4}\right)^3 = \widehat{1}$. With $n = 3$ and $\widehat{1}/(n\widehat{1}) = \widehat{5}$, we set $\rho = \widehat{2}$ in (C.3) and (C.2) to obtain

$$L = \widehat{5} \begin{bmatrix} \widehat{1} & \widehat{1} & \widehat{1} \\ \widehat{1} & \widehat{4} & \widehat{2} \\ \widehat{1} & \widehat{2} & \widehat{4} \end{bmatrix} = \begin{bmatrix} \widehat{5} & \widehat{5} & \widehat{5} \\ \widehat{5} & \widehat{6} & \widehat{3} \\ \widehat{5} & \widehat{3} & \widehat{6} \end{bmatrix} \quad \text{and} \quad M = \begin{bmatrix} \widehat{1} & \widehat{1} & \widehat{1} \\ \widehat{1} & \widehat{2} & \widehat{4} \\ \widehat{1} & \widehat{4} & \widehat{2} \end{bmatrix}.$$

To construct a 3×3 circulant matrix A over $\mathcal{F}$ that has, for example, $\widehat{3}, \widehat{5}, \widehat{6}$ as eigenvalues, we use (C.6) with $d_1 = \widehat{3}$, $d_2 = \widehat{5}$, $d_3 = \widehat{6}$ to obtain

$$\begin{bmatrix} a_1, a_2, a_3 \end{bmatrix} = \begin{bmatrix} d_1, d_2, d_3 \end{bmatrix} M^{-1} = \begin{bmatrix} \widehat{3}, \widehat{5}, \widehat{6} \end{bmatrix} L = \begin{bmatrix} \widehat{0}, \widehat{0}, \widehat{3} \end{bmatrix}$$

and

$$A = \begin{bmatrix} a_1 & a_2 & a_3 \\ a_3 & a_1 & a_2 \\ a_2 & a_3 & a_1 \end{bmatrix} = \begin{bmatrix} \widehat{0} & \widehat{0} & \widehat{3} \\ \widehat{3} & \widehat{0} & \widehat{0} \\ \widehat{0} & \widehat{3} & \widehat{0} \end{bmatrix}.$$

As a check, we note that the matrix $D = M^{-1}AM = LAM$ is given by

$$D = \begin{bmatrix} \widehat{5} & \widehat{5} & \widehat{5} \\ \widehat{5} & \widehat{6} & \widehat{3} \\ \widehat{5} & \widehat{3} & \widehat{6} \end{bmatrix} \begin{bmatrix} \widehat{0} & \widehat{0} & \widehat{3} \\ \widehat{3} & \widehat{0} & \widehat{0} \\ \widehat{0} & \widehat{3} & \widehat{0} \end{bmatrix} \begin{bmatrix} \widehat{1} & \widehat{1} & \widehat{1} \\ \widehat{1} & \widehat{2} & \widehat{4} \\ \widehat{1} & \widehat{4} & \widehat{2} \end{bmatrix} = \begin{bmatrix} \widehat{5} & \widehat{5} & \widehat{5} \\ \widehat{5} & \widehat{6} & \widehat{3} \\ \widehat{5} & \widehat{3} & \widehat{6} \end{bmatrix} \begin{bmatrix} \widehat{3} & \widehat{5} & \widehat{6} \\ \widehat{3} & \widehat{3} & \widehat{3} \\ \widehat{3} & \widehat{6} & \widehat{5} \end{bmatrix} = \begin{bmatrix} \widehat{3} & \widehat{0} & \widehat{0} \\ \widehat{0} & \widehat{5} & \widehat{0} \\ \widehat{0} & \widehat{0} & \widehat{6} \end{bmatrix}.$$

Thus, we see that A is a circulant matrix and

(C.15) $X^3 + \widehat{1} = \det\left(XI_3 - A\right) = \det\left(XI_3 - D\right) = \left(X - \widehat{3}\right)\left(X - \widehat{5}\right)\left(X - \widehat{6}\right).$

C.5. Notation for M and L in (C.2) and (C.3)

We define M^* as the $n \times n$ matrix obtained from M in (C.2) by replacing ρ with ρ^{-1}. Then, L in (C.3) is given by $L = \left(1/(n \cdot 1)\right) M^*$.

When $\mathcal{F}$ is $\mathbb{C}$ or a field extension of $\mathbb{C}$ as in [**19**], the complex number ρ has $|\rho| = 1$ and therefore ρ^{-1} is the complex conjugate of ρ. Then, M^* is the matrix obtained from $M^T = M$ by replacing each component with its complex conjugate.

Philip J. Davis expressed preferences in [**19**] where $\mathcal{F}$ is $\mathbb{C}$ or a field extension of $\mathbb{C}$. With $\rho = \cos(2\pi/n) + i\sin(2\pi/n)$ and $i^2 = -1$, he chose to replace M with F^* where

$$(C.16) \qquad F = \frac{1}{\sqrt{n}} M^* \quad \text{and} \quad F^* = \frac{1}{\sqrt{n}} M = F^{-1}.$$

We give three reasons for not preferring F^*.

(i) The key result

$$[d_1, d_2, \ldots, d_n] = [a_1, a_2, \ldots, a_n] M$$

in (C.5) would then involve an unnecessary factor.

(ii) For computations like FAF^* or F^*DF, it introduces n^2 unnecessary factors.

(iii) It can only be used where $\sqrt{n}$ has meaning. For instance, it is inappropriate for the example of Section C.4 where $n = 3$ because there is no element x in the field $\mathcal{F}$ of that example such that x^2 is equal to $\widehat{3}$.

C.6. Example to illustrate Formula (D.3) on page 217

Of course, we know that a circulant matrix is a special kind of group-pattern matrix. Nevertheless, we can apply the program of Section 2.5 on page 19 to a matrix like

$$(C.17) \qquad A = \begin{bmatrix} a_1 & a_2 & a_3 & a_4 \\ a_4 & a_1 & a_2 & a_3 \\ a_3 & a_4 & a_1 & a_2 \\ a_2 & a_3 & a_4 & a_1 \end{bmatrix}.$$

Namely, after the evaluation for `injectiveGroupPatternQ[Z_]` of page 19 with a version of *Mathematica* such as [**48**] and the evaluation of a representation `A` for A in (C.17), the evaluation of `injectiveGroupPatternQ[A]` yields output summarized as follows.

The matrix A in (C.17) is an injective group-pattern matrix with respect to the group $\mathfrak{P} = \{P_1, P_2, P_3, P_4\}$ and the list $\mathfrak{L} = (P_1, P_2, P_3, P_4)$ of its elements that are given as the four permutation matrices

$$P_1 = \begin{bmatrix} 1 & 0 & 0 & 0 \\ 0 & 1 & 0 & 0 \\ 0 & 0 & 1 & 0 \\ 0 & 0 & 0 & 1 \end{bmatrix}, \quad P_2 = \begin{bmatrix} 0 & 1 & 0 & 0 \\ 0 & 0 & 1 & 0 \\ 0 & 0 & 0 & 1 \\ 1 & 0 & 0 & 0 \end{bmatrix}, \quad P_3 = \begin{bmatrix} 0 & 0 & 1 & 0 \\ 0 & 0 & 0 & 1 \\ 1 & 0 & 0 & 0 \\ 0 & 1 & 0 & 0 \end{bmatrix}, \quad P_4 = \begin{bmatrix} 0 & 0 & 0 & 1 \\ 1 & 0 & 0 & 0 \\ 0 & 1 & 0 & 0 \\ 0 & 0 & 1 & 0 \end{bmatrix}.$$

Their table of matrix multiplications is shown to be

$$(C.18)$$

$\cdot$	P_1	P_2	P_3	P_4
P_1	P_1	P_2	P_3	P_4
P_4	P_4	P_1	P_2	P_3
P_3	P_3	P_4	P_1	P_2
P_2	P_2	P_3	P_4	P_1

Thus, $\mathfrak{P}$ is a cyclic group and each of P_2, P_4 is a generator.

For $n = 4$, this illustrates (D.3) as

$$A = a_1 P_1 + a_2 P_2 + a_3 P_3 + a_4 P_4$$
$$= a_1 (P_2)^0 + a_2 (P_2)^1 + a_3 (P_2)^2 + a_4 (P_2)^3.$$

APPENDIX D

Definition of a Circulant Matrix in [19]

D.1. Introduction

For any $m \times n$ matrix A to be defined from an algebraic viewpoint, one would expect that the component $\begin{bmatrix} A \end{bmatrix}_{r,s}$ of A in its rth row and sth column would be explicitly specified as a function of r and s, for $1 \leq r \leq m$ and $1 \leq s \leq n$. By avoiding this, Philip J. Davis was able to interestingly present non-algebraic properties of circulant matrices to a wider audience. Since the main algebraic properties of circulant matrices are ones closely related to Theorem C.2 and its consequences, we indicate how Davis characterized circulant matrices to accommodate those results.

D.2. Characterization in [19] of a circulant matrix

Philip J. Davis was interested in developing new geometric and numerical applications of circulant matrices in [19] rather than algebraic ones. Thus, for his definition of an $n \times n$ circulant matrix A, he preferred a visual representation with ellipsis such as

$$
(\text{D.1}) \qquad A = \begin{bmatrix} a_1 & a_2 & a_3 & a_4 & \cdots & a_n \\ a_n & a_1 & a_2 & a_3 & \cdots & a_{n-1} \\ a_{n-1} & a_n & a_1 & a_2 & \cdots & a_{n-2} \\ \vdots & \vdots & \vdots & \vdots & \vdots & \vdots \\ a_3 & a_4 & a_5 & a_6 & \cdots & a_2 \\ a_2 & a_3 & a_4 & a_5 & \cdots & a_1 \end{bmatrix}
$$

rather than an algebraic one like (0.9) on page xvi of the preface. For $1 \leq k \leq n$, let $P_{n,k}$ denote the $n \times n$ matrix over $\mathbb{C}$ that is obtained from A in (D.1) by setting $a_k = 1$ and $a_j = 0$ when $j \neq k$. Since $P_{n,1}$ is the $n \times n$ identity matrix I_n, we therefore have

$$
(\text{D.2}) \qquad A = a_1 I_n + a_2 P_{n,2} + a_3 P_{n,3} + \cdots + a_n P_{n,n}.
$$

We note that

$$
P_{2,2} = \begin{bmatrix} 0 & 1 \\ 1 & 0 \end{bmatrix}, \qquad P_{2,2}^2 = \begin{bmatrix} 1 & 0 \\ 0 & 1 \end{bmatrix} = P_{2,1} = I_2,
$$

$$
P_{3,2} = \begin{bmatrix} 0 & 1 & 0 \\ 0 & 0 & 1 \\ 1 & 0 & 0 \end{bmatrix}, \qquad P_{3,2}^2 = \begin{bmatrix} 0 & 0 & 1 \\ 1 & 0 & 0 \\ 0 & 1 & 0 \end{bmatrix} = P_{3,3}, \qquad P_{3,2}^3 = \begin{bmatrix} 1 & 0 & 0 \\ 0 & 1 & 0 \\ 0 & 0 & 1 \end{bmatrix} = P_{3,1} = I_3,
$$

$$
P_{4,2} = \begin{bmatrix} 0 & 1 & 0 & 0 \\ 0 & 0 & 1 & 0 \\ 0 & 0 & 0 & 1 \\ 1 & 0 & 0 & 0 \end{bmatrix}, \qquad P_{4,2}^2 = \begin{bmatrix} 0 & 0 & 1 & 0 \\ 0 & 0 & 0 & 1 \\ 1 & 0 & 0 & 0 \\ 0 & 1 & 0 & 0 \end{bmatrix} = P_{4,3}, \qquad P_{4,2}^3 = \begin{bmatrix} 0 & 0 & 0 & 1 \\ 1 & 0 & 0 & 0 \\ 0 & 1 & 0 & 0 \\ 0 & 0 & 1 & 0 \end{bmatrix} = P_{4,4}.
$$

This and similar evidence make it convincing that (D.2) can be rewritten as

$$
(\text{D.3}) \qquad A = a_1 I_n + a_2 P_{n,2} + a_3 (P_{n,2})^2 + \cdots + a_n (P_{n,2})^{n-1}.
$$

Thus, with the agreement that $(P_{n,2})^0 = I_n$, this motivates the following characterization of a circulant matrix in [19].

CHARACTERIZATION D.1. *An $n \times n$ matrix A having components in $\mathbb{C}$ (or $\mathcal{F}$) is a circulant matrix if and only if there are elements a_1, a_2, ..., a_n in $\mathbb{C}$ (or $\mathcal{F}$) such that*

$$(D.4) \qquad\qquad A = \sum_{\nu=1}^{n} a_\nu \left(P_{n,2} \right)^{\nu-1}.$$

Using M and $M^{-1} = L$ from (C.2) and (C.3) rather than F^* and F of (C.16) from [**19**] because it is advantageous to do so, we shall show how the diagonalization of circulant matrices was done in [**19**] after first establishing the following result.

LEMMA D.2. *The $n \times n$ matrix $D_{n,2}$ defined by $D_{n,2} = M^{-1} P_{n,2} M$ is a diagonal matrix and its diagonal components are given by*

$$(D.5) \qquad\qquad d_k = \left[D_{n,2} \right]_{k,k} = \rho^{k-1}, \quad for \; k = 1, 2, \dots, n.$$

PROOF. We observe that: for $1 \le r, s \le n$, the component of $P_{n,2}$ in its rth row and sth column is given by

$$(D.6) \qquad\qquad \left[P_{n,2} \right]_{r,s} = \begin{cases} 1, & \text{if } r+1 \equiv s \mod n, \\ 0, & \text{if } r+1 \not\equiv s \mod n. \end{cases}$$

We use $M^{-1} = L$, (D.6), (C.3), (C.2), and $\left[M \right]_{1,s} = 1 = \rho^{n(s-1)}$ to obtain

$$(D.7) \quad \left[D_{n,2} \right]_{r,s} = \left[M^{-1} P_{n,2} M \right]_{r,s} = \sum_{\mu=1}^{n} \sum_{\nu=1}^{n} \left[L \right]_{r,\mu} \left[P_{n,2} \right]_{\mu,\nu} \left[M \right]_{\nu,s}$$

$$= \left[\sum_{k=1}^{n-1} \sum_{\nu=1}^{n} \left[L \right]_{r,k} \left[P_{n,2} \right]_{k,\nu} \left[M \right]_{\nu,s} \right] + \sum_{\nu=1}^{n} \left[L \right]_{r,n} \left[P_{n,2} \right]_{n,\nu} \left[M \right]_{\nu,s}$$

$$= \left[\sum_{k=1}^{n-1} \left[L \right]_{r,k} \left[P_{n,2} \right]_{k,k+1} \left[M \right]_{k+1,s} \right] + \left[L \right]_{r,n} \left[P_{n,2} \right]_{n,1} \left[M \right]_{1,s}$$

$$= \left[\sum_{k=1}^{n-1} (1/n) \rho^{-(r-1)(k-1)} \rho^{k(s-1)} \right] + (1/n) \rho^{-(r-1)(n-1)} \rho^{n(s-1)}$$

$$= \sum_{k=1}^{n} (1/n) \rho^{-(r-1)(k-1)} \rho^{k(s-1)} = \rho^{r-1} \left[\frac{1}{n} \sum_{k=1}^{n} \rho^{k(s-r)} \right]$$

$$= \rho^{r-1} \left[\frac{1}{n} \sum_{k=1}^{n} \rho^{(k-1)(s-r)} \right], \quad \text{for } 1 \le r, s \le n.$$

For $1 \le r, s \le n$, we find that (D.7) and, from page 211, (C.4) yield

$$\left[D_{n,2} \right]_{r,s} = \begin{cases} \rho^{r-1}, & \text{if } r = s, \\ 0, & \text{if } r \ne s. \end{cases}$$

Consequently, $D_{n,2}$ is a diagonal matrix and its diagonal components are given by (D.5). This completes the proof. $\qquad\qquad\qquad\qquad\qquad\qquad\qquad\qquad\qquad\quad \square$

Next, we follow the proof from [**19**] for the following restatement of Theorem C.2.

THEOREM D.3. *Suppose that A is an $n \times n$ circulant matrix and the components of its first row are a_1, a_2, ..., a_n. Then, $M^{-1}AM$ is a diagonal matrix whose diagonal components d_1, d_2, ..., d_n are given by*

$$(D.8) \qquad\qquad \left[d_1, d_2, \dots, d_n \right] = \left[a_1, a_2, \dots, a_n \right] M.$$

Moreover, if D is an $n \times n$ diagonal matrix, then MDM^{-1} is a circulant matrix.

PROOF. (i) Starting with A given by (D.4), we use Lemma D.2 to verify that

$$M^{-1}AM = \sum_{\nu=1}^{n} a_\nu M^{-1}(P_{n,2})^{\nu-1}M = \sum_{\nu=1}^{n} a_\nu \big(M^{-1}P_{n,2}M\big)^{\nu-1} = \sum_{\nu=1}^{n} a_\nu (D_{n,2})^{\nu-1}$$

$$= \sum_{\nu=1}^{n} a_\nu \begin{bmatrix} \rho^0 & 0 & 0 & \cdots & 0 \\ 0 & \rho^1 & 0 & \cdots & 0 \\ 0 & 0 & \rho^2 & \cdots & 0 \\ \vdots & \vdots & \vdots & \ddots & 0 \\ 0 & 0 & 0 & \cdots & \rho^{n-1} \end{bmatrix}^{\nu-1}$$

$$= \sum_{\nu=1}^{n} a_\nu \begin{bmatrix} \rho^{0(\nu-1)} & 0 & 0 & \cdots & 0 \\ 0 & \rho^{1(\nu-1)} & 0 & \cdots & 0 \\ 0 & 0 & \rho^{2(\nu-1)} & \cdots & 0 \\ \vdots & \vdots & \vdots & \ddots & 0 \\ 0 & 0 & 0 & \cdots & \rho^{(n-1)(\nu-1)} \end{bmatrix}$$

$$= \sum_{\nu=1}^{n} \begin{bmatrix} a_\nu\rho^{0(\nu-1)} & 0 & 0 & \cdots & 0 \\ 0 & a_\nu\rho^{1(\nu-1)} & 0 & \cdots & 0 \\ 0 & 0 & a_\nu\rho^{2(\nu-1)} & \cdots & 0 \\ \vdots & \vdots & \vdots & \ddots & 0 \\ 0 & 0 & 0 & \cdots & a_\nu\rho^{(n-1)(\nu-1)} \end{bmatrix}$$

$$= \begin{bmatrix} d_1 & 0 & 0 & \cdots & 0 \\ 0 & d_2 & 0 & \cdots & 0 \\ 0 & 0 & d_3 & \cdots & 0 \\ \vdots & \vdots & \vdots & \ddots & 0 \\ 0 & 0 & 0 & \cdots & d_n \end{bmatrix},$$

where

(D.9) $$d_k = \sum_{\nu=1}^{n} \rho^{(k-1)(\nu-1)}a_\nu, \quad \text{for } k = 1,\, 2,\, \ldots,\, n.$$

Thus, $M^{-1}AM$ is a diagonal matrix and (D.9) shows that its diagonal components are given by $[d_1,\, d_2,\, \ldots,\, d_n]^T = M[a_1,\, a_2,\, \ldots,\, a_n]^T$. In view of $M^T = M$, this yields (D.8).

(ii) To complete this proof, repeat the argument at the top of page 213 for Part (ii) of the proof for Theorem C.2. $\square$

EXERCISES

1. Find a 3×3 circulant matrix A over $\mathbb{C}$ such that
$$\det(XI_3 - A) = X^3 - 6X^2 + 11X - 6.$$

2. Find a 4×4 circulant matrix B over $\mathbb{C}$ such that
$$\det(XI_4 - B) = X^4 - 10X^3 + 35X^2 - 50x + 24.$$

APPENDIX E

Result of Richard Baltzer in [2] of 1864

E.1. Introduction

As reported in [**36**, Volume 3, pages 374–375], R, Baltzer expressed the determinant

$$(\text{E.1}) \qquad D(Y) \equiv \begin{vmatrix} a_0 - Y & a_1 & a_2 & \cdots & a_{n-1} \\ a_{n-1} & a_0 - Y & a_1 & \cdots & a_{n-2} \\ a_{n-2} & a_{n-1} & a_0 - Y & \cdots & a_{n-3} \\ \vdots & \vdots & \vdots & \cdots & \vdots \\ a_1 & a_2 & a_3 & \cdots & a_0 - Y \end{vmatrix}, \quad \text{for } n \geq 2,$$

as the resultant of two polynomials and then indicated how that resultant could then be used to obtain $D(Y) \equiv P(Y)$ where, in terms of a primitive nth root ρ of unity,

$$(\text{E.2}) \qquad P(Y) \equiv \prod_{k=1}^{n} \left(a_0 - Y + a_1 \rho^{(k-1)\cdot 1} + a_2 \rho^{(k-1)\cdot 2} + \cdots + a_{n-1} \rho^{(k-1)(n-1)} \right).$$

While the identity $D(Y) \equiv P(Y)$ can be easily derived from Theorem 6.4 on page 65 by performing substitutions, Baltzer's specialized argument is instructive to follow.

E.2. The resultant of two polynomials having degrees $n - 1$ and n

Let $\mathfrak{F}$ be a field that contains the coefficients of two polynomials

$$(\text{E.3}) \qquad F(X) \equiv a_{n-1}X^{n-1} + a_{n-2}X^{n-2} + \cdots + a_1 X + a_0$$

and

$$(\text{E.4}) \qquad G(X) \equiv b_n X^n + b_{n-1}X^{n-1} + \cdots + b_1 X + b_0$$

of respectively degrees $n - 1$ and $n \geq 2$. Then, with $a_{n-1}b_n \neq 0$, the determinant

$$(\text{E.5}) \qquad R = \left. \begin{vmatrix} a_{n-1} & a_{n-2} & \cdots & a_1 & a_0 & 0 & \cdots & 0 \\ 0 & a_{n-1} & \cdots & a_2 & a_1 & a_0 & \cdots & 0 \\ \vdots & \vdots & \vdots & \vdots & \vdots & \vdots & \vdots & \vdots \\ 0 & 0 & \cdots & 0 & a_{n-1} & a_{n-2} & \cdots & a_0 \\ b_n & b_{n-1} & \cdots & b_2 & b_1 & b_0 & \cdots & 0 \\ 0 & b_n & \cdots & b_3 & b_2 & b_1 & \cdots & 0 \\ \vdots & \vdots & \vdots & \vdots & \vdots & \vdots & \vdots & \vdots \\ 0 & 0 & \cdots & b_n & b_{n-1} & b_{n-2} & \cdots & b_0 \end{vmatrix} \right\} \begin{matrix} \\ n \\ \\ \\ \\ n-1 \\ \\ \\ \end{matrix}$$

is the resultant of $F(X)$ and $G(X)$. This resultant R has the property that: *if $F(X)$ and $G(X)$ have a common root in some extension field of $\mathfrak{F}$, then $R = 0$.* For details about that, see [**31**, page 208, Proposition 10.1].

222 E. RESULT OF RICHARD BALTZER IN [2] OF 1864

E.3. Computation of R. Baltzer in a modified context

We assume that $n \geq 2$ and $a_{n-1}, a_{n-2}, \ldots, a_1, a_0, Y$ are algebraically independent variables over a field $\mathcal{F}$ that contains a primitive nth root ρ of unity having period n. To include the context for (E.3)–(E.5), let $\mathfrak{F}$ be the quotient field of the polynomial ring $\mathcal{F}[a_{n-1}, a_{n-2}, \ldots, a_1, a_0, Y]$ in the variables $a_{n-1}, a_{n-2}, \ldots, a_1, a_0, Y$ over $\mathcal{F}$.

Let $D(Y)$ be defined by (E.1) and let $R(Y)$ denote the resultant of

(E.6) $\qquad f(X, Y) \equiv a_{n-1}X^{n-1} + a_{n-2}X^{n-2} + \cdots + a_1 X + (a_0 - Y)$

and

(E.7) $\qquad g(X) \equiv X^n - 1,$

as polynomials with respect to the variable X.

PROPOSITION E.1. *As polynomials in Y, $D(Y)$ and $R(Y)$ satisfy $D(Y) \equiv R(Y)$.*

PROOF. We use (E.3), (E.4), and (E.5) to see that the determinant

$$
\left.\begin{array}{ccccc|ccccc}
a_{n-1} & a_{n-2} & \cdots & a_2 & a_1 & a_0 - Y & 0 & 0 & \cdots & 0 & 0 \\
0 & a_{n-1} & \cdots & a_3 & a_2 & a_1 & a_0 - Y & 0 & \cdots & 0 & 0 \\
\vdots & \vdots & \cdots & \vdots & \vdots & \vdots & \vdots & \vdots & \cdots & \vdots & \vdots \\
0 & 0 & \cdots & 0 & a_{n-1} & a_{n-2} & a_{n-3} & a_{n-4} & \cdots & a_0 - Y & 0 \\
\hline
0 & 0 & \cdots & 0 & 0 & a_{n-1} & a_{n-2} & a_{n-3} & \cdots & a_1 & a_0 - Y \\
\hline
1 & 0 & \cdots & 0 & 0 & 0 & -1 & 0 & \cdots & 0 & 0 \\
0 & 1 & \cdots & 0 & 0 & 0 & 0 & -1 & \cdots & 0 & 0 \\
\vdots & \vdots & \cdots & \vdots & \vdots & \vdots & \vdots & \vdots & \cdots & \vdots & \vdots \\
0 & 0 & \cdots & 0 & 1 & 0 & 0 & 0 & \cdots & 0 & -1
\end{array}\right\}
\begin{array}{c} \\ \\ n \\ \\ \\ \\ \\ \\ n-1 \\ \\ \end{array}
$$

$$\underbrace{\qquad}_{n-1} \quad \underbrace{\quad}_{1} \quad \underbrace{\qquad}_{n-1}$$

is the resultant $R(Y)$ of $f(X, Y)$ and $g(X)$ when they are considered as polynomials in X. The preceding formula expresses $R(Y)$ as the determinant of a $(2n-1)\times(2n-1)$ matrix H. For $k = 1, 2, \ldots, n-1$, we add the kth column of H to the $(n+k)$th column of H and leave the nth column unchanged to see that the determinant

$$
\left.\begin{array}{ccccc|ccccc}
a_{n-1} & a_{n-2} & \cdots & a_2 & a_1 & a_0 - Y & a_{n-1} & a_{n-2} & \cdots & a_2 & a_1 \\
0 & a_{n-1} & \cdots & a_3 & a_2 & a_1 & a_0 - Y & a_{n-1} & \cdots & a_3 & a_2 \\
\vdots & \vdots & \cdots & \vdots & \vdots & \vdots & \vdots & \vdots & \cdots & \vdots & \vdots \\
0 & 0 & \cdots & 0 & a_{n-1} & a_{n-2} & a_{n-3} & a_{n-4} & \cdots & a_0 - Y & a_{n-1} \\
\hline
0 & 0 & \cdots & 0 & 0 & a_{n-1} & a_{n-2} & a_{n-3} & \cdots & a_1 & a_0 - Y \\
\hline
1 & 0 & \cdots & 0 & 0 & 0 & 0 & 0 & \cdots & 0 & 0 \\
0 & 1 & \cdots & 0 & 0 & 0 & 0 & 0 & \cdots & 0 & 0 \\
\vdots & \vdots & \cdots & \vdots & \vdots & \vdots & \vdots & \vdots & \cdots & \vdots & \vdots \\
0 & 0 & \cdots & 0 & 1 & 0 & 0 & 0 & \cdots & 0 & 0
\end{array}\right\}
\begin{array}{c} \\ \\ n \\ \\ \\ \\ \\ \\ n-1 \\ \\ \end{array}
$$

$$\underbrace{\qquad}_{n-1} \quad \underbrace{\quad}_{1} \quad \underbrace{\qquad}_{n-1}$$

is equal to $R(Y)$. We expand the preceding determinant according to the elements in its $(n+1)$th row and after repeating that action an additional $n-2$ times, we obtain

$$(E.8) \quad R(Y) \equiv \left((-1)^{n+2}\right)^{n-1} \begin{vmatrix} a_0 - Y & a_{n-1} & a_{n-2} & \cdots & a_2 & a_1 \\ a_1 & a_0 - Y & a_{n-1} & \cdots & a_3 & a_2 \\ \vdots & \vdots & \vdots & \cdots & \vdots & \vdots \\ a_{n-2} & a_{n-3} & a_{n-4} & \cdots & a_0 - Y & a_{n-1} \\ a_{n-1} & a_{n-2} & a_{n-3} & \cdots & a_1 & a_0 - Y \end{vmatrix}.$$

Since $(n+2)(n-1)$ is an even integer, we have $\left((-1)^{n+2}\right)^{n-1} = 1$. A transposition of the determinant in (E.8) yields

$$(E.9) \quad R(Y) \equiv \begin{vmatrix} a_0 - Y & a_1 & a_2 & \cdots & a_{n-2} & a_{n-1} \\ a_{n-1} & a_0 - Y & a_1 & \cdots & a_{n-3} & a_{n-2} \\ a_{n-2} & a_{n-1} & a_0 - Y & \cdots & a_{n-4} & a_{n-3} \\ \vdots & \vdots & \vdots & \cdots & \vdots & \vdots \\ a_1 & a_2 & a_3 & \cdots & a_{n-1} & a_0 - Y \end{vmatrix}.$$

Thus, in view of (E.1), we have $R(Y) = D(Y)$. This completes the proof. $\quad\square$

THEOREM E.2. *For $n \geq 2$, $D(Y)$ in (E.1) and $P(Y)$ in (E.2) satisfy $D(Y) \equiv P(Y)$.*

PROOF. We set

$$(E.10) \quad y_k = a_{n-1}\left(\rho^{k-1}\right)^{n-1} + a_{n-2}\left(\rho^{k-1}\right)^{n-2} + \cdots + a_1\left(\rho^{k-1}\right) + a_0, \quad \text{when } 1 \leq k \leq n.$$

For $1 \leq k \leq n$, the resultant of $f(X, y_k)$ and $g(X)$ is equal to $R(y_k)$. By using (E.10) and substituting y_k for Y in (E.6), we see that $f(X, y_k)$ and $g(X)$ have ρ^{k-1} as a common root, for $1 \leq k \leq n$. Therefore, the corresponding resultants satisfy $R(y_k) = 0$, for $1 \leq k \leq n$. In view of Proposition E.1, this yields $D(y_k) = 0$, for $1 \leq k \leq n$. Consequently, the polynomial $D(Y)$ is divisible $Y - y_k$, for $1 \leq k \leq n$.

Suppose that $1 \leq i < j \leq n$. Then, we have $y_j - y_i = a_1\left(\rho^{j-1} - \rho^{i-1}\right) + T$, where T does not involve a_1. We use $\rho^{i-1} \neq \rho^{j-1}$ and the algebraic independence of the variables $a_{n-1}, a_{n-2}, \ldots, a_1, a_0$ over $\mathcal{F}$ to deduce $y_j \neq y_i$ and conclude that $D(Y)$ has n distinct factors given by $(Y - y_k)$, for $1 \leq k \leq n$. Since the coefficient of Y^n in $(-1)^n D(Y)$ is 1, we therefore have

$$(-1)^n D(Y) \equiv \prod_{k=1}^{n} (Y - y_k) \quad \text{and} \quad D(Y) \equiv \prod_{k=1}^{n} (y_k - Y).$$

We uses this with (E.10) and (E.2) to obtain $D(Y) \equiv P(Y)$ and completes the proof. $\quad\square$

REMARK E.3. We observe that $D(Y) \equiv P(Y)$ is a polynomial identity in the $n+1$ variables $a_0, a_1, \ldots, a_{n-1}, Y$. Consequently, it yields a valid polynomial identity in Y when the n variables $a_0, a_1, \ldots, a_{n-1}$ are replaced by specific elements in $\mathcal{F}$. In particular, when $\mathcal{F} \equiv \mathbb{C}$ and $a_0, a_1, \ldots, a_{n-1}$ in $(-1)^n D(Y)$ are replaced by complex numbers, the resulting expression is a monic polynomial of degree n in Y having complex coefficients. It is natural to inquire: Does any monic polynomial of degree $n \geq 1$ in Y over $\mathbb{C}$ have a representation of that kind? That motivated [7] and therefore, as indicated on page 209 and in Corollary C.3 on page 213, the answer is yes.

APPENDIX F

Automorphisms
for Developers of Computer Algebra

A program to find all of the automorphisms for a group G of order n is presented on page 34. It functions efficiently when the computer is not hindered by the storage of representations for the $(n-1)!$ permutation matrices of size $n \times n$ having 1 as their $(1, 1)$-component.

Fortunately, the algorithm merely needs the permutation matrices to be generated one-at-a-time. To reduce memory requirements when n exceeds some value, developers can make the program one that is desirable for inclusion in *Mathematica*. Namely, there has been considerable research about efficient methods to generate $n \times n$ permutation matrices. That specialized knowledge would naturally be modified to generate $n \times n$ permutation matrices having 1 as their $(1, 1)$-component. When each such permutation matrix is generated, it would be checked as on page 34 to see if it specifies an automorphism. Memory requirements are reduced by immediately deleting each generated permutation matrix that does not specify an automorphism.

The method in Section 3.4 to compute the automorphisms of G was a consequence of the plan to compute $\mathfrak{N}_\mathfrak{G}(G)$ for G. Here, we focus on just the argument to obtain the automorphisms of G.

F.1. The automorphisms for a group G of order n

Let $\mathscr{L}_0 = (g_1, g_2, \ldots, g_n)$ be a list of the elements for a group G of order n in which g_1 is the identity element e of G. Let Z_0 be the standard group-pattern matrix for G and $\mathscr{L}_0$ as explained in Definition 3.1 on page 27. Thus, we have

(F.1) $$\left[Z_0\right]_{r,s} = \sigma\!\left(g_r^{-1} g_s\right) \quad \text{for } r, s = 1, 2, \ldots, n,$$

where σ is selected as the one-to-one function from G to $\mathbb{Q}[X_1, X_2, \ldots, X_n]$ defined by

(F.2) $$\sigma(g_k) = X_k, \quad \text{for } k = 1, 2, \ldots, n.$$

Let $\mathfrak{P}$ denote the set of $n \times n$ permutation matrices P having $\left[P\right]_{1,1} = 1$.
Let Π denote the set of permutations π of $S = \{1, 2, \ldots, n\}$ with $\pi(1) = 1$.
Let F_1 denote the function from $\mathfrak{P}$ to Π defined, for P in $\mathfrak{P}$, by $F_1(P) = \pi$ where

(F.3) $$\pi(k) = \left[[1, 2, \ldots, n]P\right]_{1,k}, \quad \text{for } k = 1, 2, \ldots, n.$$

Let $\mathscr{F}$ denote the set of one-to-one functions ϕ from G onto G such that $\phi(g_1) = g_1$.
Let F_2 denote the function from Π to $\mathscr{F}$ defined, for π in Π, by $F_2(\pi) = \phi$ where

(F.4) $$\phi(g_k) = g_{\pi(k)}, \quad \text{for } k = 1, 2, \ldots, n.$$

Let $\mathfrak{L}$ be the set of those lists for G that have $g_1 = e$ as their first element.
Let F_3 denote the function from Π to $\mathfrak{L}$ defined by

(F.5) $$F_3(\pi) = \left(g_{\pi(1)}, g_{\pi(2)}, \ldots, g_{\pi(n)}\right), \quad \text{for each } \pi \text{ in } \Pi.$$

PROPOSITION F.1. *One-to-one correspondences of $\mathfrak{P}$ onto Π, Π onto $\mathscr{F}$, and Π onto $\mathfrak{L}$ are respectively provided by F_1, F_2, and F_3.*

225

PROOF. To prove that F_1 is one-to-one, suppose that P, Q in $\mathfrak{P}$ yield $F_1(P) = F_1(Q)$. Then, we have $[1, 2, \ldots, n] P = [\pi(1), \pi(2), \ldots, \pi(n)] = [1, 2, \ldots, n] Q$. We write

$$P = [p_1 \,|\, p_2 \,|\, \cdots \,|\, p_n] \quad \text{and} \quad Q = [q_1 \,|\, q_2 \,|\, \cdots \,|\, q_n]$$

where, for $k = 1, 2, \ldots, n$, p_k and q_k are $n \times 1$ matrices having a single component equal to 1 and $n - 1$ components equal to 0. Since we must have

$$[1, 2, \ldots, n] \, q_k = [1, 2, \ldots, n] \, p_k, \quad \text{for } k = 1, 2, \ldots, n,$$

we find that $q_k = p_k$, for $k = 1, 2, \ldots, n$, and $P = Q$. Thus, F_1 is one-to-one.

To prove that F_1 is onto, let π be an element of $\varPi$. We define P in $\mathfrak{P}$ through

(F.6) $$[P]_{r,s} = \begin{cases} 1, & \text{if } r = \pi(s), \\ 0, & \text{if } r \neq \pi(s), \end{cases} \quad \text{for } r, s = 1, 2, \ldots, n.$$

Then, for $k = 1, 2, \ldots, n$, we have

$$\left[[1, 2, \ldots, n] P\right]_{1,k} = \sum_{\mu=1}^{n} \left[[1, 2, \ldots, n]\right]_{1,\mu} [P]_{\mu,k} = \left[[1, 2, \ldots, n]\right]_{1,\pi(k)} = \pi(k).$$

This gives $F_1(P) = \pi$ and shows that F_1 is onto.

Clearly, F_2 and F_3 are one-to-one.

To prove F_2 is onto, let ϕ be an element of $\mathscr{F}$. The list $(\phi(g_1), \phi(g_2), \ldots, \phi(g_n))$ is a permutation of $(g_1, g_2, \ldots, g_n)$ with $\phi(g_1) = g_1$. Thus, there is a π in $\varPi$ such that

(F.7) $$(g_{\pi(1)}, g_{\pi(2)}, \ldots, g_{\pi(n)}) = (\phi(g_1), \phi(g_2), \ldots, \phi(g_n)).$$

This shows that $F_2(\pi) = \phi$. Thus, F_2 is onto.

For any $\mathscr{L}$ in $\mathfrak{L}$, there is a π in $\varPi$ such that $\mathscr{L} = (g_{\pi(1)}, g_{\pi(2)}, \ldots, g_{\pi(n)})$. Then, we have $F_3(\pi) = \mathscr{L}$ and note that F_3 is onto. This completes the proof $\qquad \square$

For each P in $\mathfrak{P}$, there is π_P in $\varPi$ defined by $\pi_P = F_1(P)$; there is ϕ_P in $\mathscr{F}$ defined by $\phi_P = (F_2 \circ F_1)(P) = F_2(\pi_P)$ for which (F.4) yields

(F.8) $$\phi_P(g_k) = g_{\pi_P(k)}, \quad \text{for } k = 1, 2, \ldots, n;$$

and there is a list $\mathscr{L}_P$ in $\mathfrak{L}$ given by

(F.9) $$\mathscr{L}_P = (F_3 \circ F_1)(P) = F_3(\pi_P) = \left(g_{\pi_P(1)}, g_{\pi_P(2)}, \ldots, g_{\pi_P(n)}\right).$$

For each P in $\mathfrak{P}$, the standard group-pattern matrix for G and $\mathscr{L}_P$, named $\mathcal{S}_P$, has

(F.10) $$[\mathcal{S}_P]_{r,s} = f_P\big(g_{\pi_P(r)}^{-1} g_{\pi_P(s)}\big), \quad \text{for } r, s = 1, 2, \ldots, n,$$

where f_P is the one-to-one function from G to $\mathbb{Q}[X_1, X_2, \ldots, X_n]$ defined by

(F.11) $$f_P(g_{\pi_P(k)}) = X_k, \quad \text{for } k = 1, 2, \ldots, n.$$

We use $\pi_P(1) = 1$, $g_1 = e$, (F.10), and (F.11) to check that

(F.12) $$(\text{The first row of } \mathcal{S}_P) = (X_1, X_2, \ldots, X_n).$$

THEOREM F.2. *For each P in $\mathfrak{P}$, the corresponding standard group-pattern matrix $\mathcal{S}_P$ is equal to the matrix obtained from the $n \times n$ matrix $P^T Z_0 P$, with $(b_1, b_2, \ldots, b_n)$ as its first row, when each b_k in $P^T Z_0 P$ is replaced with X_k, for $k = 1, 2, \ldots, n$. Moreover, each standard group-pattern matrix is obtained in this manner for some P in $\mathfrak{P}$.*

PROOF. Since any list $\mathscr{L}$ in $\mathfrak{L}$ is expressible as $\mathscr{L}_P$ in (F.9) for some P in $\mathfrak{P}$, each standard group-pattern matrix for G is given by $\mathcal{S}_P$ in (F.10) for some P in $\mathfrak{P}$.

For any P in $\mathfrak{P}$, with $\pi_P = F_1(P)$ and $P = F_1^{-1}(\pi_P)$, we use (F.6) to see that

(F.13) $$[P]_{r,s} = \begin{cases} 1, & \text{if } r = \pi_P(s), \\ 0, & \text{if } r \neq \pi_P(s), \end{cases} \quad \text{for } r, s = 1, 2, \ldots, n.$$

For r, $s = 1, 2, \ldots, n$, we use (F.13) and (F.1) to obtain

$$(F.14) \qquad \left[P^T Z_0 P\right]_{r,s} = \sum_{\mu=1}^{n} \sum_{\nu=1}^{n} \left[P^T\right]_{r,\mu} \left[Z_0\right]_{\mu,\nu} \left[P\right]_{\nu,s} = \sum_{\mu=1}^{n} \sum_{\nu=1}^{n} \left[P\right]_{\mu,r} \left[Z_0\right]_{\mu,\nu} \left[P\right]_{\nu,s}$$

$$= \left[Z_0\right]_{\pi_P(r),\pi_P(s)} = \sigma\left(g_{\pi_P(r)}^{-1} g_{\pi_P(s)}\right).$$

We use $\pi_P(1) = 1$, $g_1 = e$, (F.14), and (F.2), to find that

$$(F.15) \qquad \text{(The first row of } P^T Z_0 P) = \left(X_{\pi_P(1)},\, X_{\pi_P(2)},\, \ldots,\, X_{\pi_P(n)}\right).$$

Comparing (F.10) and (F.12) with (F.14) and (F.15), we see that $\mathcal{S}_P$ is equal to the matrix obtained from $P^T Z_0 P$ by replacing each $X_{\pi_P(k)}$ in $P^T Z_0 P$ with X_k, for $1 \leq k \leq n$.

To describe this for computer algebra applications, let the first row of $P^T Z_0 P$ be denoted by $(b_1, b_2, \ldots, b_n)$. Then, $\mathcal{S}_P$ is equal to the matrix obtained from $P^T Z_0 P$ by replacing each b_k in $P^T Z_0 P$ with X_k, for $k = 1, 2, \ldots, n$. This completes the proof. $\square$

THEOREM F.3. *For P in $\mathfrak{P}$, the standard group-pattern matrix $\mathcal{S}_P$ satisfies $\mathcal{S}_P = Z_0$ if and only if the function $\phi_P = (F_2 \circ F_1)(P)$ is an automorphism of G.*

PROOF. We use (F.8) to rewrite $\mathscr{L}_P$ in (F.9) for $\mathcal{S}_P$ as

$$(F.16) \qquad \mathscr{L}_P == \left(\phi_P(g_1),\, \phi_P(g_2),\, \ldots,\, \phi_P(g_n)\right).$$

In view of (F.10) and (F.16), the standard group-pattern matrix $\mathcal{S}_P$ for G and $\mathscr{L}_P$ has

$$(F.17) \qquad \left[\mathcal{S}_P\right]_{r,s} = f_P\left(\left(\phi_P(g_r)\right)^{-1}\phi_P(g_s)\right), \quad \text{for } r, s = 1, 2, \ldots, n,$$

where f_P is defined in (F.11). We apply (F.8), (F.11), and (F.2) to verify that

$$f_P(\phi_P(g_k)) = f_P(g_{\pi_P(k)}) = X_k = \sigma(g_k), \quad \text{for } k = 1, 2, \ldots, n.$$

This yields $\sigma = f_P \circ \phi_P$.

(i) Suppose that $\mathcal{S}_P = Z_0$. Then, (F.1) with $\sigma = f_P \circ \phi_P$, $Z_0 = \mathcal{S}_P$, and (F.17) yield

$$f_P\left(\phi_P(g_r^{-1}g_s)\right) = \sigma(g_r^{-1}g_s) = \left[Z_0\right]_{r,s} = \left[\mathcal{S}_P\right]_{r,s} = f_P\left(\left(\phi_P(g_r)\right)^{-1}\phi_P(g_s)\right),$$

for $1 \leq r, s \leq n$. Since f_P is one-to-one, we therefore have

$$\phi_P(g_r^{-1}g_s) = \left(\phi_P(g_r)\right)^{-1}\phi_P(g_s), \quad \text{for } r, s = 1, 2, \ldots, n.$$

With $s = 1$ and $g_1 = e = \phi_P(g_1)$, this gives $\phi_P(g_r^{-1}) = \left(\phi_P(g_r)\right)^{-1}$, when $1 \leq r \leq n$. Thus, we find that $\phi_P(g_r^{-1}g_s) = \phi_P(g_r^{-1})\,\phi_P(g_s)$, for $1 \leq r, s \leq n$. We rewrite this in the form $\phi_P(xy) = \phi_P(x)\,\phi_P(y)$, for each x, y in G, and therefore conclude that ϕ_P is an automorphism of G.

(ii) Suppose that ϕ_P is an automorphism of G. Then, for $r, s = 1, 2, \ldots, n$, we use (F.17), properties of ϕ_P as an automorphism, $f_P \circ \phi_P = \sigma$, and (F.1) to obtain

$$\left[\mathcal{S}_P\right]_{r,s} = f_P\left(\left(\phi_P(g_r)\right)^{-1}\phi_P(g_s)\right) = f_P\left(\phi_P(g_r^{-1}g_s)\right) = \sigma(g_r^{-1}g_s) = \left[Z_0\right]_{r,s}.$$

This yields $\mathcal{S}_P = Z_0$ and completes the proof. $\square$

OBSERVATION F.4. The program of page 34 to obtain the automorphisms of a group runs correctly and efficiently when the input statement

```
Do[( B[j] = Transpose[P[j]].Z0.P[j];
```

is replaced by the input statement

```
Do[( B[j] = P[j].Z0.Transpose[P[j]];
```

and no other alterations are made. In this regard, see Explanation F.6.

THEOREM F.5. *For P in $\mathfrak{P}$, the function $\phi = (F_2 \circ F_1)(P)$ in $\mathscr{F}$ is an automorphism of G if and only if the function $\psi = (F_2 \circ F_1)(P^T)$ in $\mathscr{F}$ is an automorphism of G.*

PROOF. We use P to define the permutation $\pi = F_1(P)$ of S subject to $\pi(1) = 1$. Thus, we have $[1, 2, \ldots, n]P = [\pi(1), \pi(2), \ldots, \pi(n)]$. We use (F.6) to obtain

$$(\text{F.18}) \quad [P]_{r,s} = \begin{cases} 1, & \text{if } r = \pi(s), \\ 0, & \text{if } r \neq \pi(s), \end{cases} \quad \text{and} \quad [P^T]_{r,s} = [P]_{s,r} = \begin{cases} 1, & \text{if } r = \pi^{-1}(s), \\ 0, & \text{if } r \neq \pi^{-1}(s). \end{cases}$$

For $k = 1, 2, \ldots, n$, this yields $\phi(g_k) = g_{\pi(k)}$, $\psi(g_k) = g_{\pi^{-1}(k)}$,

$$(\phi \circ \psi)(g_k) = \phi(g_{\pi^{-1}(k)}) = g_{\pi(\pi^{-1}(k))} = g_k,$$

$$(\psi \circ \phi)(g_k) = \psi(g_{\pi(k)}) = g_{\pi^{-1}(\pi(k))} = g_k,$$

and $\phi \circ \psi = \psi \circ \phi = id_G$. Thus, ϕ is an automorphism of G if and only if ψ is an automorphism of G. This completes the proof. $\square$

EXPLANATION F.6. Theorem F.3 shows that: ϕ is an automorphism if and only if $\mathcal{S}_P = Z_0$; and, ψ is an automorphism if and only if $\mathcal{S}_{P^T} = Z_0$. We use this with Theorem F.5 to verify that the condition $\mathcal{S}_P = Z_0$ is satisfied if and only if $\mathcal{S}_{P^T} = Z_0$. The modification described in Observation F.4 about the program on page 34 to compute automorphisms merely amounts to replacing the condition $\mathcal{S}_P = Z_0$ by the equivalent condition $\mathcal{S}_{P^T} = Z_0$.

APPENDIX G

Computer Algebra for Transformations
of Homogeneous Linear Differential Equations

Appendix F presents the finding of automorphisms in a focused manner to encourage preparation of programs that greatlly reduce memory requirements.

Here, we show that the significant subject of transformations for homogeneous linear differential equations is ready for adoption by systems of computer algebra based on the fully developed presentation in [**12**].

G.1. The two basic types of transformations

Throughout, we let $c_{m,1}(z)$, $c_{m,2}(z)$, $\ldots$, $c_{m,m}(z)$ denote meromorphic functions on a region Ω, we define $c_{m,0}(z) \equiv 1$ on Ω, we let ρ denote a meromorphic function on Ω that satisfies $\rho(z) \not\equiv 0$, and we assume that $z = f(\zeta)$ is a univalent analytic function on a region Ω^{**} such that $f(\Omega^{**}) = \Omega$. Then, we have $f'(\zeta) \neq 0$, for each ζ in Ω^{**}.

The monic mth-order homogeneous linear differential equation

$$(G.1) \qquad y^{(m)}(z) + \sum_{i=1}^{m} c_{m,i}(z)\, y^{(m-i)}(z) = 0, \quad \text{on } \Omega \text{ with } c_{m,0}(z) \equiv 1,$$

for $y(z)$ has two basic types of transformations that can be applied to it.

(i) A *transformation of the first kind* for (G.1) is specified by $y(z) = \rho(z)\,v(z)$. It changes the function from $y(z)$ to $v(z)$. Namely, after first using $y(z) = \rho(z)\,v(z)$ to obtain

$$(G.2) \qquad y^{(i)}(z) = \sum_{k=0}^{i} \binom{i}{k} \rho^{(i-k)}(z)\, v^{(k)}(z), \quad \text{for } i \geq 0,$$

we replace $y^{(0)}(z)$, $y^{(1)}(z)$, $\ldots, y^{(m)}(z)$ in (G.1) with the corresponding expressions in the right member of (G.2). Then, when the resulting expression is rewritten as

$$(G.3) \qquad v^{(m)}(z) + \sum_{i=1}^{m} c_{m,i}^{*}(z)\, v^{(m-i)}(z) = 0, \quad \text{on } \Omega \text{ with } c_{m,0}^{*}(z) \equiv 1,$$

the coefficients are easily found to be given by

$$(G.4) \qquad c_{m,i}^{*}(z) \equiv \sum_{j=0}^{i} \binom{m-j}{i-j} \frac{\rho^{(i-j)}(z)}{\rho(z)}\, c_{m,j}(z), \quad \text{for } i = 0,\, 1,\, 2,\, \ldots,\, m.$$

(ii) A *transformation of the second kind* for (G.1) is specified by $z = f(\zeta)$. It changes the independent variable from z in Ω to ζ in Ω^{**}. It is made by first replacing each z in (G.1) with $f(\zeta)$ to obtain

$$(G.5) \quad y^{(m)}\big(f(\zeta)\big) + \sum_{i=1}^{m} c_{m,i}\big(f(\zeta)\big)\, y^{(m-i)}\big(f(\zeta)\big) = 0, \quad \text{on } \Omega^{**} \text{ with } c_{m,0}\big(f(\zeta)\big) \equiv 1.$$

With $u(\zeta) = y\big(f(\zeta)\big)$ on Ω^{**}, we see that: in order to rewrite (G.5) as a monic mth-order homogeneous linear differential equation in a form analogous to (G.1), the challenge is to express each of $y^{(i)}\big(f(\zeta)\big)$, for $0 \leq i \leq m$, in terms of $u^{(j)}(\zeta)$, for $0 \leq j \leq m$.

Progress about rewriting (G.5) did not occur prior to [**11**] of 1989. Our key discovery was the result presented in [**12**, page 135, Proposition A.2] that: for any meromorphic function $y(z)$ on Ω, the functions $y^{(i)}\big(f(\zeta)\big)$ are given on Ω^{**} in terms of $u(\zeta) = y\big(f(\zeta)\big)$ on Ω^{**} and $f'(\zeta) \neq 0$, for each ζ in Ω^{**}, by

$$(\text{G.6}) \qquad y^{(i)}\big(f(\zeta)\big) \equiv \sum_{j=0}^{i} \big(f'(\zeta)\big)^{-i} \alpha_{i-j,j}(\zeta)\, u^{(j)}(\zeta), \quad \text{for } i \geq 0,$$

where the functions $\alpha_{i,j}(\zeta)$ are defined as meromorphic functions on Ω^{**} by

$$(\text{G.7}) \quad \alpha_{0,j}(\zeta) \equiv 1, \quad \text{for any } j,$$

and

$$(\text{G.8}) \quad \alpha_{i,j}(\zeta) \equiv \sum_{k=1}^{j} \left[\alpha_{i-1,k}^{(1)}(\zeta) - (i+k-1)\frac{f''(\zeta)}{f'(\zeta)}\, \alpha_{i-1,k}(\zeta) \right], \quad \text{for } i \geq 1 \text{ and any } j.$$

When each $y^{(i)}\big(f(\zeta)\big)$ in (G.5) is replaced by the corresponding right member of (G.6) and the resulting expression is rewritten in the form

$$(\text{G.9}) \qquad u^{(m)}(\zeta) + \sum_{i=1}^{m} c_{m,i}^{**}(\zeta)\, u^{(m-i)}(\zeta) = 0, \quad \text{with } c_{m,0}^{**}(\zeta) \equiv 1,$$

we use [**12**, page 136, Theorem A.3] to see that the coefficients $c_{m,i}^{**}(\zeta)$ are given by

$$(\text{G.10}) \quad c_{m,i}^{**}(\zeta) \equiv \sum_{j=0}^{i} \alpha_{i-j,m-i}(\zeta) \big(f'(\zeta)\big)^{j} c_{m,j}\big(f(\zeta)\big), \quad \text{on } \Omega^{**} \text{ for } i = 0,\, 1,\, 2,\, \ldots,\, m.$$

The remarkable feature about (G.4) and (G.10) is that: for any integer $i \geq 0$, these formulas give the meromorphic coefficients $c_{m,i}^{*}(z)$ and $c_{m,i}^{**}(\zeta)$ for all of the differential equations (G.3) and (G.9) of order m for any $m \geq i$.

When James Cockle began the study of invariants for homogeneous linear differential equations in the early 1860's, he inserted binomial coefficients in the writing of (G.1), (G.3), and (G.9). That same binomial notation was employed in each of the numerous research papers on that subject prior to 1989. For details about how that notation impeded progress, see [**15**, Chapter 15].

G.2. Corresponding transformation formulas for *Mathematica*

The meromorphic functions have derivatives of all orders. Thus, there is no need for restrictions on the *Mathematica* representations `c[m,i][z]` for $c_{m,i}(z)$, or `cS[m,i][z]` for $c_{m,i}^{*}(z)$, or `cSS[m,i][zet]` for $c_{m,i}^{**}(\zeta)$, or `rho[z]` for $\rho(z)$, or `f[zet]` for $f(\zeta)$, ...

With $c_{0,m}(z) \equiv 1$, the key transformation formulas (G.4), (G.7), (G.8), and (G.10) are represented in a version of *Mathematica* such as: [**48**] by the input statements

```
c[m_,0][z_] = 1;

cS[m_,i_][z_] := Sum[ Binomial[m-j,i-j]*(D[rho[z],{z,i-j}]/rho[z])*
                        c[m,j][z], {j,0,i}]

alpha[0,j_][zet_] := 1

alpha[i_,j_][zet_] := Sum[( D[alpha[i-1,k][zet],zet]
    - (i+k-1)(f''[zet]/f'[zet])*alpha[i-1,k][zet] ), {k,1,j}] /; i >= 1

cSS[m_,i_][zet_] := Sum[ alpha[i-j,m-i][zet]*(f'[zet]^j)*
                        c[m,j][f[zet]], {j,0,i} ]
```

that we have entered and evaluated in a notebook that can be downloaded as indicated on page 235. Then, we entered and evaluated

```
eq[m_,z_] := ( D[y[z],{z,m}]
                + Sum[ c[m, j][z]*D[y[z],{z,m-j}], {j,1,m}] )

eqS[m_,z_] := ( D[v[z], {z,m}]
                + Sum[ cS[m,j][z]*D[v[z],{z,m-j}], {j,1,m}] )

eqSS[m_,zet_] := ( D[u[zet], {zet, m}]
                + Sum[ cSS[m,j][zet]*D[u[zet],{zet,m-j}], {j,1,m}] )
```

to represent the respective left members of (G.1), (G.3), and (G.9). As a check when $y(z) = \rho(z)\, v(z)$, the output for entering and evaluating

```
y[z_] = rho[z]*v[z];

first[m_] := FullSimplify[ rho[z]*eqS[m,z] - eq[m,z] ]

Do[ Print[ "first[",m,"] = ",first[m] ], {m,1,30} ]
```

shows that: for $1 \le m \le 30$, the left member of (G.1) is equal to the product of the left member of (G.3) and the nonzero factor $\rho(z)$. When $z = f(\zeta)$ and $u(\zeta) = y\big(f(\zeta)\big)$, the output for entering and evaluating

```
Clear[y];    z = f[zet];    u[zet_] := y[f[zet]];

second[m_] := FullSimplify[ eqSS[m,zet] - (f'[zet])^m*eq[m,f[zet]] ]

Do[ Print[ "second[",m,"] = ", second[m] ], {m,1,10} ]

Clear[y, z, v, u]
```

provides the check that: for $1 \le m \le 10$, the left member of (G.9) is equal to the product of the left member of (G.5) and the nonzero factor $\big(f'(\zeta)\big)^m$.

G.3. Identities for several of the principal invariants

The principal invariants for monic homogeneous linear differential equations have been called *relative invariants* since Edmund Laguerre introduced that terminology in [**29, 30**] of 1879. For their history, see [**15**, Chapter 1 and Chapters 15–18].

Our discovery that the coefficients of (G.9) are given by (G.10) enabled formulas to be developed in [**12**] that yield all of the basic relative invariants. An mth-order equation has relative invariants if and only if $m \ge 3$; and when $m \ge 3$, the mth-order equation has $m - 2$ basic relative invariants that we have designated by $\mathcal{I}_{m,3}, \mathcal{I}_{m,4}, \ldots, \mathcal{I}_{m,m}$.

For any integers k, m that satisfy $3 \le k \le m$, a *Mathematica* program to compute $\mathcal{I}_{m,k}$ is presented in [**15**, pages 53–54]. The downloadable notebook for this page uses that program to obtain, for each integer $m \ge 3$,

$$(G.11) \qquad \mathcal{I}_{m,3} \equiv w_3 - \frac{(m-2)}{m} w_1\, w_2 - \frac{m-2}{2} w_2^{(1)} + \frac{(m-1)(m-2)}{3m^2} (w_1)^3$$
$$+ \frac{(m-1)(m-2)}{2m} w_1\, w_1^{(1)} + \frac{(m-1)(m-2)}{12} w_1^{(2)}$$

as the basic relative invariant of weight 3 for the equations (G.1) having $m \ge 3$.

To illustrate the key properties of (G.11), we evaluate the *Mathematica* input

```
inv[m_,3][z_] := ( c[m,3][z]
                 - ((m-2)/m)c[m,1][z]*c[m,2][z]
                 - ((m-2)/2)D[c[m,2][z],z]
                 + ((m-1)(m-2)/(3m^2))c[m,1][z]^3
                 + ((m-1)(m-2)/(2m))c[m,1][z]*D[c[m,1][z],z]
                 + ((m-1)(m-2)/12)D[c[m,1][z],{z,2}] )

invS[m_,3][z_] := ( cS[m,3][z]
                  - ((m-2)/m)cS[m,1][z]*cS[m,2][z]
                  - ((m-2)/2)D[cS[m,2][z],z]
                  + ((m-1)(m-2)/(3m^2))cS[m, 1][z]^3
                  + ((m-1)(m-2)/(2m))cS[m,1][z]*D[cS[m,1][z],z]
                  + ((m-1)(m-2)/12)D[cS[m,1][z],{z,2}] )

invSS[m_,3][zet_] := ( cSS[m,3][zet]
                     - ((m-2)/m)cSS[m,1][zet]*cSS[m,2][zet]
                     - ((m-2)/2)D[cSS[m,2][zet],zet]
                     + ((m-1)(m-2)/(3m^2))cSS[m, 1][zet]^3
                     + ((m-1)(m-2)/(2m))cSS[m,1][zet]*D[cSS[m,1][zet],zet]
                     + ((m-1)(m-2)/12)D[cSS[m,1][zet],{zet,2}] )
```

and then find that the evaluation of

```
FullSimplify[ invS[m,3][z] - inv[m,3][z] ]
```

is zero and the evaluation of

```
FullSimplify[ invSS[m,3][zet] - (f'[zet])^3*inv[m,3][f[zet]] ]
```

is zero. Thus, for the functions $I_{m,3}(z)$ on Ω, $I_{m,3}^*(z)$ on Ω, and $I_{m,3}^{**}(\zeta)$ on Ω^{**} obtained from $\mathcal{I}_{m,3}$ by respectively replacing each $w_i^{(j)}$ in $\mathcal{I}_{m,3}$ with the corresponding $c_{m,i}^{(j)}(z)$ from (G.1) or $c_{m,i}^{*(j)}(z)$ from (G.3) or $c_{m,i}^{**(j)}(\zeta)$ from (G.9), the conditions

$$I_{m,3}^*(z) \equiv I_{m,3}(z), \quad \text{on } \Omega, \quad \text{and} \quad I_{m,3}^{**}(\zeta) \equiv \big(f'(\zeta)\big)^3 I_{m,3}\big(f(\zeta)\big), \quad \text{on } \Omega^{**},$$

are satisfied. This shows that $\mathcal{I}_{m,3}$ is an invariant for both transformations of the first kind and transformations of the second kind for each (G.1) having order $m \geq 3$.

Similarly, for each (G.1) of order $m \geq 4$, the basic relative invariant of weight 4 is

$$\begin{aligned}
\mathcal{I}_{m,4} \equiv\ & w_4 - \frac{(m-3)}{m}w_1 w_3 - \frac{(m-3)}{2}\mathfrak{w}_3^{(1)} - \frac{(m-2)(m-3)(5m+7)}{10(m+1)m(m-1)}w_2^2 \\
& + \frac{(m-2)(m-3)(5m+6)}{5m^2(m+1)}w_1^2 w_2 + \frac{(m-2)(m-3)(5m+7)}{10m(m+1)}w_1^{(1)} w_2 \\
& + \frac{(m-2)(m-3)}{2m}w_1 w_2^{(1)} + \frac{(m-2)(m-3)}{10}w_2^{(2)} \\
& - \frac{(m-1)(m-2)(m-3)(5m+6)}{20m^3(m+1)}w_1^4 - \frac{(m-1)(m-2)(m-3)(5m+6)}{10m^2(m+1)}w_1^2 w_1^{(1)} \\
& - \frac{(m-1)(m-2)(m-3)(2m+3)}{20m(m+1)}\big(w_1^{(1)}\big)^2 - \frac{(m-1)(m-2)(m-3)}{10m}w_1 w_1^{(2)} \\
& - \frac{(m-1)(m-2)(m-3)}{120}w_1^{(3)}.
\end{aligned}$$

It serves as a pattern to specify functions $I_{m,4}(z)$ on Ω, $I_{m,4}^*(z)$ on Ω, and $I_{m,4}^{**}(\zeta)$ on Ω^{**} respectively obtained from the coefficients of (G.1), (G.3), and (G.9) such that

$$I_{m,4}^*(z) \equiv I_{m,4}(z) \quad \text{on } \Omega \quad \text{and} \quad I_{m,4}^{**}(\zeta) \equiv \big(f'(\zeta)\big)^4 I_{m,4}\big(f(\zeta)\big), \quad \text{on } \Omega^{**}.$$

In particular, $I_{m,4}(z)$ is obtained from (G.1) by the replacement of each $\boldsymbol{w}_i^{(j)}$ in $\boldsymbol{\mathcal{I}}_{m,4}$ with the corresponding $c_{m,i}^{(j)}(z)$ from (G.1). Similarly, $I_{m,4}^*(z)$ and $I_{m,4}^{**}(\zeta)$ are obtained from (G.3) and (G.9) by using $\boldsymbol{\mathcal{I}}_{m,4}$ as a pattern.

For each pair of integers m, k that satisfy $3 \leq k \leq m$, there is a basic relative invariant $\boldsymbol{\mathcal{I}}_{m,k}$ of weight k for the equations (G.1) of order m. Then, the corresponding functions $I_{m,k}(z)$ on $\varOmega$, $I_{m,k}^*(z)$ on $\varOmega$, and $I_{m,k}^{**}(\zeta)$ on $\varOmega^{**}$ satisfy

$$I_{m,k}^*(z) \equiv I_{m,k}(z) \quad \text{on } \varOmega \text{ and} \quad I_{m,k}^{**}(\zeta) \equiv \big(f'(\zeta)\big)^k I_{m,k}\big(f(\zeta)\big), \quad \text{on } \varOmega^{**}.$$

Two completely independent methods to compute $\boldsymbol{\mathcal{I}}_{m,k}$ are presented in [**12**, pages 4–6].

G.4. Brief observations

Mathematica acquired its built-in capability of efficiently evaluating symbolic sums like the one in (G.8) after Version 3 was used in 2002 for [**12**]. That capability is included in Version 7.0 and later versions. In particular, the *Mathematica* evaluation of `alpha[i,j]` on the lower portion of page 230 is somewhat simpler than the corresponding one in [**12**].

The two independent methods of evaluating $\boldsymbol{\mathcal{I}}_{m,k}$, for $3 \leq k \leq m$, in [**12**] provide a useful check on the computations. By using a later version, the two corresponding *Mathematica* computations for $\boldsymbol{\mathcal{I}}_{m,k}$ in [**15**] as $\boldsymbol{\mathcal{I}}_{m,1;k}$, for $3 \leq k \leq m$, are a bit simpler.

To explain why the notation $\boldsymbol{\mathcal{I}}_{m,1;k}$ was employed in [**15**] for $\boldsymbol{\mathcal{I}}_{m,k}$ in [**12**], let $\mathfrak{F}$ denote the field of meromorphic functions defined on a region $\varOmega$ of the complex plain and observe that the left members of the differential equations in (G.1) can be described as homogeneous polynomial combinations of $y(z)$, $y^{(1)}(z)$, ..., $y^{(m)}(z)$ over $\mathfrak{F}$ having degree 1 such that the coefficient of $y^{(m)}(z)$ is 1. In [**13, 14, 15**], the results of [**12**] are included as the case $n = 1$ of the differential equations whose left members are homogeneous polynomial combinations of $y(z)$, $y^{(1)}(z)$, ..., $y^{(m)}(z)$ over $\mathfrak{F}$ having degree n such that the coefficient of $\big(y^{(m)}(z)\big)^n$ is 1. Just as for (G.1), the substitutions $y(z) = \rho(z)\,v(z)$ and $z = f(\zeta)$ transform those equations into ones having a similar form. All of their basic relative invariants are explicitly given in [**13, 14, 15**] with notation $\boldsymbol{\mathcal{I}}_{m,n;k_1,k_2,...,k_n}$. We advise any interested reader to first see the development in [**13**] from $n = 1$ to $n = 2$. Then, the special manner of writing the equations in [**15**] for any positive integers m and n is well motivated by the need to have simple formulas for the transformations.

The subject of differential algebra was initiated with the research of Joseph Ritt; e.g., see [**46**] of 1950 and [**28**] of 1973. Relative invariants for differential equations are naturally viewed as differential polynomials into which substitutions can be made. Prior to 1988, that had been overlooked. In particular, the differential polynomials that represent relative invariants have a weight defined for them. As examples, $\boldsymbol{\mathcal{I}}_{m,3}$ in (G.11) has weight 3 and $\boldsymbol{\mathcal{I}}_{m,4}$ has weight 4. For the differential equations given by m and n, an algorithm in [**14, 15**] uses the basic relative invariants to specify all of the relative invariants having a given weight.

Other contexts are available for the preceding results. For example, the coefficients of (G.1) as well as $\rho(z)$ and $f(\zeta)$ can be real-valued functions of a real variable when care is taken to provide suitable additional hypotheses about them.

G.5. Infinitesimal transformations were a great hindrance

Suitable formulas for transforming (G.1) under a change $z = f(\zeta)$ of the independent variable were not available for another 100 years when Andrew Forsyth wrote [**22**] in 1888. In their place, Forsyth introduced infinitesimal transformations as a means to deduce necessary details about several relative invariants whose existence was assumed. For 100 years, that technique of Forsyth was mistakenly made to seem essential. Thus, numerous elaborations about applying infinitesimal transformations were published. That was accompanied by wide acceptance of the completely absurd view that the subject of relative invariants was merely a detail in a general study of infinitesimal transformations.

APPENDIX H

Reference to *Mathematica* Notebooks

Mathematica notebooks have been prepared to supplement various pages throughout this monograph. Any particular notebook can be downloaded by first using the web browser Google Chrome or the web browser Microsoft Edge to visit the web page

`http://homepages.uc.edu/~chalklr/group-pattern.htm`

and then click the hyperactive text on that web page for the notebook in question. This places the name of the selected notebook in a small rectangular area at the lower portion of the web page. A click on that rectangular area will then download the *Mathematica* notebook that opens automatically if the computer has a version of *Mathematica* installed.

We have named the notebooks after the numbers of the pages whose content they supplement. For example, page 19 presents a computer program to determine whether any given matrix is an injective group-pattern matrix and, when so, to provide details. For page 19, the corresponding notebook is named P-019.nb and it contains the program of that page ready for application.

The pages that have supplemental *Mathematica* notebooks are provided by the names of the pages for each chapter in the following list.

Preface.
Chapter 1.
Chapter 2: P-019, P-020, P-021, P-025, P-026.
Chapter 3: P-034, P-035, P-036-037, P-038, P-039, P-039-040.
Chapter 4: P-047.
Chapter 5: P-055-056-and-P-061, P-057-060, P-062.
Chapter 6.
Chapter 7: P-077.
Chapter 8: P-082-083.
Chapter 9.
Chapter 10: P-092, P-093, P-095, P-096, P-097, P-098.
Chapter 11: P-105, P-107-108, P-108-(not-in-monograph), P-110.
Chapter 12: P-113-114, P-118-121.
Chapter 13: P-130-131.
Chapter 14.
Chapter 15: P-145-D3-D4-D5.
Chapter 16: P-155-157, P-157-159, P-159-162.
Chapter 17: P-172, P-173-174, P-175.
Chapter 18: P-181.
Chapter 19: P-185, P-186, P-187.
Chapters 20-22.
Chapter 23: P-201-202, P-203-204, P-203-204-(not-in-monograph), P-206.
Appendices A-B.
Appendix C: P-215.
Appendix D: P-219.
Appendices E-F.
Appendix G: P-230-232.

Bibliography

1. C. M. Ablow and J. L. Brenner, *Roots and canonical forms for circulant matrices*, Trans. Amer. Math. Soc. **107** (1963) 360–376.
2. R. Baltzer, *Theorie und Anwendung der Determinanten*, 2nd edition, Leipzig, 1864.
3. M. Bôcher, *Introduction to Higher Algebra*, Macmillan, New York, 1907.
4. E. C. Catalan, *Recherches sur les déterminants*, Bull. de l'Acad. royal de Belgique **13** (1846) 534–555.
5. A. Cayley, *A solvable case of the quintic equation*, Quart. J. Pure Appl. Math. **18** (1882) 154–157.
6. R. Chalkley, *Cardan's formulas and biquadratic equations*, Math. Mag. **47** (1974) 8–14.
7. _____, *Circulant matrices and algebraic equations*, Math. Mag. **48** (1975) 73–80.
8. _____, *Quartic equations and tetrahedral symmetries*, Math. Mag. **48** (1975) 211–215.
9. _____, *Matrices derived from finite abelian groups*, Math. Mag. **49** (1976) 121–129.
10. _____, *Information about group matrices*, Linear Algebra Appl. **38** (1981) 121–133.
11. _____, *Relative invariants for homogeneous linear differential equations*, J. Differential Equations, **80** (1989), 107–153.
12. _____, *Basic Global Relative Invariants for Homogeneous Linear Differential Equations*, no. 744, Memoirs Amer. Math. Soc., Providence, 2002, 1–204.
13. _____, *Basic Global Relative Invariants for Nonlinear Differential Equations*, no. 888, Memoirs Amer. Math. Soc., Providence, 2007, 1–365.
14. _____, *Relative Invariants from 1879 Onward: Their Evolution for Differential Equations*, Lumina Press, Plantation, Florida, 2014, 1 – 145 + xviii.
15. _____, *The Research about Invariants of Ordinary Differential Equations*, Available from Amazon.com and other retail outlets, Cincinnati, 2018, 1–190.
16. L. Cremona, *Intorno ad un theorema di Abel*, Annali di Sci. mat. e fis., **7** (1856) 99-105.
17. C. W. Curtis, *Pioneers of Representation Theory: Frobenius, Burnside, Schur, and Brauer*, Amer. Math. Soc., Providence, 1999.
18. C. W. Curtis and I. Reiner, *Representation Theory of Finite Groups and Associative Algebras*, Wiley-Interscience, New York, 1962.
19. P. J. Davis, *Circulant Matrices*, Wiley-Interscience, New York, 1979.
20. J. W. R. Dedekind, *Gesammelte mathematische Werke, Zweiter Band*, Friedrich Vieweg & Sohn, Braunschweig, 1931.
21. G. J. Dostor, *Éléments de la Théorie des Déterminants*, Gauthier-Villars, Paris, 1877.
22. A. R. Forsyth, *Invariants, covariants and quotient derivatives associated with linear differential equations*, Philosophical Transactions of the Royal Society of London, **179** (1888) 377–489.
23. F. G. Frobenius, *Über Gruppencharaktere*, Sitzungsberichte Akad. Wiss. Berlin (1896), Zweiter Halbband, 985–1021.
24. _____, *Über die Primfactoren der Gruppendeterminante*, Sitzungsberichte Akad. Wiss. Berlin (1896), Zweiter Halbband, 1343–1382.
25. _____, *Über die Darstellung der endlichen Gruppen durch lineare Substitutionen*, Sitzungsberichte Akad. Wiss. Berlin (1897), Zweiter Halbband, 994–1015.
26. _____, *Gesammelte Abhandlungen III*, Springer-Verlag, Berlin, 1968, pp. 1–37, 38–77, and 82–103.
27. D. Kalman and J. E. White, *Polynomial equations and circulant matrices*, Amer. Math. Monthly **108** (2001) 821-840.
28. E. R. Kolchin, *Differential Algebra and Algebraic Groups*, Academic Press, New York, 1973.
29. E. Laguerre, *Sur les équations différentielles linéaires du troisième ordre*, C. R. Acad. Sci. Paris, **88** (1879) 116–119.

30. _____, *Sur quelques invariants des équations différentielles linéaires*, C. R. Acad. Sci. Paris, **88** (1879) 224–227.

31. S. Lang, *Algebra,* Second edition, Addison-Wesley, New York, 1984.

32. E. A. Legoux, *Sur une application d'un déterminant*, Quar. J. Pure Appl. Math. **19** (1883) 41–43.

33. C. Le Paige, *Sur l'équation du quatrième degré*, Časopis pro pěstování math. a fys., **14** (1884) 26–28.

34. A. Lodge, *On the solution of the general equation of the fourth degree*, Quar. J. Pure Appl. Math. **19** (1883) 257–262.

35. C. C. MacDuffee, *An Introduction to Abstract Algebra,* Wiley, New York, 1950.

36. T. Muir, *The Theory of Determinants in the Historical Order of Development*, Vol. 1 (1906), Vol. 2 (1911), Vol. 3 (1920), Vol. 4 (1923), Macmillan, London. (Reprinted by Dover, New York.)

37. M. Newman, *Matrix Representations of Groups*, National Bureau of Standards, Applied Mathematics Series 60, Washington, D. C., 1968.

38. I. M. Niven, H. S. Zuckerman, and H. L. Montgomery, *An Introduction to the Theory of Numbers,* 5th edition, Wiley, New York, 1991.

39. M. Noether, *Zur Theorie der Thetafunctionen von beliebig vielen Argumenten* (§ 15, pp. 322–325), Math. Annalen, **16** (1880), pp. 270–344.

40. _____, *Notiz über eine Classe symmetrischer Determinanten*, Math. Annalen, **16** (1880), pp. 551–555.

41. D. S. Passman, *What is a group ring?*, Amer. Math. Monthly **83** (1976), 173-184.

42. A. Puchta, *Ein Determinantensatz und seine Umkehrung*, Denkschriften Akad. d. Wiss. (Wien), **38** (1877), pp. 215-221.

43. A, Puchta, *Ein neuer Satz aus der Theorie der Determinanten*, Denkschriften Akad. d. Wiss. (Wien), **44** (1881), pp. 277–282.

44. L. Rédei, *Algebra*, Akademische Verlagsgesellschaft Geest & Portig K.-G., Leipzig, 1959.

45. L. Clariana y Ricart, *Aplicación de las determinantes a la resolución de las ecuaciones de cuarto grado*, Crónica Científica Revista Internacional de Ciencias (Barcelona) **3** (1880) 425–429.

46. J. F. Ritt, *Differential Algebra*, Amer. Math. Soc. Colloq. Publ., vol. 33, Amer. Math. Soc., New York, 1950.

47. W. Spottiswoode, *Elementary theorems relating to determinants. Rewritsten and much enlarged by the author*, J. Reine Angew. Math., **51** (1853) 209–271, 328–381.

48. S. Wolfram, *Mathematica*, Version 11.2, Wolfram Research Inc. Champaign, Illinois, 2017.

Index

$Aut(G)$, 30
$\mathfrak{A}$, 123
$\mathfrak{B}$, 123
$\mathbb{C}$, xvi, 14
$\mathbb{C}^*$, 86
$\mathbb{C}[X_1, X_2, \ldots, X_n]$, 137
$\delta(x, y)$, 124
$\mathbb{F}$, xvi, 47, 123
$\mathcal{F}$, xviii, 63
$\mathcal{F}^*$, 63, 86
$\mathfrak{F}$, 13, 42
$\mathfrak{F}[G]$, 44
$G_1 \times G_2 \times \cdots \times G_d$, 36
$\widehat{g}$, 42
$\Gamma(\lambda, \mu, \nu)$, 104
I_n, 42
$\mathscr{L}$, 11
$\lambda(s)$, 105
M, 53, 63, 64, 73, 79, 124
$\mathcal{M}[\mathbb{C}, G, \mathscr{L}]$, 14
$\mathcal{M}[\mathbb{F}, G, \mathscr{L}]$, 47
$\mathcal{M}[\mathcal{F}, G, \mathscr{L}]$, 84
$\mathcal{M}[\mathfrak{F}, G, \mathscr{L}]$, 42
$\mathcal{M}[\mathfrak{R}, G, \mathscr{L}]$, 41
$\mathcal{M}[S, G, \mathscr{L}]$, 12
$\mu(s)$, 105
N, 68, 85, 166
$\mathfrak{N}_{\mathfrak{A}}(G)$, 180
$\mathfrak{N}_{\mathfrak{G}}(G)$, 30
$\nu(s)$, 105
$\mathbb{Q}[X_1, X_2, \ldots, X_n]$, 33
$\mathfrak{R}$, 13, 41, 137
$\sigma * \tau$, 41, 44
$Stan_{\mathfrak{A}}(G, \mathscr{L})$, 177
$Stan_{\mathfrak{G}}(G, \mathscr{L})$, 27
Θ, 91, 92, 103
abelian group, 11
automorphisms of a group G
  algorithm for, 33, 183, 227
  computation of, 34, 185, 227
  computer program for, 34, 185
  definition of, 30
  number of, 30, 180

auxiliary-pattern matrix
  definition of, 166
  displayed example for the group
    $\mathcal{D}_3$, 167
  injective, 166
  multiplication table for, 165
  standard, 177

circulant matrix, 24, 53, 63, 67
complete set of matrix representations, 99
conjugacy classes for a group, 99
conjugate elements in a group, 99

degree of a matrix representation, 96
diagonalization, 65, 66, 74, 75, 81, 89
dihedral group $\mathcal{D}_n$ of order $2n$, 139
direct product of groups, 36

equivalent matrix representations, 97

Frobenius block-diagonal matrix
  definition of, 123
  displayed example for the group
    $\mathcal{D}_3$, 156
    $\mathcal{D}_4$, 159
    $\mathcal{D}_5$, 162
    Q_8, 202, 204
    $\mathfrak{S}_3 \cong \mathcal{D}_3$, 96, 108
    $\mathfrak{T}$, as the twelve rotational symmetries
      of a regular tetrahedron, 116
  historical introduction of, 96
Frobenius block-diagonalization, 125

generator for cyclic group, 12
group, 11
  abelian (i.e. commutative), 11
  cyclic, $\mathcal{C}_n$ of order n, 12
  definition of, 11
  dihedral, $\mathcal{D}_n$ of order $2n$, 139–141
  direct product of groups, 36
  multiplication table for, 11, 165
  order of, 11
  quaternion Q_8, 21, 35, 199, 203
  $\mathfrak{T}$, as the twelve rotational symmetries of
    a regular tetrahedron, 111

group character (abelian group), 86
group homomorphism, 96
group ring and group algebra, 44
group-matrix, xiii, 91, 103, 197
group-pattern matrix
    definition of, 12
    displayed example for the group
        $\mathcal{C}_2$, 76
        $\mathcal{C}_3$, 1, 214
        $\mathcal{C}_4$, 2, 76
        $\mathcal{C}_5$, 39
        $\mathcal{C}_6$, 3, 4, 76
        $\mathcal{C}_n$, 12, 54, 209, 217
        $\mathcal{C}_2 \times \mathcal{C}_2$, 3, 55
        $\mathcal{C}_3 \times \mathcal{C}_3$, 20, 58, 83
        $\mathcal{C}_4 \times \mathcal{C}_2$, 8
        $\mathcal{C}_2 \times \mathcal{C}_2 \times \mathcal{C}_2$, 7, 56
        $\mathcal{C}_2 \times \mathcal{C}_2 \times \mathcal{C}_2 \times \mathcal{C}_2$, 25
        $\mathcal{D}_3$, 5, 47, 155, 167
        $\mathcal{D}_4$, 6, 157
        $\mathcal{D}_5$, 25, 160
        $\mathcal{D}_n$, primary, 142
        $\mathcal{D}_n$, secondary, 146
        Q_8, 21, 199, 203
        $\mathfrak{S}_3 \cong \mathcal{D}_3$, 94, 107
        $\mathfrak{T}$, as the twelve rotational symmetries
            of a regular tetrahedron, 115
    injective, 12
    multiplication table for, 11
    standard, 27
Gruppendeterminante, xiv, 91, 197, 198
Gruppenmatrix, xiv, 91, 197, 198

identity element of a group, 11
inequivalent matrix representations, 97
injective
    auxiliary-pattern matrix, 166
    group-pattern matrix, 12
inverse element in a group, 11
irreducible matrix representation, 98

leftward-circulant matrix, 24, 53, 67
list $\mathscr{L}$ for group G, 11

matrix representations for a group
    complete set of, 99
    definition of, 97
    degree of, 97
    equivalent, 97
    irreducible, 98
    pairwise inequivalent, 97
    reducible, 98
multiplication table for G and $\mathscr{L}$, 11

number q of conjugacy classes, 99

period (or order) of element in a group, 40
primitive nth root of unity
    having period n, 63

reducible matrix representation, 98
reference to work of
    C. M. Ablow, 24
    Heinrich Richard Baltzer, 221
    Maxime Bôcher, 137
    Francisco Brioschi, 54, 61
    Eugène Charles Catalan, 53
    Arthur Cayley, 207
    Roger Chalkley, xv, xvii, 12, 24, 47, 98,
        107, 207, 209, 210, 213, 229–231, 233
    James Cockle, 230
    Luigi Cremona, 53, 54, 61, 64
    Charles W. Curtis, 96
    Philip J. Davis, xv, 24, 210, 215, 217, 218
    Richard Dedekind
        (October 6,1831–February 12, 1916), 2,
        57, 61, 86, 87, 91, 93–96, 106, 193,
        197, 199, 203
    Georges Dostor, 207
    Andrew Forsyth, 233
    Georg Frobenius
        (October 26, 1849–August 3, 1917), 2,
        57, 61, 86, 91, 92, 96, 103, 106, 193,
        197
    Emilie V. Haynsworth, 210
    Dan Kalman, 207
    Ellis Robert Kolchin, 233
    Constantin Le Paige, 207
    Edmond Laguerre, 231
    Serge Lang, 45, 63, 96
    Alphonse Legoux, 207
    Alfred Lodge, 207
    Cyrus Colton MacDuffee, 79
    Thomas Muir, 24, 53, 54, 57, 207, 209,
        221
    Morris Newman, 99, 100, 112
    Max Noether, 57, 59, 61
    Donald S. Passman, 46
    Anton Puchta, 55–59, 61
    László Rédei, 209
    Irving Reiner, 96
    L. Clariana y Ricart, 207
    Joseph Fels Ritt, 233
    William H. Spottiswode, 53
    James E. White, 207
    Wolfram Research Inc., 19, 38, 47, 61,
        83, 109, 110, 114, 118, 121, 173, 201,
        230, 232
    Chao-Hui Yang, 207
reproduce Cardano's formula, 207

standard auxiliary-pattern matrices for G
    definition of, 177
    number of, 179
standard group-pattern matrices for G
    definition of, 27
    number of, 30

www.ingramcontent.com/pod-product-compliance
Lightning Source LLC
Chambersburg PA
CBHW061324190326
41458CB00011B/3886